Patrick Moore's Yearbook of Astronomy 2013

Patrick Moore's Yearbook of Astronomy 2013

EDITED BY

Patrick Moore

AND

John Mason

MACMILLAN

First published 2012 by Macmillan
an imprint of Pan Macmillan, a division of Macmillan Publishers Limited
Pan Macmillan, 20 New Wharf Road, London N1 9RR
Basingstoke and Oxford
Associated companies throughout the world
www.panmacmillan.com

ISBN 978-0-230-76750-8

One article from Part II is republished from a previous edition:

'The Supernova That Won't Go Away' first published in the *Yearbook of Astronomy 1991*.

1 3 5 7 9 8 6 4 2

A CIP catalogue record for this book is available from
the British Library.

Typeset by Ellipsis Digital Limited
Printed and bound by CPI Group (UK) Ltd, Croydon, CR0 4YY

Contents

Part Two

Article Section

Part Three

Miscellaneous

Editors' Foreword

This latest edition of the *Yearbook* follows the long-established pattern. Wil Tirion, who produced our stars maps for the Northern and Southern Hemispheres, has again drawn all of the line diagrams showing the positions and movements of the planets to accompany the Monthly Notes. These provide a detailed guide to what's happening in the night sky throughout the year, on a month-by-month basis. Martin Mobberley has supplied the notes on eclipses, comets and minor planets, and Nick James has produced the data for the phases of the Moon, longitudes of the Sun, Moon and planets, and details of lunar occultations. As always, John Isles and Bob Argyle have provided the information on variable stars and double stars, respectively.

We also have a fine selection of previously unpublished longer articles, both from our regular contributors and from those new to the *Yearbook* this year. As usual, we have done our best to give you a wide choice, both of subject and of technical level. On the eightieth anniversary of the discovery of a Great White Spot on Saturn by the stage and screen comedian and amateur astronomer Will Hay, Martin Mobberley takes a fascinating in-depth look at Will Hay, the man, while Paul Abel examines these great storms on Saturn, both past and present. David Harland describes the *Dawn* spacecraft's ongoing mission to the main belt asteroids Vesta and Ceres, Stephen Webb looks at the ALMA, the remarkable telescope now being assembled on the plains of the Atacama Desert in Chile, while Fred Watson takes us on a tour of the subatomic universe, from the Northern Lights of the high Arctic to the underground tunnels of the Large Hadron Collider. For the amateur observer, Nick James gives some important advice on imaging meteors using digital cameras. We also have Richard Baum's fascinating tale of the arguments for the existence of Vulcan, a small planetary object at one time proposed to exist in orbit between Mercury and the Sun. Finally, historian Allan Chapman gives an absorbing account of the seventeenth-century instrument maker, lunar cartographer and surveyor of the heavens – Johannes Hevelius.

In last year's special fiftieth anniversary publication, we picked out a selection of previously published articles that we thought would be of interest to readers today. This proved so popular that we have decided to republish one article from the past in each *Yearbook* from now on. This time, from the 1991 *Yearbook*, we have the late David Allen's contribution about the remarkable supernova SN1987A which burst on to the astronomical scene in spectacular fashion in February 1987. David was one of the *Yearbook*'s most regular and welcome contributors until his untimely death in 1994, a few days before his forty-eighth birthday. We are indebted to Professor Fred Watson for carrying out the necessary updating of David's original article.

We hope you enjoy reading the range of articles we have selected for you this year.

Patrick Moore
John Mason
Selsey, August 2012

Preface

New readers will find that all the information in this *Yearbook* is given in diagrammatic or descriptive form; the positions of the planets may easily be found from the specially designed star charts, while the Monthly Notes describe the movements of the planets and give details of other astronomical phenomena visible in both the Northern and Southern Hemispheres. Two sets of star charts are provided. The **Northern Star Charts** (pp. 7 to 31) are designed for use at latitude 52° N, but may be used without alteration throughout the British Isles, and (except in the case of eclipses and occultations) in other countries of similar northerly latitude. The **Southern Star Charts** (pp. 33 to 57) are drawn for latitude 35°S, and are suitable for use in South Africa, Australia and New Zealand, and other locations in approximately the same southerly latitude. The reader who needs more detailed information will find *Norton's Star Atlas* an invaluable guide, while more precise positions of the planets and their satellites, together with predictions of occultations, meteor showers and periodic comets, may be found in the *Handbook of the British Astronomical Association.* Readers will also find details of forthcoming events given in the American monthly magazine *Sky & Telescope* and the British periodicals *The Sky at Night, Astronomy Now* and *Astronomy and Space.*

Important note

The times given on the star charts and in the Monthly Notes are generally given as local times, using the twenty-four-hour clock, the day beginning at midnight. All the dates, and the times of a few events (e.g. eclipses), are given in Greenwich Mean Time (GMT), which is related to local time by the formula:

Local Mean Time = GMT – west longitude

In practice, small differences in longitude are ignored, and the observer will use local clock time, which will be the appropriate Standard (or Zone) Time. As the formula indicates, places in west longitude will have a Standard Time slow on GMT, while places in east longitude will have a Standard Time fast on GMT. As examples we have:

Standard Time in	
New Zealand	GMT + 12 hours
Victoria, NSW	GMT + 10 hours
Western Australia	GMT + 8 hours
South Africa	GMT + 2 hours
British Isles	GMT
Eastern ST	GMT – 5 hours
Central ST	GMT – 6 hours, etc.

If Summer Time is in use, the clocks will have been advanced by one hour, and this hour must be subtracted from the clock time to give Standard Time.

Part One

Monthly Charts and Astronomical Phenomena

Notes on the Star Charts

The stars, together with the Sun, Moon and planets, seem to be set on the surface of the celestial sphere, which appears to rotate about the Earth from east to west. Since it is impossible to represent a curved surface accurately on a plane, any kind of star map is bound to contain some form of distortion.

Most of the monthly star charts which appear in the various journals and some national newspapers are drawn in circular form. This is perfectly accurate, but it can make the charts awkward to use. For the star charts in this volume, we have preferred to give two hemispherical maps for each month of the year, one showing the northern aspect of the sky and the other showing the southern aspect. Two sets of monthly charts are provided, one for observers in the Northern Hemisphere and one for those in the Southern Hemisphere.

Unfortunately the constellations near the overhead point (the zenith) on these hemispherical charts can be rather distorted. This would be a serious drawback for precision charts, but what we have done is to give maps which are best suited to star recognition. We have also refrained from putting in too many stars, so that the main patterns stand out clearly. To help observers with any distortions near the zenith, and the lack of overlap between the charts of each pair, we have also included two circular maps, one showing all the constellations in the northern half of the sky and one showing those in the southern half. Incidentally, there is a curious illusion that stars at an altitude of 60° or more are actually overhead, and beginners may often feel that they are leaning over backwards in trying to see them.

The charts show all stars down to the fourth magnitude, together with a number of fainter stars which are necessary to define the shapes of constellations. There is no standard system for representing the outlines of the constellations, and triangles and other simple figures have been used to give outlines which are easy to trace with the naked eye. The names of the constellations are given, together with the proper names of the brighter stars. The apparent magnitudes of the stars are indicated roughly by using different sizes of dot, the larger dots representing the brighter stars.

The two sets of star charts – one each for Northern and Southern Hemisphere observers – are similar in design. At each opening there is a single circular chart which shows all the constellations in that hemisphere of the sky. (These two charts are centred on the North and South Celestial Poles, respectively.) Then there are twelve double-page spreads, showing the northern and southern aspects for each month of the year for observers in that hemisphere. In the **Northern Charts** (drawn for latitude 52°N) the left-hand chart of each spread shows the northern half of the sky (lettered 1N, 2N, 3N . . . 12N), and the corresponding right-hand chart shows the southern half of the sky (lettered 1S, 2S, 3S . . . 12S). The arrangement and lettering of the charts is exactly the same for the **Southern Charts** (drawn for latitude 35°S).

Because the sidereal day is shorter than the solar day, the stars appear to rise and set about four minutes earlier each day, and this amounts to two hours in a month. Hence the twelve pairs of charts in each set are sufficient to give the appearance of the sky throughout the day at intervals of two hours, or at the same time of night at monthly intervals throughout the year. For example, charts 1N and 1S here are drawn for 23 hours on 6 January. The view will also be the same on 6 October at 05 hours; 6 November at 03 hours; 6 December at 01 hours and 6 February at 21 hours. The actual range of dates and times when the stars on the charts are visible is indicated on each page. Each pair of charts is numbered in bold type, and the number to be used for any given month and time may be found from the following table:

Local Time	**18h**	**20h**	**22h**	**0h**	**2h**	**4h**	**6h**
January	11	12	1	2	3	4	5
February	12	1	2	3	4	5	6
March	1	2	3	4	5	6	7
April	2	3	4	5	6	7	8
May	3	4	5	6	7	8	9
June	4	5	6	7	8	9	10
July	5	6	7	8	9	10	11
August	6	7	8	9	10	11	12
September	7	8	9	10	11	12	1

Local Time	18h	20h	22h	0h	2h	4h	6h
October	8	9	10	11	12	1	2
November	9	10	11	12	1	2	3
December	10	11	12	1	2	3	4

On these charts, the ecliptic is drawn as a broken line on which longitude is marked every 10°. The positions of the planets are then easily found by reference to the table on p. 64. It will be noticed that on the **Southern Charts** the ecliptic may reach an altitude in excess of 62.5° on the star charts showing the northern aspect (5N to 9N). The continuations of the broken line will be found on the corresponding charts for the southern aspect (5S, 6S, 8S and 9S).

Northern Star Charts

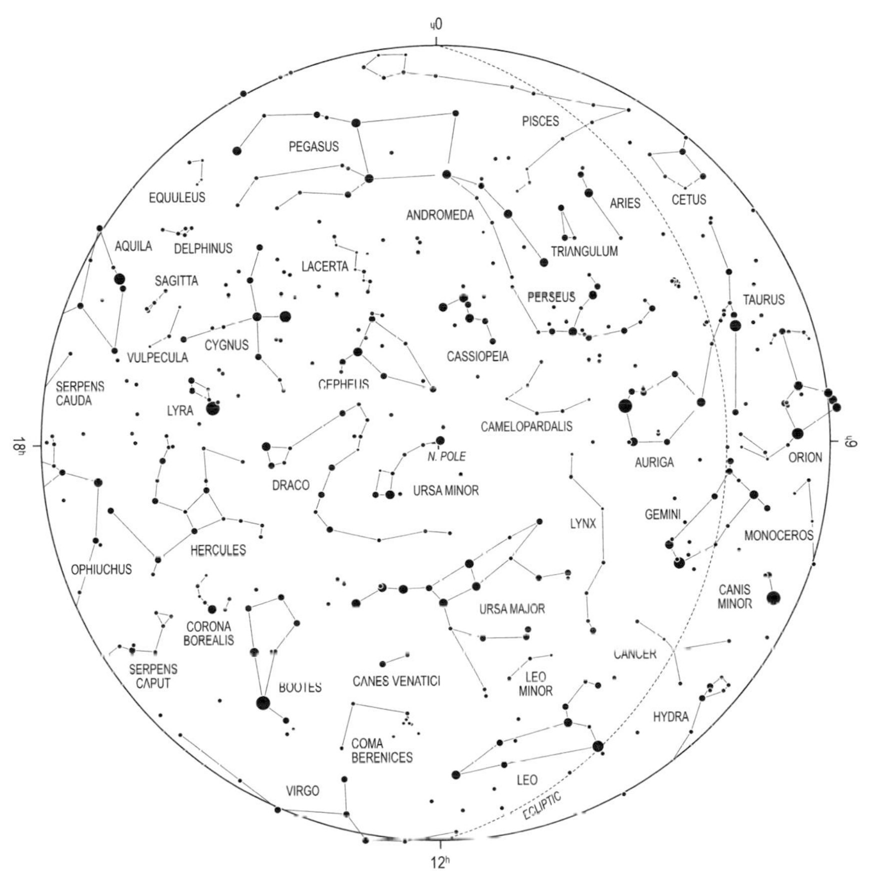

Northern Hemisphere

Note that the markers at 0h, 6h, 12h and 18h indicate hours of Right Ascension.

1N

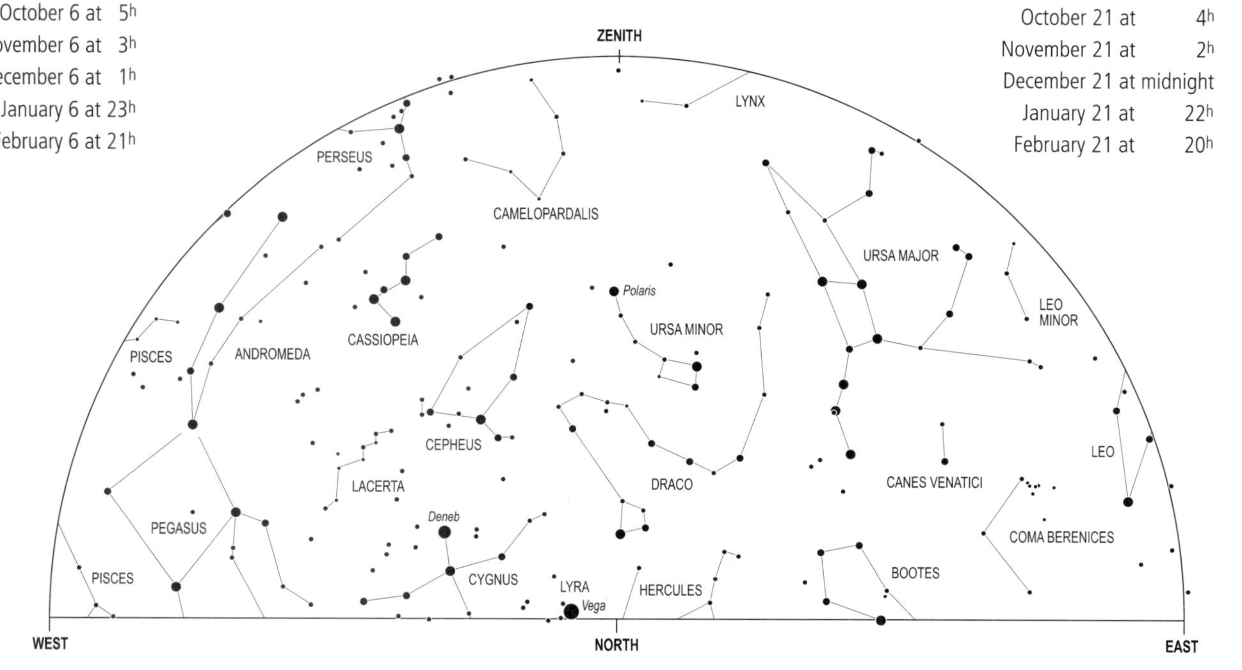
October 6 at 5h
November 6 at 3h
December 6 at 1h
January 6 at 23h
February 6 at 21h
ZENITH
LYNX
PERSEUS
CAMELOPARDALIS
URSA MAJOR
Polaris
URSA MINOR
LEO MINOR
PISCES
ANDROMEDA
CASSIOPEIA
CEPHEUS
LEO
LACERTA
DRACO
CANES VENATICI
Deneb
PEGASUS
COMA BERENICES
PISCES
CYGNUS
LYRA
Vega
HERCULES
BOOTES
WEST
NORTH
EAST
October 21 at 4h
November 21 at 2h
December 21 at midnight
January 21 at 22h
February 21 at 20h

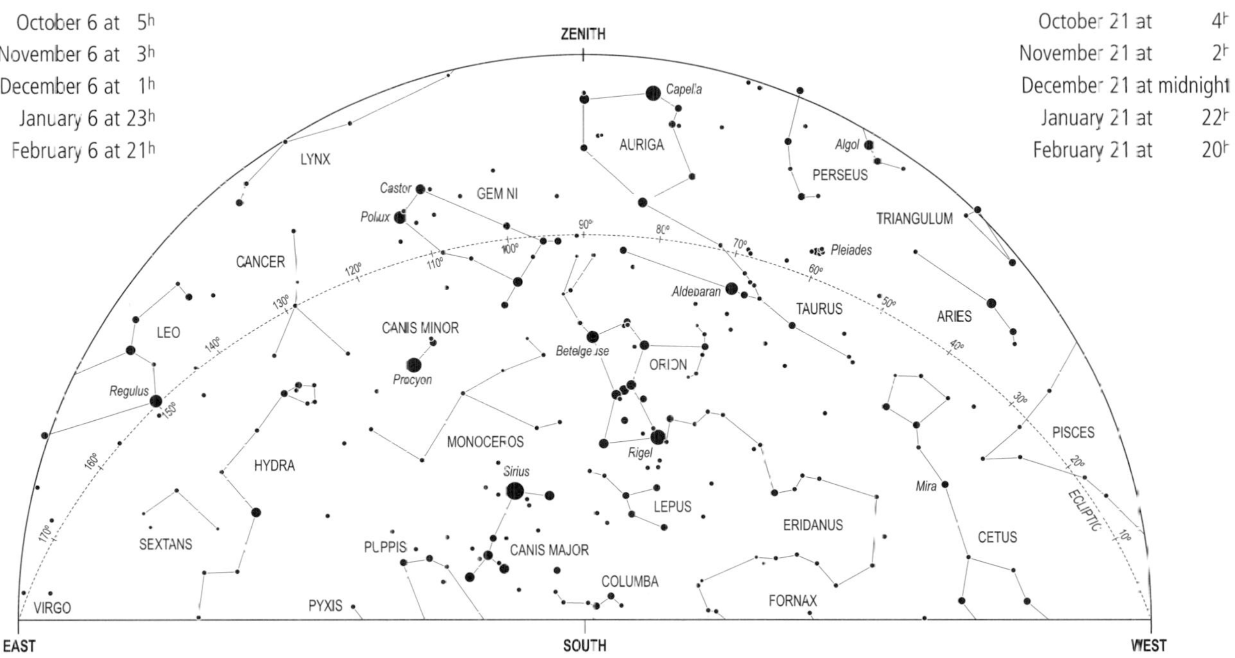
October 6 at 5h
November 6 at 3h
December 6 at 1h
January 6 at 23h
February 6 at 21h
October 21 at 4h
November 21 at 2h
December 21 at midnight
January 21 at 22h
February 21 at 20h
ZENITH
EAST
SOUTH
WEST
ECLIPTIC
LYNX
Castor
Pollux
GEMINI
CANCER
LEO
Regulus
CANIS MINOR
Procyon
HYDRA
SEXTANS
VIRGO
PYXIS
PUPPIS
MONOCEROS
Sirius
CANIS MAJOR
COLUMBA
AURIGA
Capella
Betelgeuse
ORION
Rigel
LEPUS
Aldebaran
TAURUS
Pleiades
PERSEUS
Algol
TRIANGULUM
ARIES
ERIDANUS
FORNAX
Mira
CETUS
PISCES

1S

2N

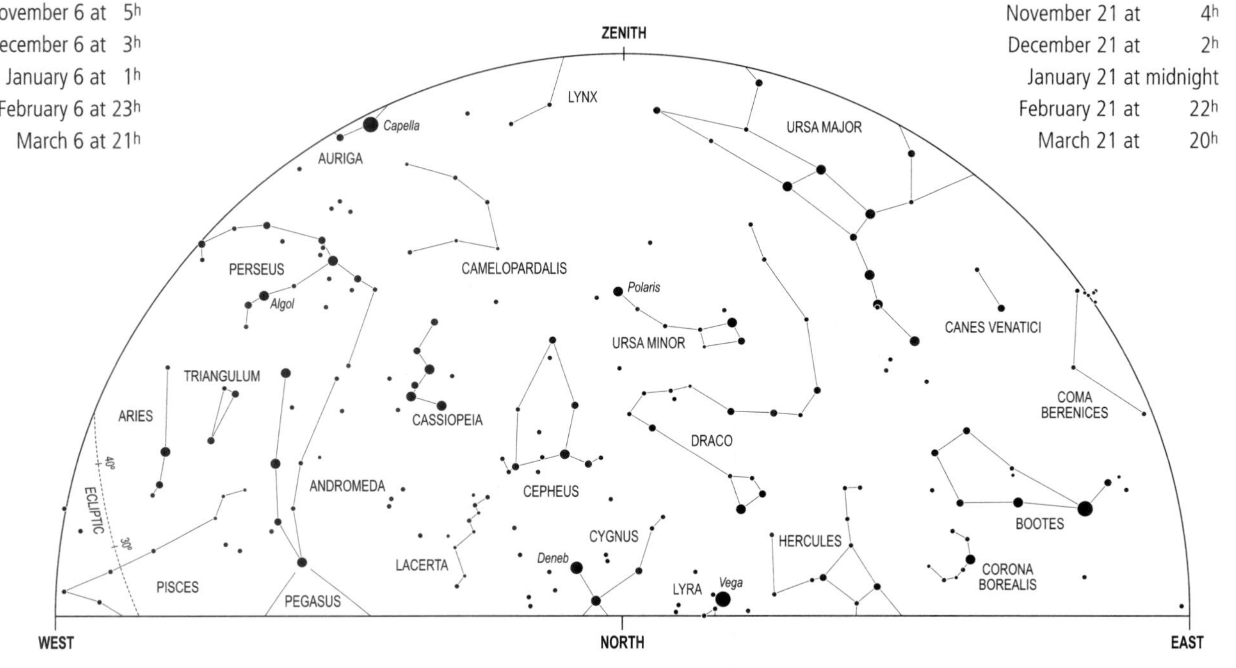
November 6 at 5h
December 6 at 3h
January 6 at 1h
February 6 at 23h
March 6 at 21h
November 21 at 4h
December 21 at 2h
January 21 at midnight
February 21 at 22h
March 21 at 20h
ZENITH
WEST
NORTH
EAST
ECLIPTIC
40°
30°
LYNX
Capella
AURIGA
URSA MAJOR
PERSEUS
Algol
CAMELOPARDALIS
Polaris
URSA MINOR
CANES VENATICI
TRIANGULUM
ARIES
CASSIOPEIA
COMA
BERENICES
DRACO
ANDROMEDA
CEPHEUS
BOOTES
CYGNUS
HERCULES
LACERTA
Deneb
CORONA
BOREALIS
PISCES
PEGASUS
LYRA
Vega

2S

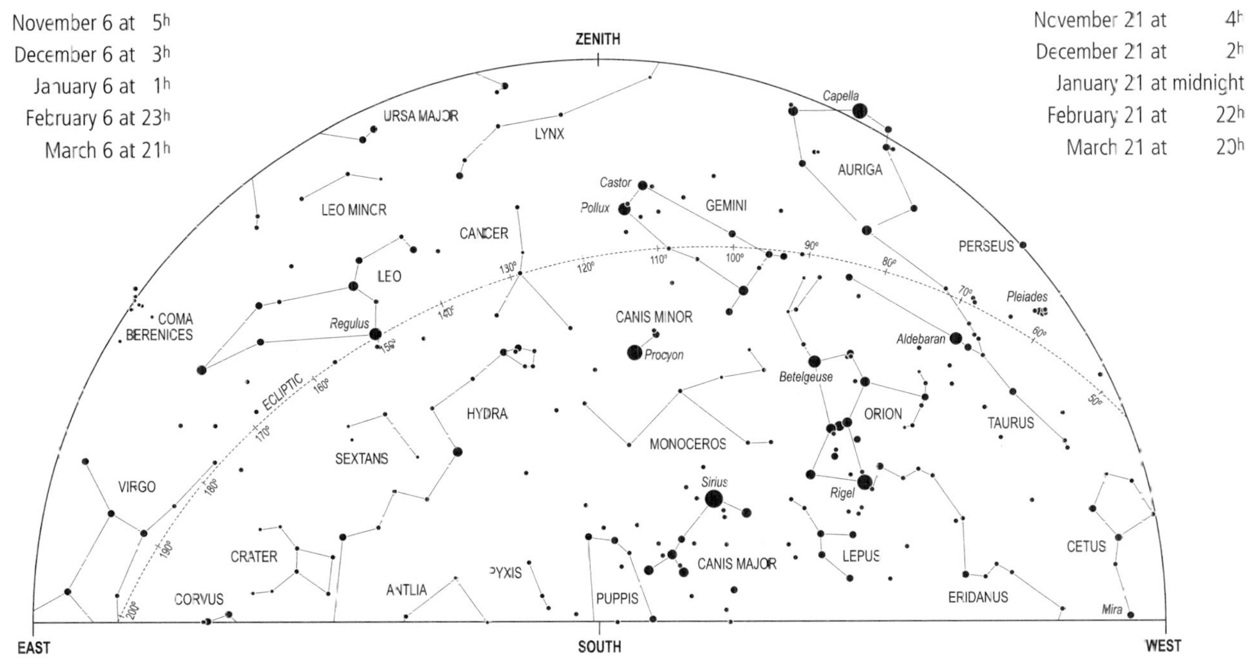
November 6 at 5h
December 6 at 3h
January 6 at 1h
February 6 at 23h
March 6 at 21h
November 21 at 4h
December 21 at 2h
January 21 at midnight
February 21 at 22h
March 21 at 20h
ZENITH
EAST
SOUTH
WEST
URSA MAJOR
LYNX
LEO MINOR
CANCER
LEO
Regulus
COMA BERENICES
Castor
Pollux
GEMINI
CANIS MINOR
Procyon
AURIGA
Capella
PERSEUS
Pleiades
Aldebaran
TAURUS
Betelgeuse
ORION
Rigel
HYDRA
MONOCEROS
Sirius
CANIS MAJOR
LEPUS
CETUS
Mira
ERIDANUS
PUPPIS
PYXIS
ANTLIA
SEXTANS
CRATER
CORVUS
VIRGO
ECLIPTIC
50°
60°
70°
80°
90°
100°
110°
120°
130°
140°
150°
160°
170°
180°
190°
200°

3N

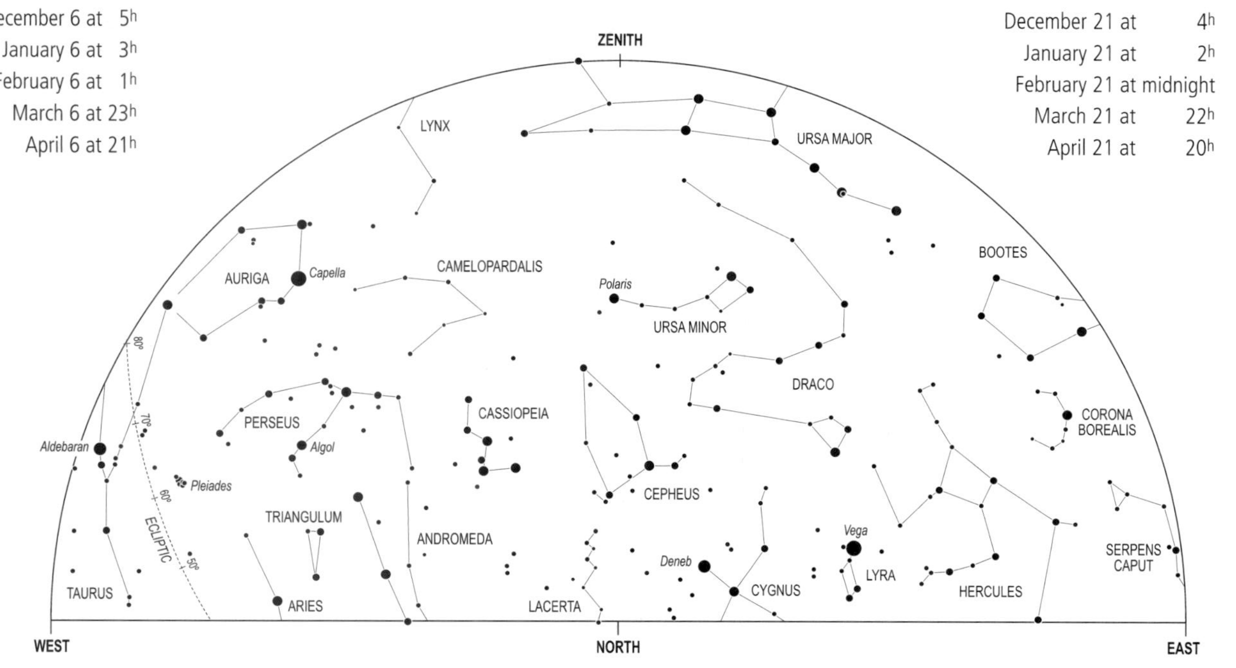
December 6 at 5h
January 6 at 3h
February 6 at 1h
March 6 at 23h
April 6 at 21h
December 21 at 4h
January 21 at 2h
February 21 at midnight
March 21 at 22h
April 21 at 20h
ZENITH
LYNX
URSA MAJOR
AURIGA
Capella
CAMELOPARDALIS
Polaris
URSA MINOR
BOOTES
80°
70°
60°
50°
ECLIPTIC
DRACO
PERSEUS
Algol
CASSIOPEIA
Aldebaran
Pleiades
CORONA BOREALIS
CEPHEUS
TRIANGULUM
ANDROMEDA
Vega
SERPENS CAPUT
Deneb
LYRA
TAURUS
ARIES
LACERTA
CYGNUS
HERCULES
WEST
NORTH
EAST

3S

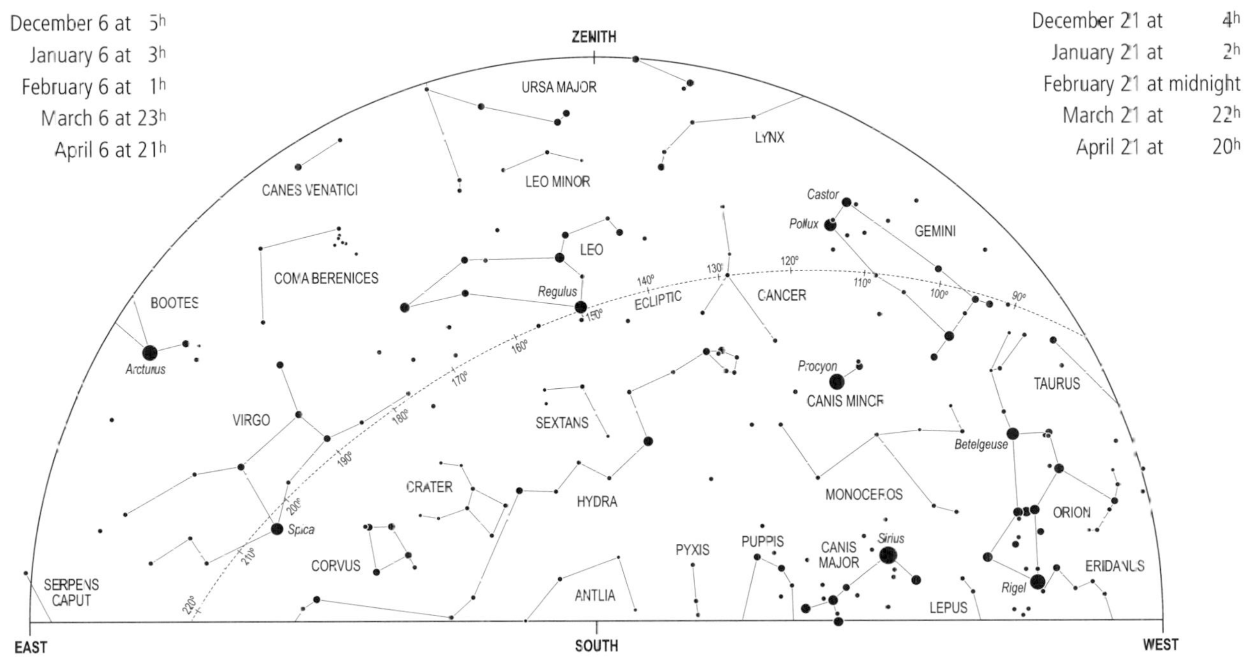
December 6 at 5h
January 6 at 3h
February 6 at 1h
March 6 at 23h
April 6 at 21h
December 21 at 4h
January 21 at 2h
February 21 at midnight
March 21 at 22h
April 21 at 20h
ZENITH
URSA MAJOR
LYNX
LEO MINOR
CANES VENATICI
Castor
Pollux
GEMINI
LEO
COMA BERENICES
BOOTES
Regulus
ECLIPTIC
CANCER
Arcturus
Procyon
CANIS MINOR
TAURUS
VIRGO
SEXTANS
Betelgeuse
CRATER
HYDRA
MONOCEROS
ORION
Spica
CORVUS
PYXIS
PUPPIS
CANIS MAJOR
Sirius
ERIDANUS
Rigel
SERPENS CAPUT
ANTLIA
LEPUS
EAST
SOUTH
WEST
90°
100°
110°
120°
130°
140°
150°
160°
170°
180°
190°
200°
210°
220°

4N

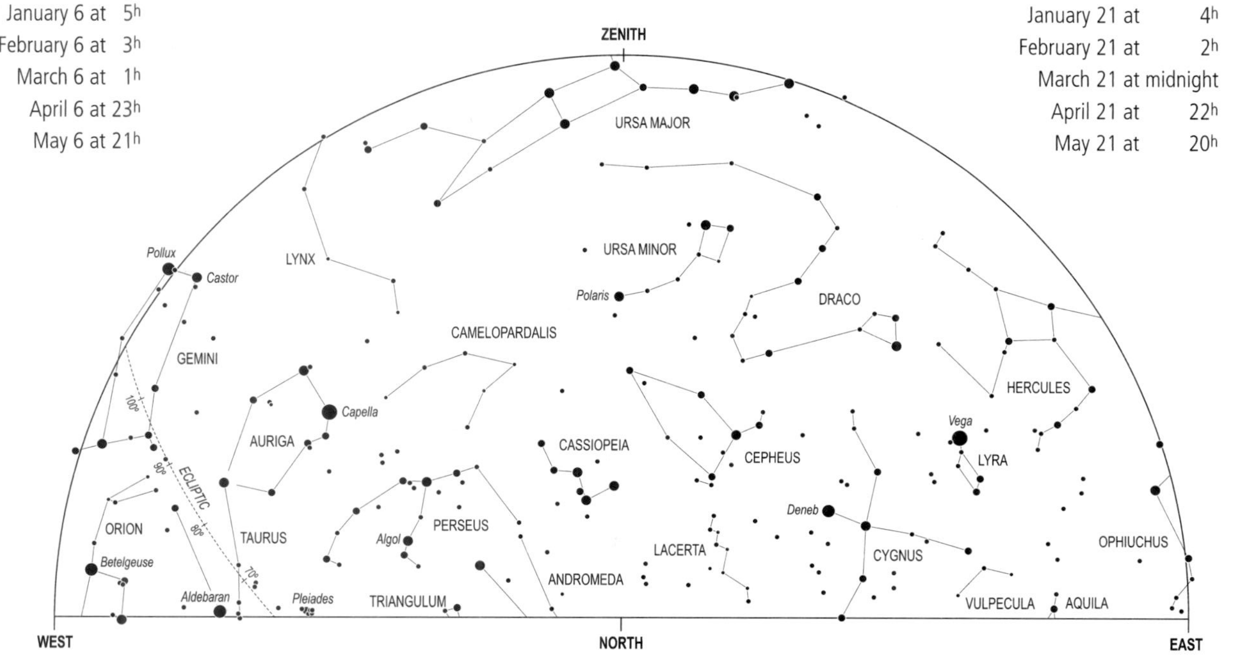
January 6 at 5h
February 6 at 3h
March 6 at 1h
April 6 at 23h
May 6 at 21h
January 21 at 4h
February 21 at 2h
March 21 at midnight
April 21 at 22h
May 21 at 20h
ZENITH
WEST
NORTH
EAST
URSA MAJOR
URSA MINOR
Polaris
LYNX
Pollux
Castor
GEMINI
CAMELOPARDALIS
DRACO
HERCULES
100°
90°
ECLIPTIC
80°
70°
Capella
AURIGA
CASSIOPEIA
CEPHEUS
Vega
LYRA
ORION
TAURUS
Algol
PERSEUS
Deneb
LACERTA
CYGNUS
OPHIUCHUS
Betelgeuse
Aldebaran
Pleiades
TRIANGULUM
ANDROMEDA
VULPECULA
AQUILA

4S

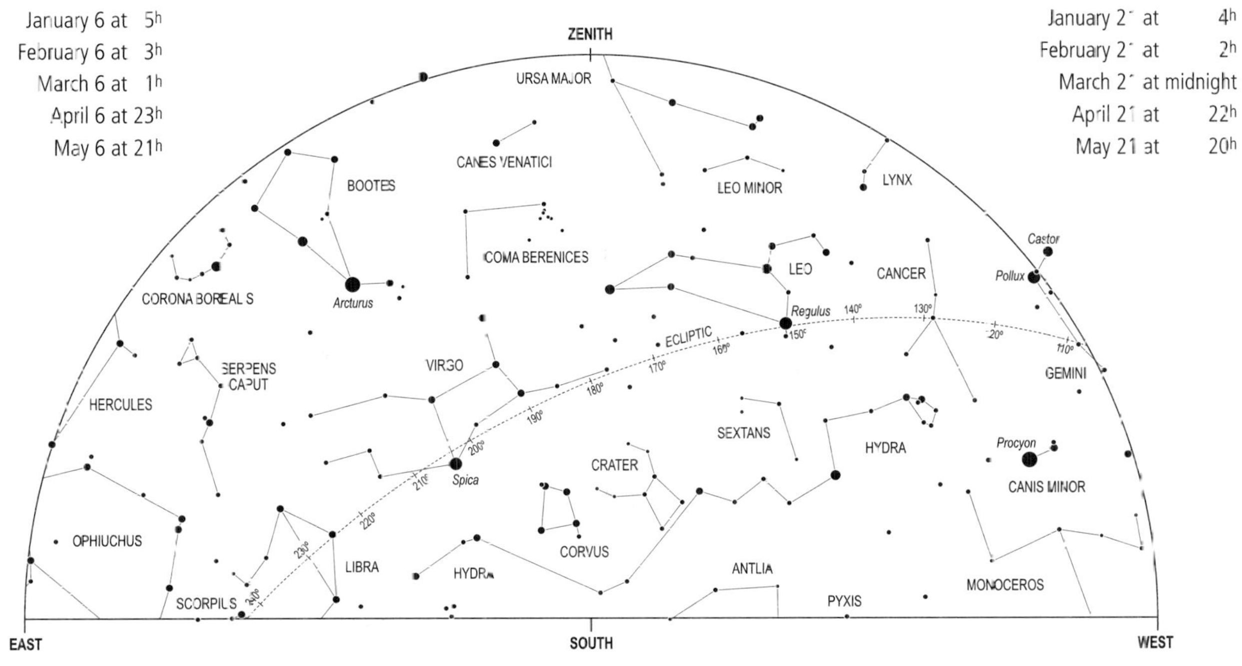
January 6 at 5h
February 6 at 3h
March 6 at 1h
April 6 at 23h
May 6 at 21h
January 21 at 4h
February 21 at 2h
March 21 at midnight
April 21 at 22h
May 21 at 20h
ZENITH
URSA MAJOR
CANES VENATICI
BOOTES
LEO MINOR
LYNX
COMA BERENICES
LEO
CANCER
Castor
Pollux
CORONA BOREALIS
Arcturus
Regulus
ECLIPTIC
GEMINI
VIRGO
SERPENS CAPUT
HERCULES
SEXTANS
HYDRA
Procyon
CRATER
Spica
CANIS MINOR
OPHIUCHUS
CORVUS
LIBRA
HYDRA
ANTLIA
MONOCEROS
SCORPIUS
PYXIS
EAST
SOUTH
WEST

5N

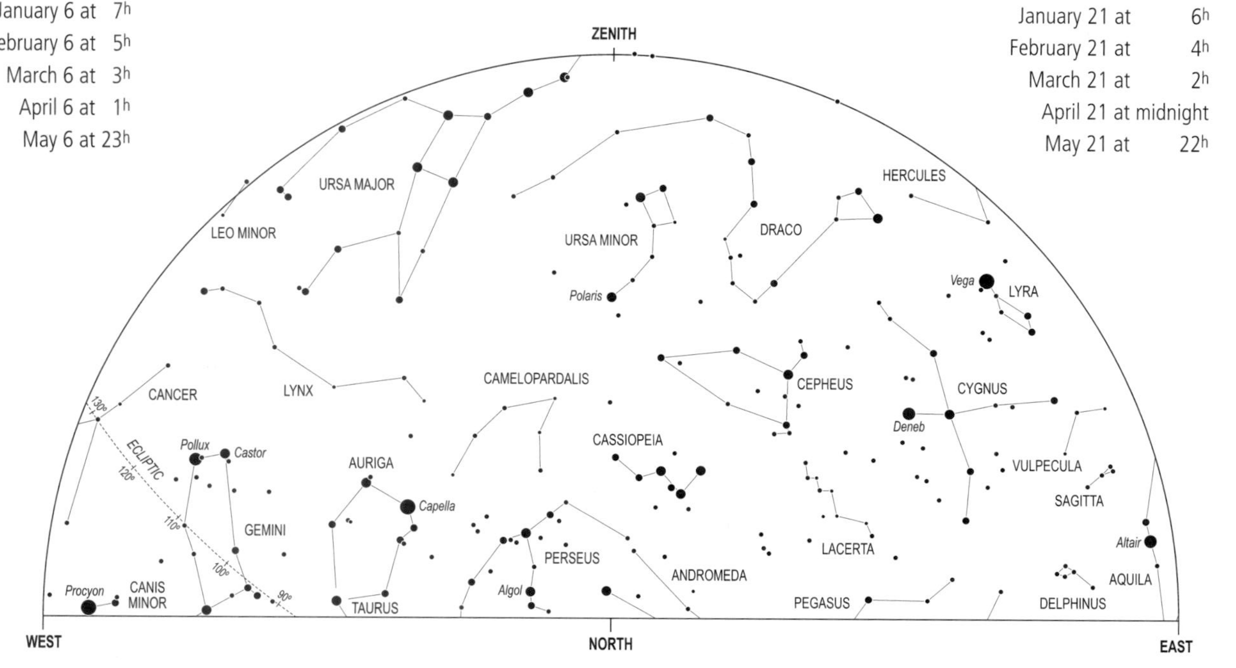
January 6 at 7h
February 6 at 5h
March 6 at 3h
April 6 at 1h
May 6 at 23h
January 21 at 6h
February 21 at 4h
March 21 at 2h
April 21 at midnight
May 21 at 22h
ZENITH
WEST
NORTH
EAST
ECLIPTIC
130°
120°
110°
100°
90°
URSA MAJOR
LEO MINOR
URSA MINOR
Polaris
DRACO
HERCULES
Vega
LYRA
CANCER
LYNX
CAMELOPARDALIS
CEPHEUS
CYGNUS
Deneb
Pollux
Castor
AURIGA
CASSIOPEIA
VULPECULA
SAGITTA
Capella
GEMINI
PERSEUS
LACERTA
Altair
ANDROMEDA
AQUILA
Procyon
CANIS MINOR
Algol
TAURUS
PEGASUS
DELPHINUS

5S

January 6 at 7h
February 6 at 5h
March 6 at 3h
April 6 at 1h
May 6 at 23h

January 21 at 6h
February 21 at 4h
March 21 at 2h
April 21 at midnight
May 21 at 22h

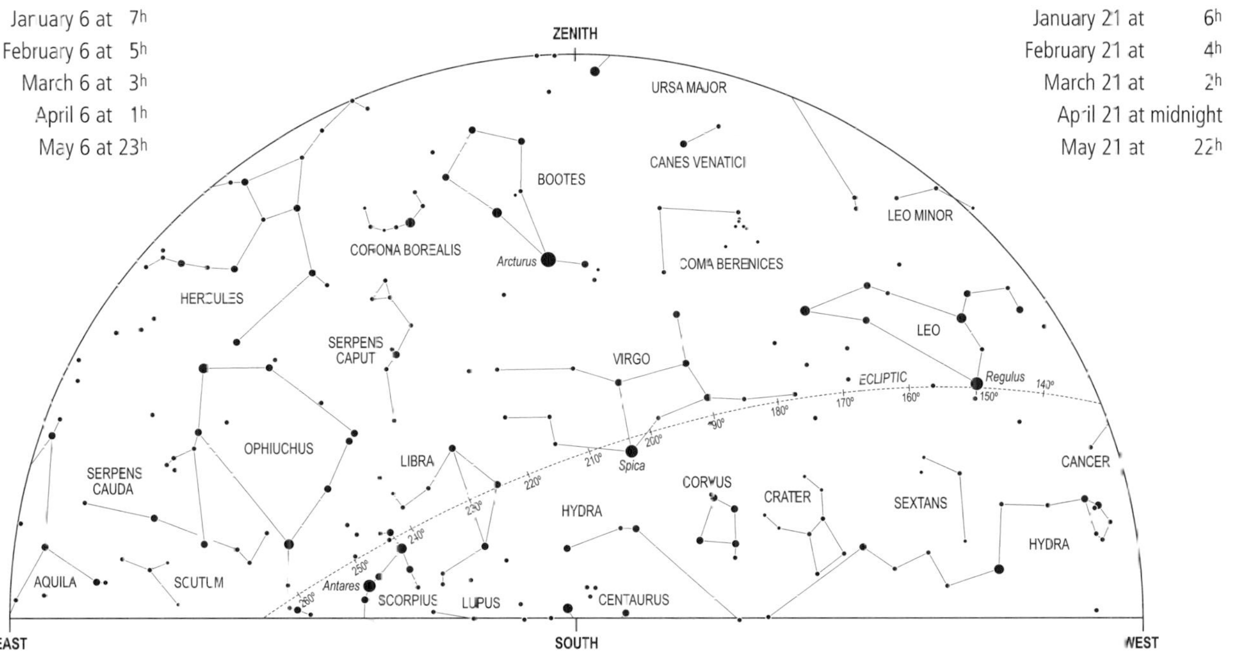

6N

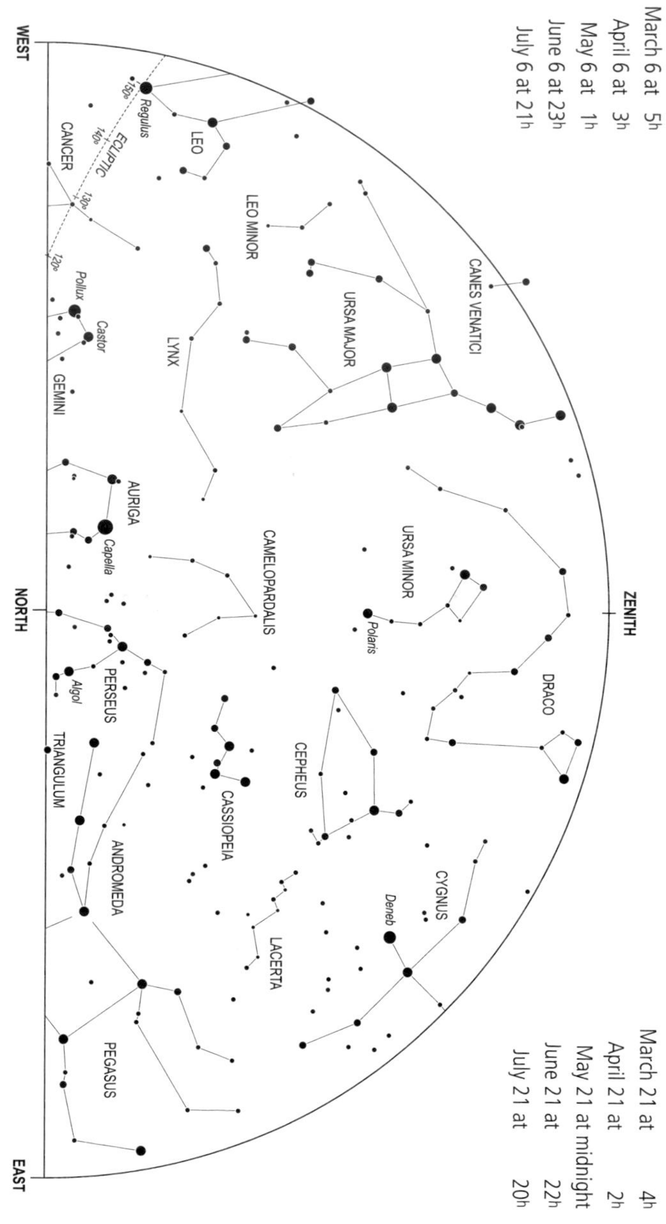
March 6 at 5h
April 6 at 3h
May 6 at 1h
June 6 at 23h
July 6 at 21h
March 21 at 4h
April 21 at 2h
May 21 at midnight
June 21 at 22h
July 21 at 20h
WEST
NORTH
EAST
ZENITH
ECLIPTIC
150°
140°
130°
120°
Regulus
LEO
CANCER
LEO MINOR
CANES VENATICI
URSA MAJOR
LYNX
Pollux
Castor
GEMINI
AURIGA
Capella
CAMELOPARDALIS
URSA MINOR
Polaris
DRACO
PERSEUS
Algol
TRIANGULUM
CEPHEUS
CASSIOPEIA
ANDROMEDA
CYGNUS
Deneb
LACERTA
PEGASUS

9S

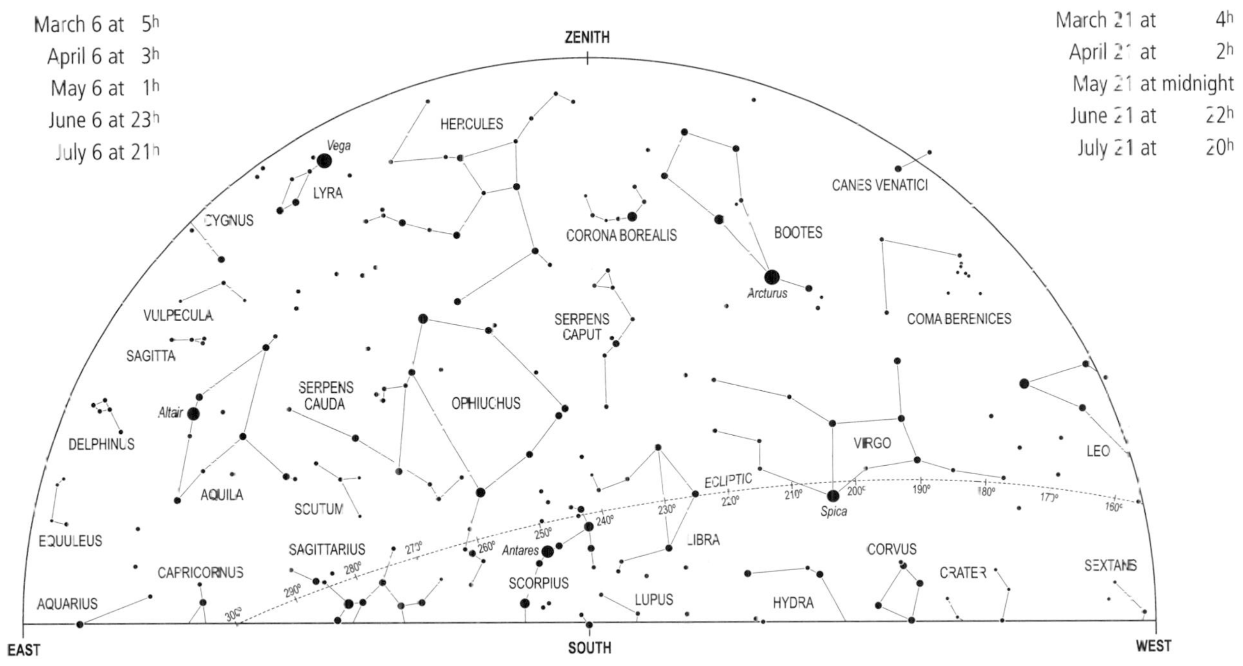
March 6 at 5h
April 6 at 3h
May 6 at 1h
June 6 at 23h
July 6 at 21h
March 21 at 4h
April 21 at 2h
May 21 at midnight
June 21 at 22h
July 21 at 20h
ZENITH
HERCULES
Vega
LYRA
CYGNUS
CORONA BOREALIS
BOOTES
CANES VENATICI
Arcturus
VULPECULA
SAGITTA
SERPENS CAPUT
COMA BERENICES
SERPENS CAUDA
OPHIUCHUS
Altair
DELPHINUS
VIRGO
LEO
AQUILA
SCUTUM
ECLIPTIC
Spica
EQUULEUS
SAGITTARIUS
Antares
LIBRA
CORVUS
CRATER
SEXTANS
CAPRICORNUS
SCORPIUS
LUPUS
HYDRA
AQUARIUS
EAST
SOUTH
WEST
300°
290°
280°
270°
260°
250°
240°
230°
220°
210°
200°
190°
180°
170°
160°

7N

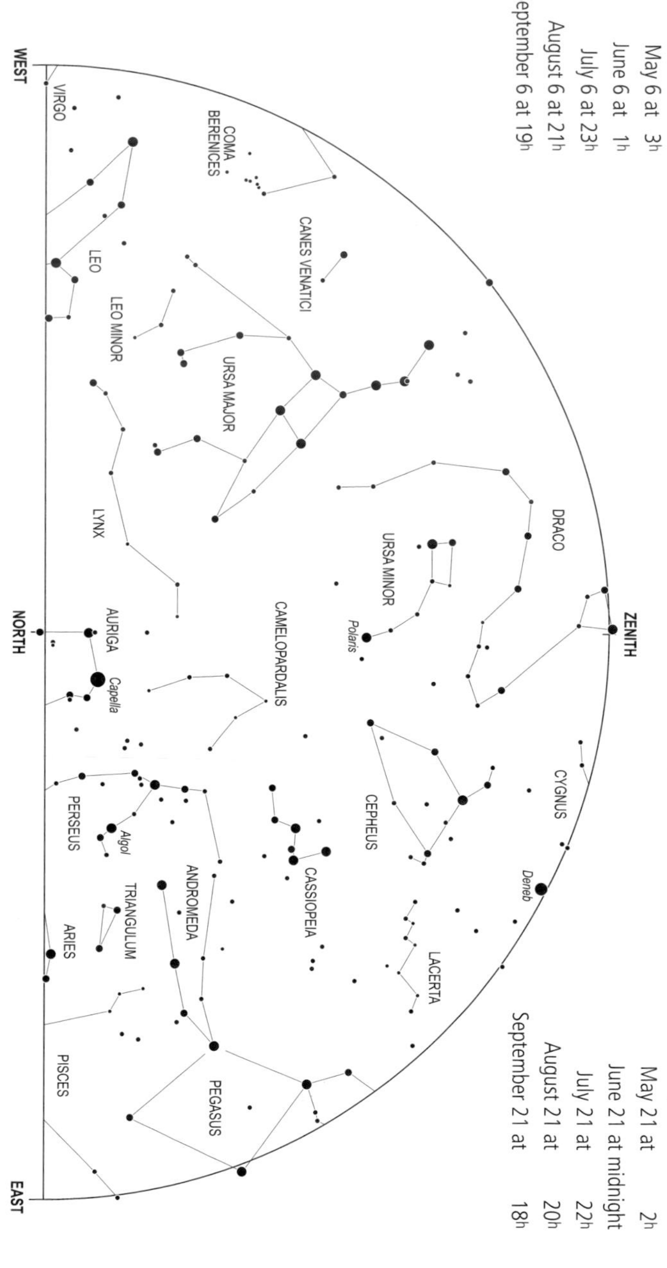
May 6 at 3h
June 6 at 1h
July 6 at 23h
August 6 at 21h
September 6 at 19h
May 21 at 2h
June 21 at midnight
July 21 at 22h
August 21 at 20h
September 21 at 18h
WEST
NORTH
EAST
ZENITH
VIRGO
COMA BERENICES
CANES VENATICI
LEO
LEO MINOR
URSA MAJOR
LYNX
DRACO
URSA MINOR
Polaris
CAMELOPARDALIS
AURIGA
Capella
CYGNUS
Deneb
CEPHEUS
CASSIOPEIA
PERSEUS
Algol
ANDROMEDA
TRIANGULUM
ARIES
LACERTA
PISCES
PEGASUS

7S

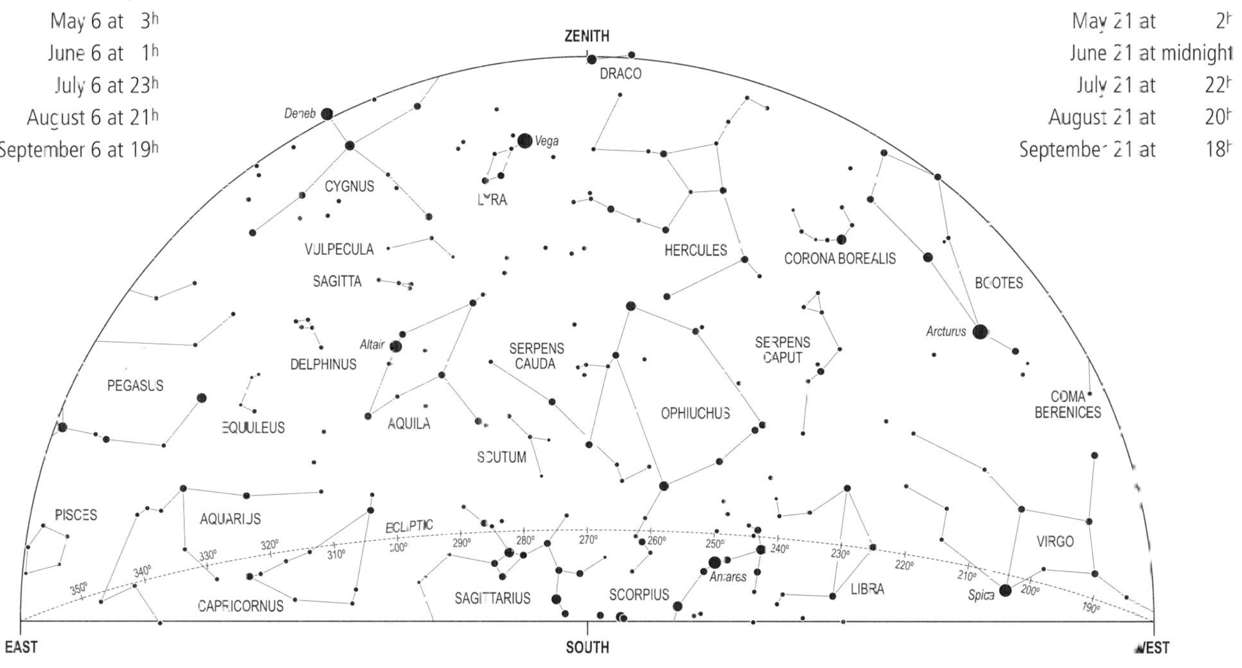
May 6 at 3h
June 6 at 1h
July 6 at 23h
August 6 at 21h
September 6 at 19h
May 21 at 2h
June 21 at midnight
July 21 at 22h
August 21 at 20h
September 21 at 18h
ZENITH
DRACO
Deneb
Vega
CYGNUS
LYRA
HERCULES
CORONA BOREALIS
BOOTES
VULPECULA
SAGITTA
Altair
DELPHINUS
SERPENS CAUDA
SERPENS CAPUT
Arcturus
PEGASUS
EQUULEUS
AQUILA
OPHIUCHUS
COMA BERENICES
SCUTUM
PISCES
AQUARIUS
ECLIPTIC
VIRGO
SAGITTARIUS
SCORPIUS
Antares
LIBRA
Spica
CAPRICORNUS
350°
340°
330°
320°
310°
300°
290°
280°
270°
260°
250°
240°
230°
220°
210°
200°
190°
EAST
SOUTH
WEST

8N

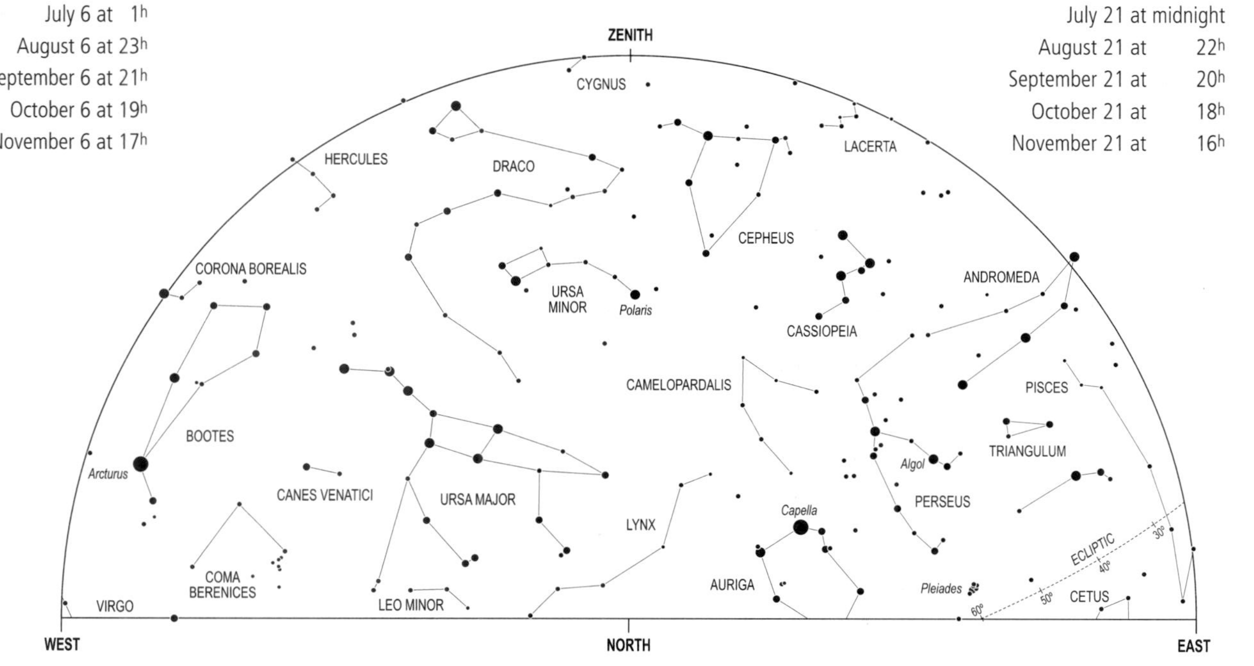
July 6 at 1h
August 6 at 23h
September 6 at 21h
October 6 at 19h
November 6 at 17h
July 21 at midnight
August 21 at 22h
September 21 at 20h
October 21 at 18h
November 21 at 16h
ZENITH
CYGNUS
HERCULES
DRACO
LACERTA
CEPHEUS
CORONA BOREALIS
URSA MINOR
Polaris
ANDROMEDA
CASSIOPEIA
CAMELOPARDALIS
PISCES
BOOTES
Arcturus
TRIANGULUM
Algol
CANES VENATICI
URSA MAJOR
PERSEUS
Capella
LYNX
ECLIPTIC
30°
40°
50°
60°
COMA BERENICES
AURIGA
Pleiades
CETUS
VIRGO
LEO MINOR
WEST
NORTH
EAST

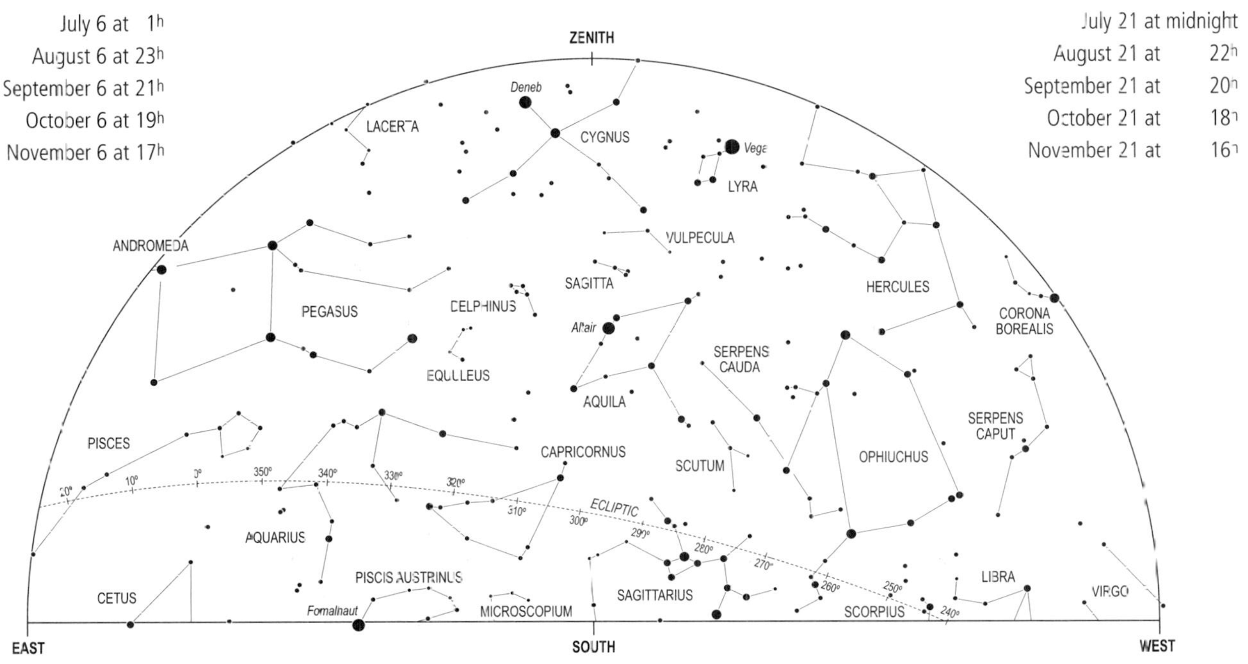
July 6 at 1^h
August 6 at 23^h
September 6 at 21^h
October 6 at 19^h
November 6 at 17^h
July 21 at midnight
August 21 at 22^h
September 21 at 20^h
October 21 at 18^h
November 21 at 16^h
ZENITH
Deneb
LACERTA
CYGNUS
Vega
LYRA
ANDROMEDA
VULPECULA
SAGITTA
PEGASUS
DELPHINUS
Altair
HERCULES
CORONA BOREALIS
EQUULEUS
SERPENS CAUDA
AQUILA
SERPENS CAPUT
PISCES
CAPRICORNUS
SCUTUM
OPHIUCHUS
ECLIPTIC
AQUARIUS
PISCIS AUSTRINUS
LIBRA
VIRGO
CETUS
Fomalhaut
MICROSCOPIUM
SAGITTARIUS
SCORPIUS
EAST
SOUTH
WEST
20°
10°
0°
350°
340°
330°
320°
310°
300°
290°
280°
270°
260°
250°
240°

8S

9N

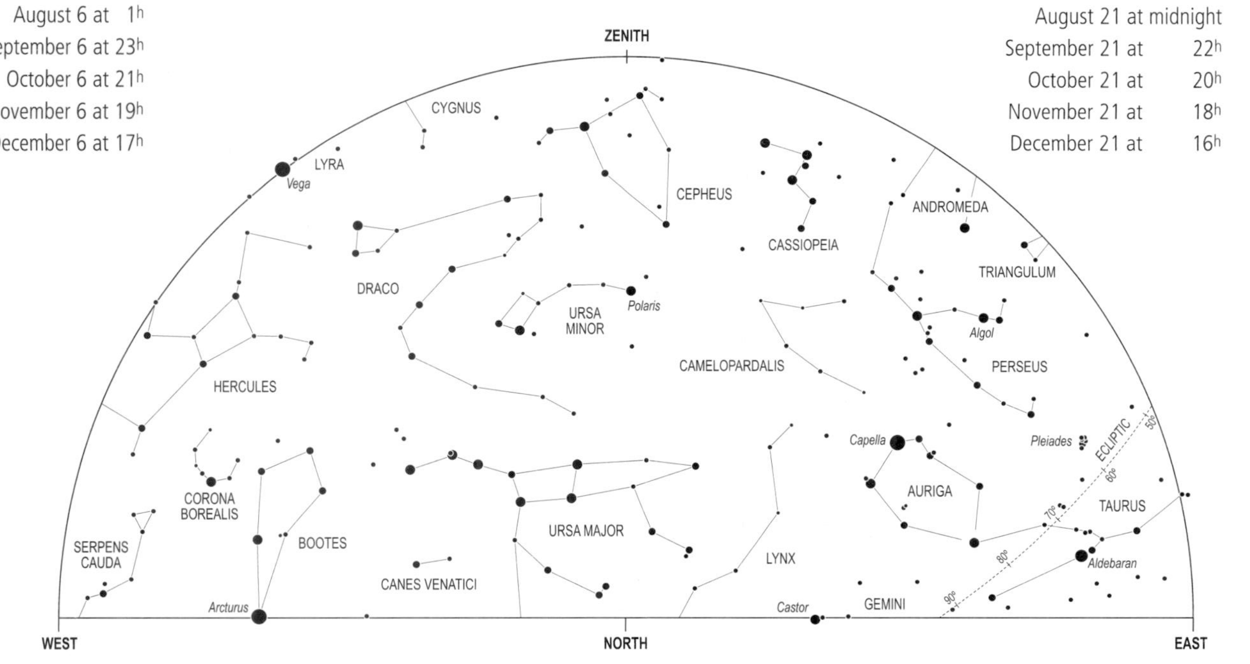
August 6 at 1h
September 6 at 23h
October 6 at 21h
November 6 at 19h
December 6 at 17h
ZENITH
August 21 at midnight
September 21 at 22h
October 21 at 20h
November 21 at 18h
December 21 at 16h
CYGNUS
LYRA
Vega
CEPHEUS
CASSIOPEIA
ANDROMEDA
TRIANGULUM
DRACO
URSA MINOR
Polaris
Algol
CAMELOPARDALIS
PERSEUS
HERCULES
Capella
Pleiades
ECLIPTIC
50°
60°
70°
80°
90°
AURIGA
TAURUS
CORONA BOREALIS
BOOTES
URSA MAJOR
SERPENS CAUDA
LYNX
CANES VENATICI
Aldebaran
Arcturus
Castor
GEMINI
WEST
NORTH
EAST

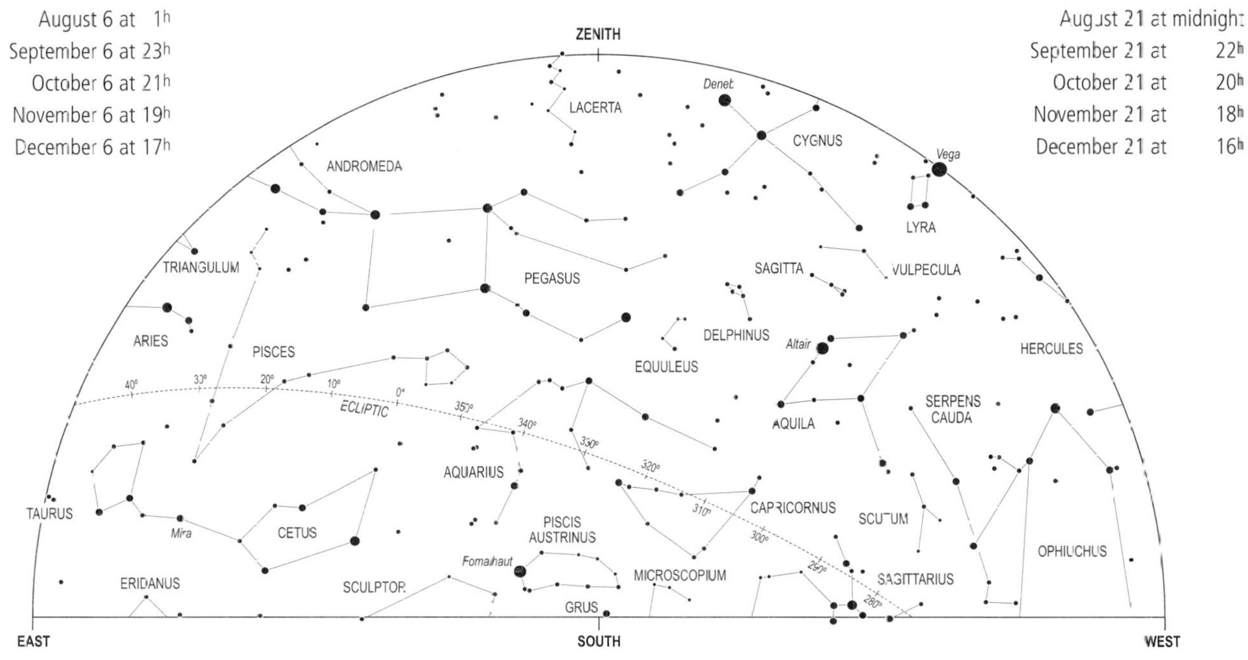
August 6 at 1h
September 6 at 23h
October 6 at 21h
November 6 at 19h
December 6 at 17h
August 21 at midnight
September 21 at 22h
October 21 at 20h
November 21 at 18h
December 21 at 16h
ZENITH
EAST
SOUTH
WEST
LACERTA
Deneb
CYGNUS
Vega
ANDROMEDA
LYRA
TRIANGULUM
PEGASUS
SAGITTA
VULPECULA
ARIES
PISCES
DELPHINUS
Altair
HERCULES
EQUULEUS
40°
30°
20°
10°
0°
ECLIPTIC
350°
340°
330°
320°
310°
300°
290°
280°
AQUILA
SERPENS CAUDA
AQUARIUS
TAURUS
CAPRICORNUS
Mira
CETUS
PISCIS AUSTRINUS
SCUTUM
OPHIUCHUS
Fomalhaut
ERIDANUS
SCULPTOR
MICROSCOPIUM
SAGITTARIUS
GRUS

S6

10N

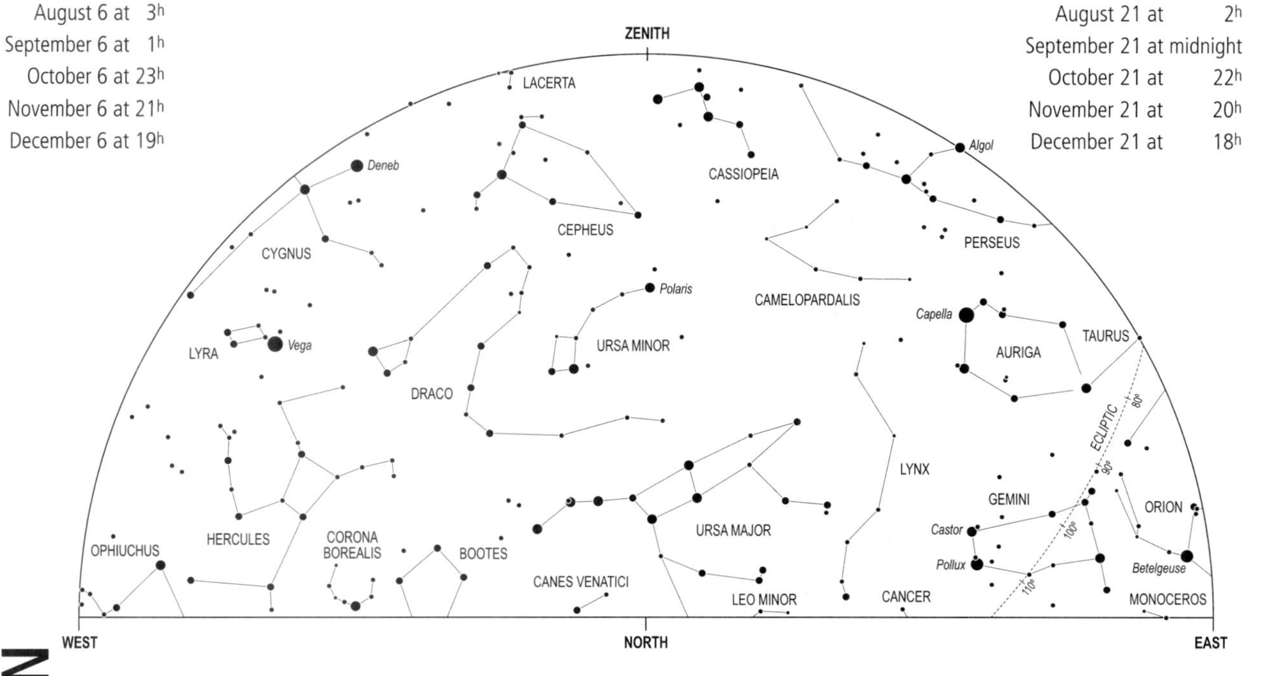
August 6 at 3h
September 6 at 1h
October 6 at 23h
November 6 at 21h
December 6 at 19h
August 21 at 2h
September 21 at midnight
October 21 at 22h
November 21 at 20h
December 21 at 18h
ZENITH
LACERTA
CASSIOPEIA
Algol
Deneb
CEPHEUS
PERSEUS
CYGNUS
Polaris
CAMELOPARDALIS
Capella
TAURUS
LYRA
Vega
URSA MINOR
AURIGA
DRACO
ECLIPTIC
80°
90°
LYNX
GEMINI
ORION
Castor
100°
URSA MAJOR
OPHIUCHUS
HERCULES
CORONA BOREALIS
BOOTES
Pollux
Betelgeuse
110°
CANES VENATICI
LEO MINOR
CANCER
MONOCEROS
WEST
NORTH
EAST

10S

August 6 at 3h	August 21 at 2h
September 6 at 1h	September 21 at midnight
October 6 at 23h	October 21 at 22h
November 6 at 21h	November 21 at 20h
December 6 at 19h	December 21 at 18h

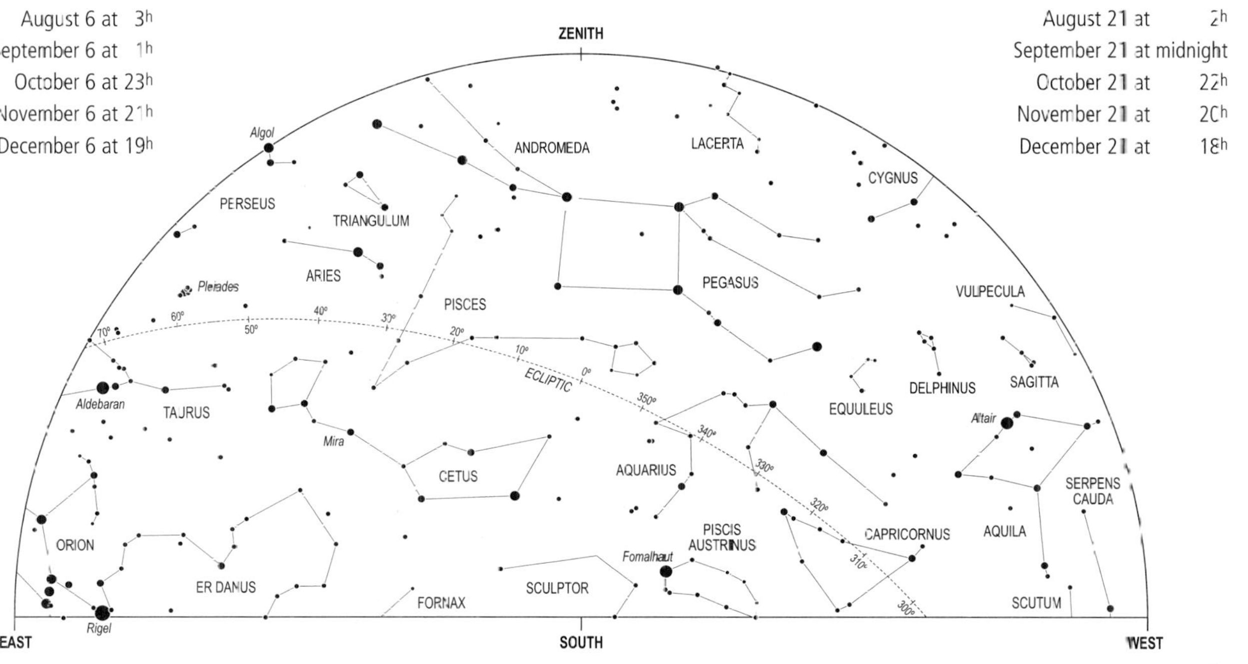

11N

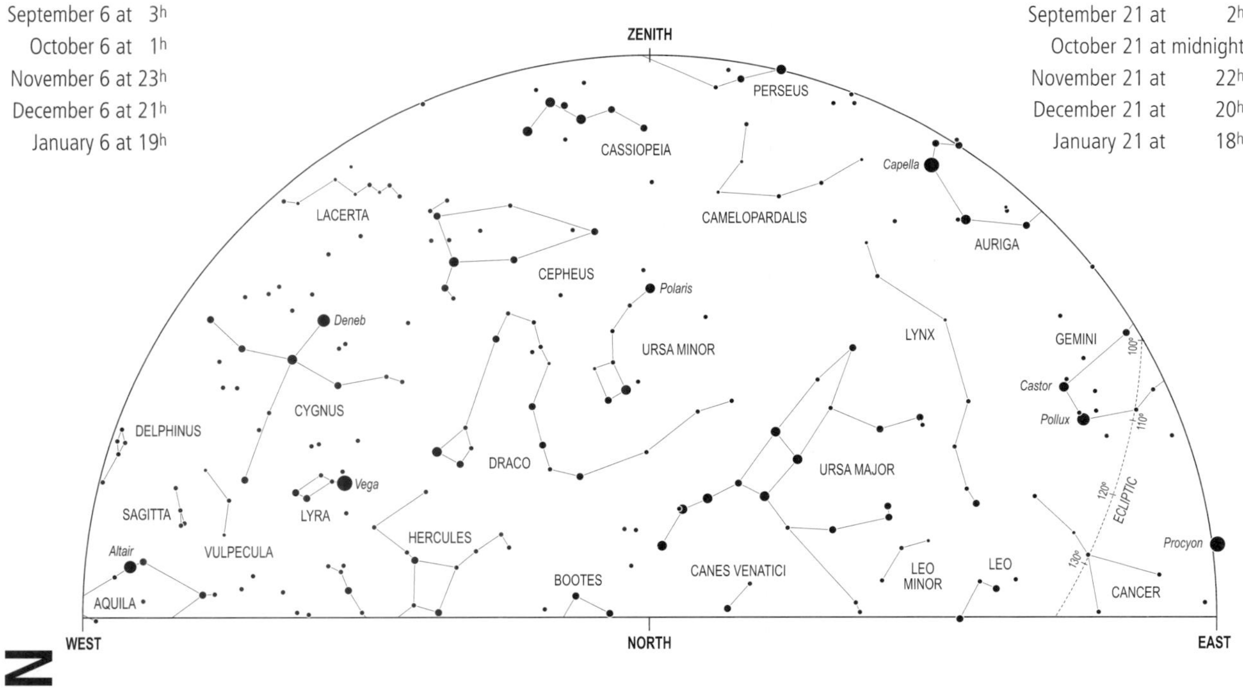
September 6 at 3h
October 6 at 1h
November 6 at 23h
December 6 at 21h
January 6 at 19h
September 21 at 2h
October 21 at midnight
November 21 at 22h
December 21 at 20h
January 21 at 18h
ZENITH
WEST
NORTH
EAST
PERSEUS
CASSIOPEIA
CAMELOPARDALIS
Capella
AURIGA
LACERTA
CEPHEUS
Polaris
URSA MINOR
Deneb
CYGNUS
LYNX
GEMINI
Castor
Pollux
DELPHINUS
DRACO
URSA MAJOR
Vega
SAGITTA
LYRA
HERCULES
Altair
VULPECULA
AQUILA
BOOTES
CANES VENATICI
LEO MINOR
LEO
CANCER
Procyon
ECLIPTIC
100°
110°
120°
130°

11S

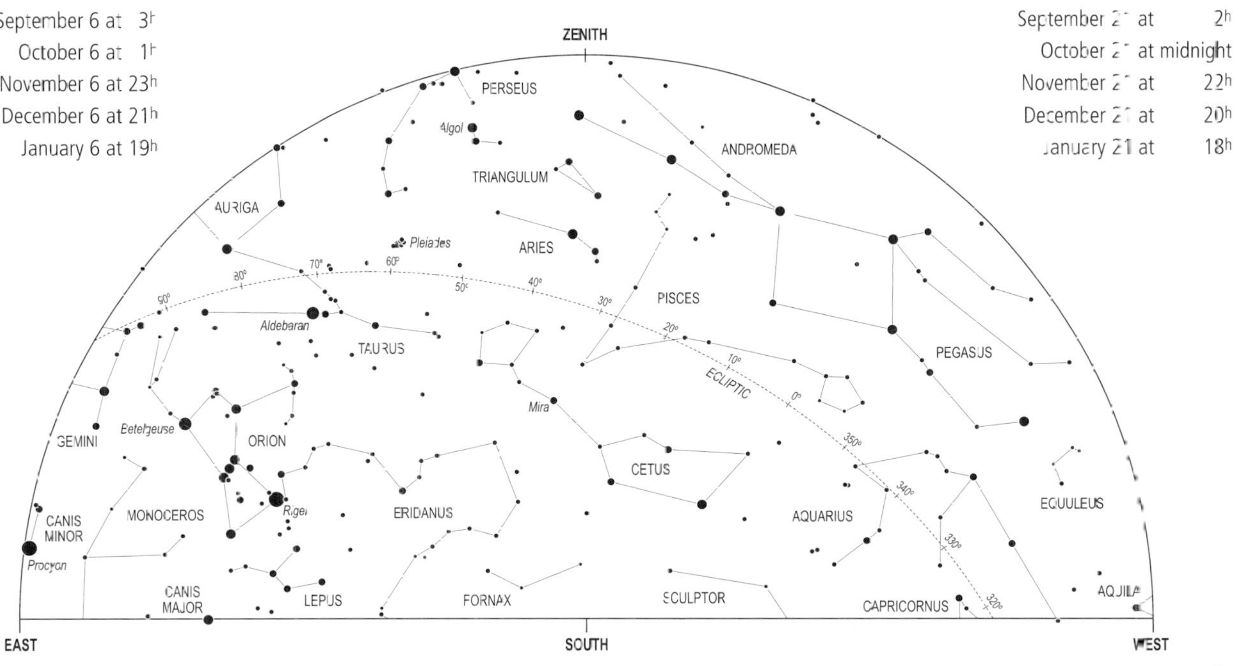
September 6 at 3h
October 6 at 1h
November 6 at 23h
December 6 at 21h
January 6 at 19h
September 21 at 2h
October 21 at midnight
November 21 at 22h
December 21 at 20h
January 21 at 18h
ZENITH
EAST
SOUTH
WEST
PERSEUS
Algol
TRIANGULUM
ANDROMEDA
AURIGA
Pleiades
ARIES
PISCES
PEGASUS
Aldebaran
TAURUS
Mira
ECLIPTIC
GEMINI
Betelgeuse
ORION
CETUS
CANIS MINOR
Procyon
MONOCEROS
Rigel
ERIDANUS
AQUARIUS
EQUULEUS
CANIS MAJOR
LEPUS
FORNAX
SCULPTOR
CAPRICORNUS
AQUILA
90°
80°
70°
60°
50°
40°
30°
20°
10°
0°
350°
340°
330°
320°

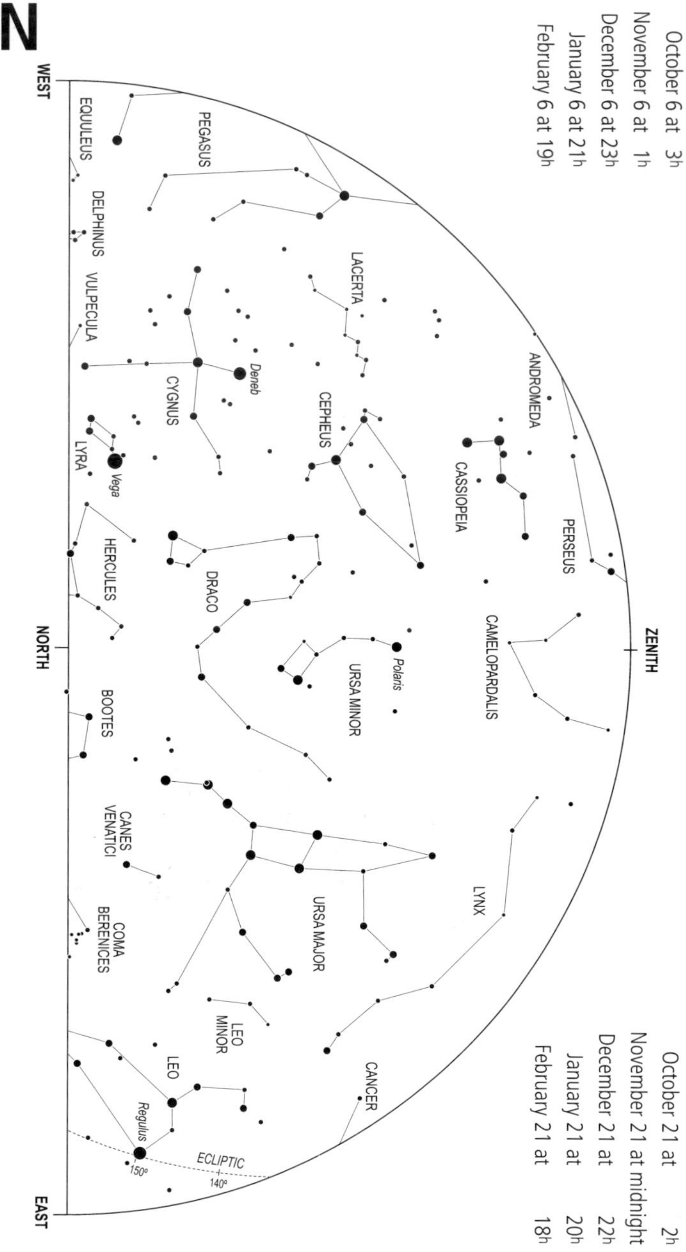
12N
October 6 at 3h
November 6 at 1h
December 6 at 23h
January 6 at 21h
February 6 at 19h
October 21 at 2h
November 21 at midnight
December 21 at 22h
January 21 at 20h
February 21 at 18h
WEST
NORTH
EAST
ZENITH
EQUULEUS
PEGASUS
DELPHINUS
VULPECULA
LACERTA
ANDROMEDA
CYGNUS
Deneb
CEPHEUS
CASSIOPEIA
PERSEUS
LYRA
Vega
HERCULES
DRACO
CAMELOPARDALIS
Polaris
URSA MINOR
BOOTES
CANES VENATICI
URSA MAJOR
LYNX
COMA BERENICES
LEO MINOR
LEO
Regulus
CANCER
ECLIPTIC
150°
140°

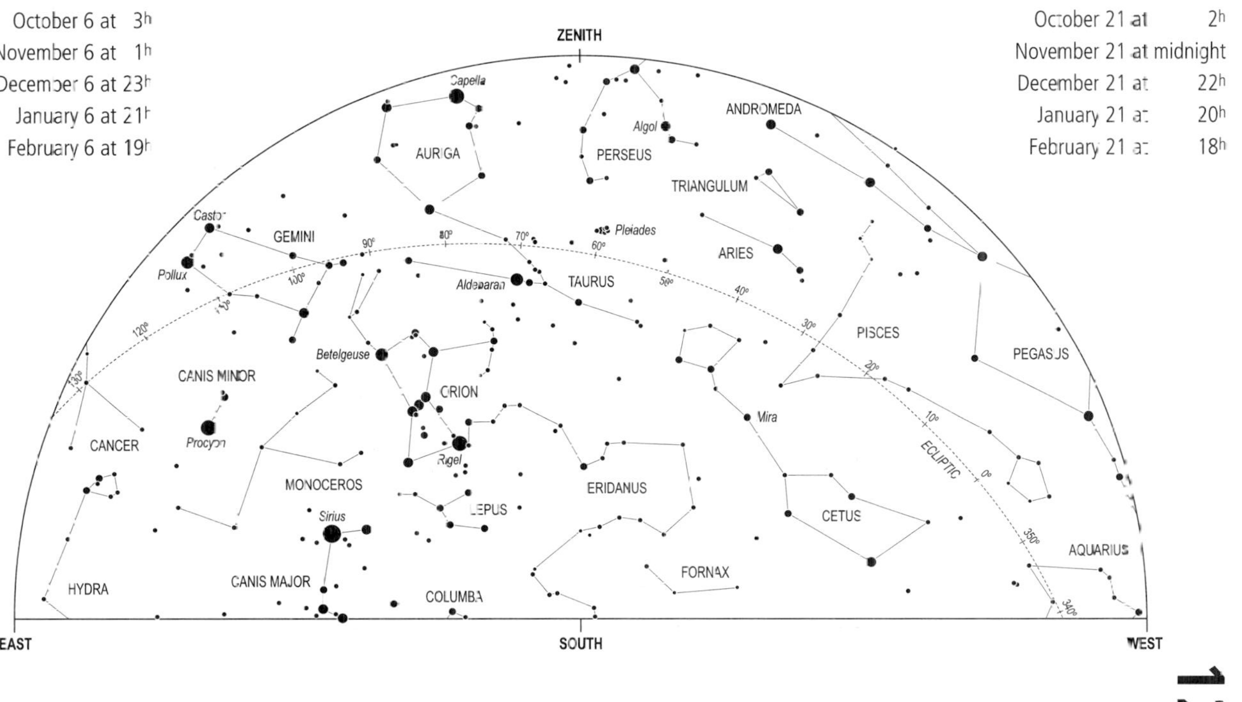
October 6 at 3h
November 6 at 1h
December 6 at 23h
January 6 at 21h
February 6 at 19h
October 21 at 2h
November 21 at midnight
December 21 at 22h
January 21 at 20h
February 21 at 18h
ZENITH
Capella
AURIGA
Algol
PERSEUS
ANDROMEDA
TRIANGULUM
Castor
GEMINI
Pleiades
ARIES
Pollux
Aldebaran
TAURUS
PISCES
PEGASUS
Betelgeuse
CANIS MINOR
ORION
Mira
CANCER
Procyon
Rigel
ECLIPTIC
MONOCEROS
ERIDANUS
LEPUS
CETUS
Sirius
AQUARIUS
FORNAX
HYDRA
CANIS MAJOR
COLUMBA
EAST
SOUTH
WEST
12S

Southern Star Charts

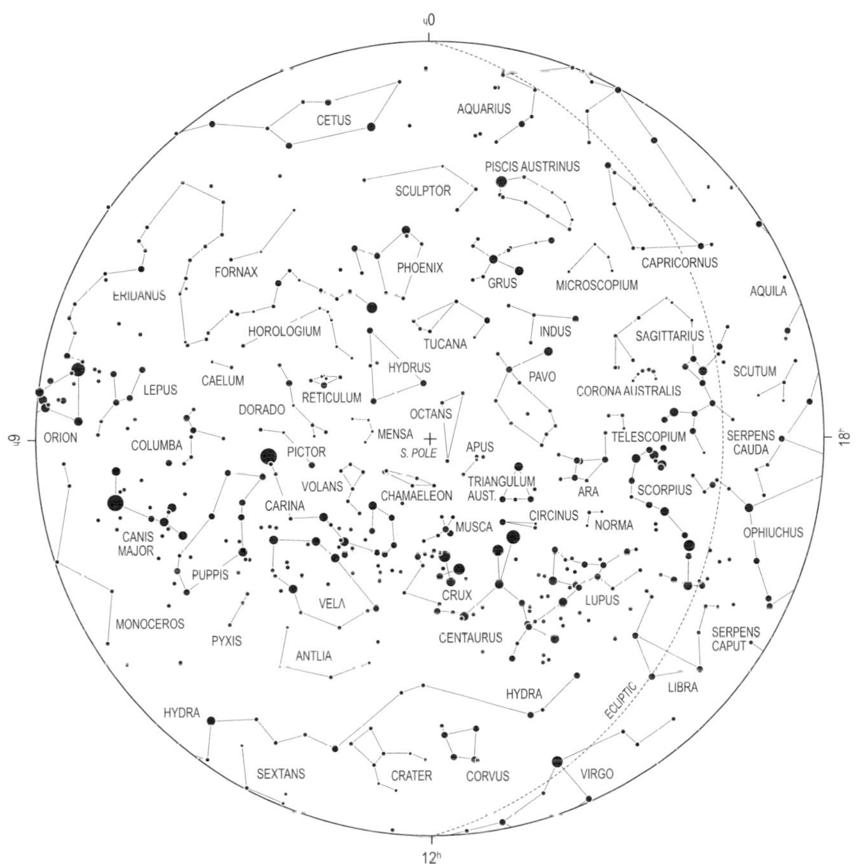

Southern Hemisphere

Note that the markers at 0^h, 6^h, 12^h and 18^h indicate hours of Right Ascension.

1N

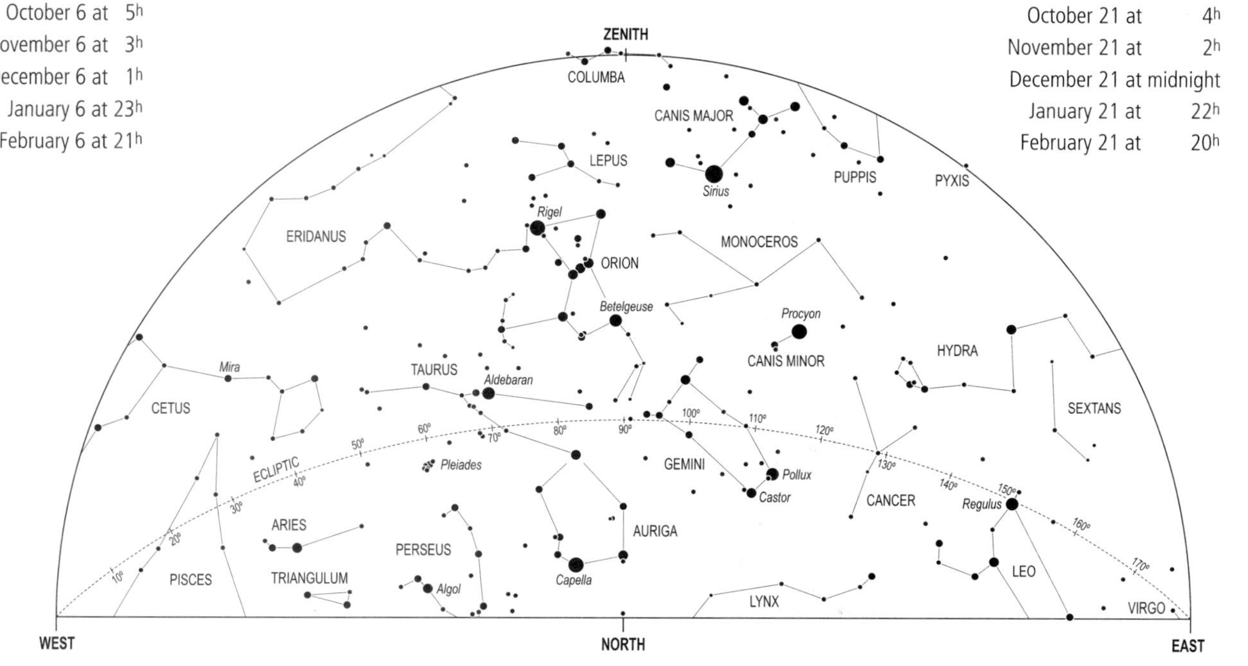
October 6 at 5h
November 6 at 3h
December 6 at 1h
January 6 at 23h
February 6 at 21h
October 21 at 4h
November 21 at 2h
December 21 at midnight
January 21 at 22h
February 21 at 20h
ZENITH
COLUMBA
CANIS MAJOR
LEPUS
Sirius
PUPPIS
PYXIS
ERIDANUS
Rigel
ORION
MONOCEROS
Betelgeuse
Procyon
CANIS MINOR
HYDRA
Mira
CETUS
TAURUS
Aldebaran
SEXTANS
ECLIPTIC
Pleiades
GEMINI
Pollux
Castor
CANCER
Regulus
ARIES
PERSEUS
AURIGA
PISCES
TRIANGULUM
Algol
Capella
LEO
LYNX
VIRGO
WEST
NORTH
EAST
10°
20°
30°
40°
50°
60°
70°
80°
90°
100°
110°
120°
130°
140°
150°
160°
170°

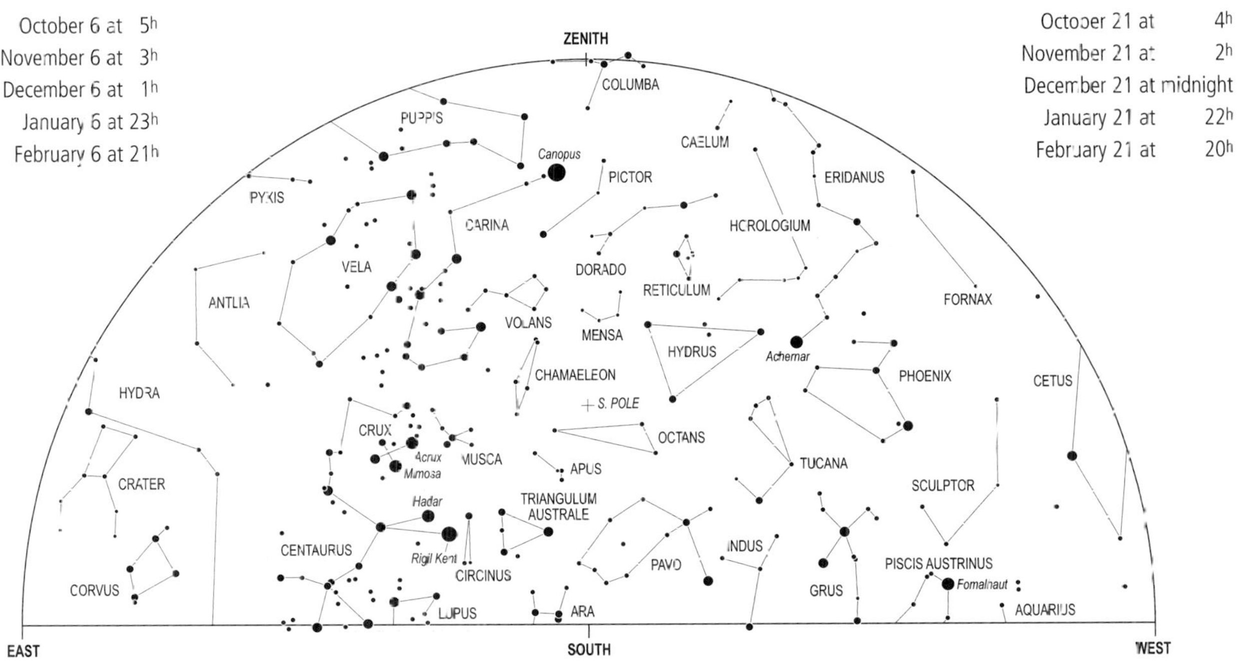
October 6 at 5h
November 6 at 3h
December 6 at 1h
January 6 at 23h
February 6 at 21h
October 21 at 4h
November 21 at 2h
December 21 at midnight
January 21 at 22h
February 21 at 20h
ZENITH
COLUMBA
PUPPIS
CAELUM
Canopus
PICTOR
ERIDANUS
PYXIS
CARINA
HOROLOGIUM
VELA
DORADO
RETICULUM
FORNAX
ANTLIA
VOLANS
MENSA
HYDRUS
Achernar
CHAMAELEON
PHOENIX
CETUS
HYDRA
S. POLE
CRUX
OCTANS
Acrux
Mimosa
MUSCA
APUS
TUCANA
CRATER
SCULPTOR
Hadar
TRIANGULUM AUSTRALE
CENTAURUS
INDUS
Rigil Kent
PAVO
PISCIS AUSTRINUS
CIRCINUS
Fomalhaut
CORVUS
GRUS
LUPUS
ARA
AQUARIUS
EAST
SOUTH
WEST

1S

2N

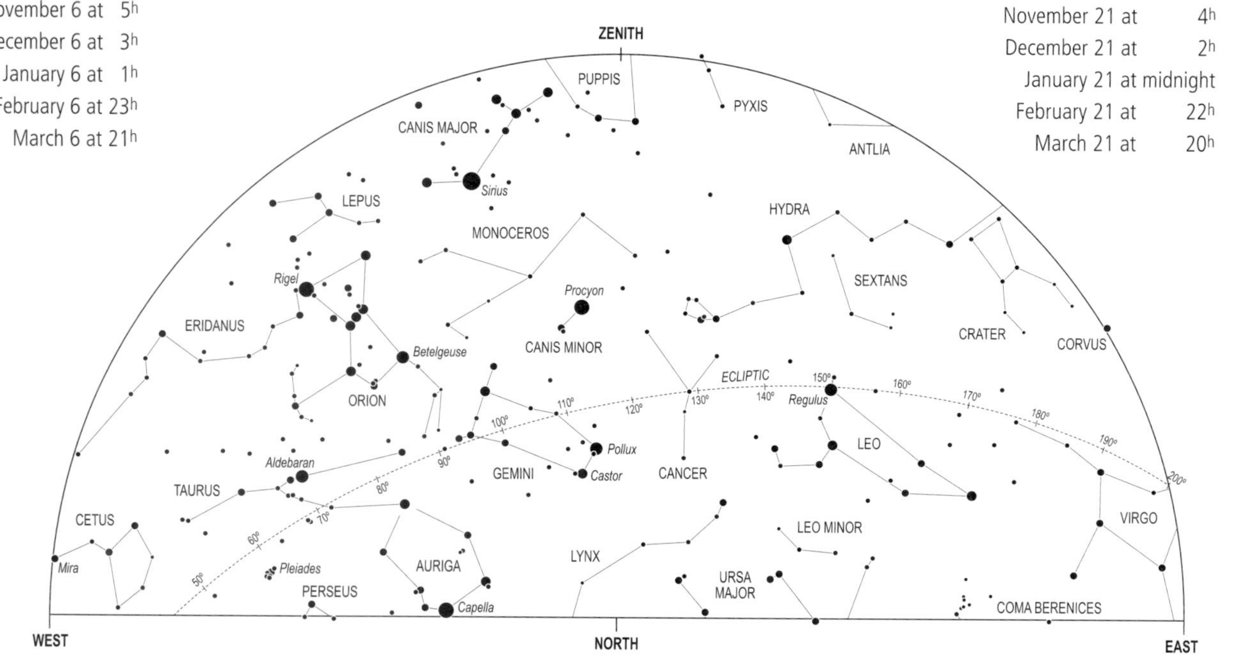
November 6 at 5h
December 6 at 3h
January 6 at 1h
February 6 at 23h
March 6 at 21h
November 21 at 4h
December 21 at 2h
January 21 at midnight
February 21 at 22h
March 21 at 20h
ZENITH
PUPPIS
PYXIS
CANIS MAJOR
ANTLIA
Sirius
LEPUS
HYDRA
MONOCEROS
SEXTANS
Rigel
Procyon
ERIDANUS
CRATER
CORVUS
Betelgeuse
CANIS MINOR
ECLIPTIC
ORION
Regulus
Pollux
LEO
Aldebaran
GEMINI
Castor
CANCER
TAURUS
CETUS
LEO MINOR
VIRGO
LYNX
Mira
Pleiades
AURIGA
URSA MAJOR
PERSEUS
Capella
COMA BERENICES
WEST
NORTH
EAST
50°
60°
70°
80°
90°
100°
110°
120°
130°
140°
150°
160°
170°
180°
190°
200°

2S

November 6 at 5h
December 6 at 3h
January 6 at 1h
February 6 at 23h
March 6 at 21h

November 21 at 4h
December 21 at 2h
January 21 at midnight
February 21 at 22h
March 21 at 20h

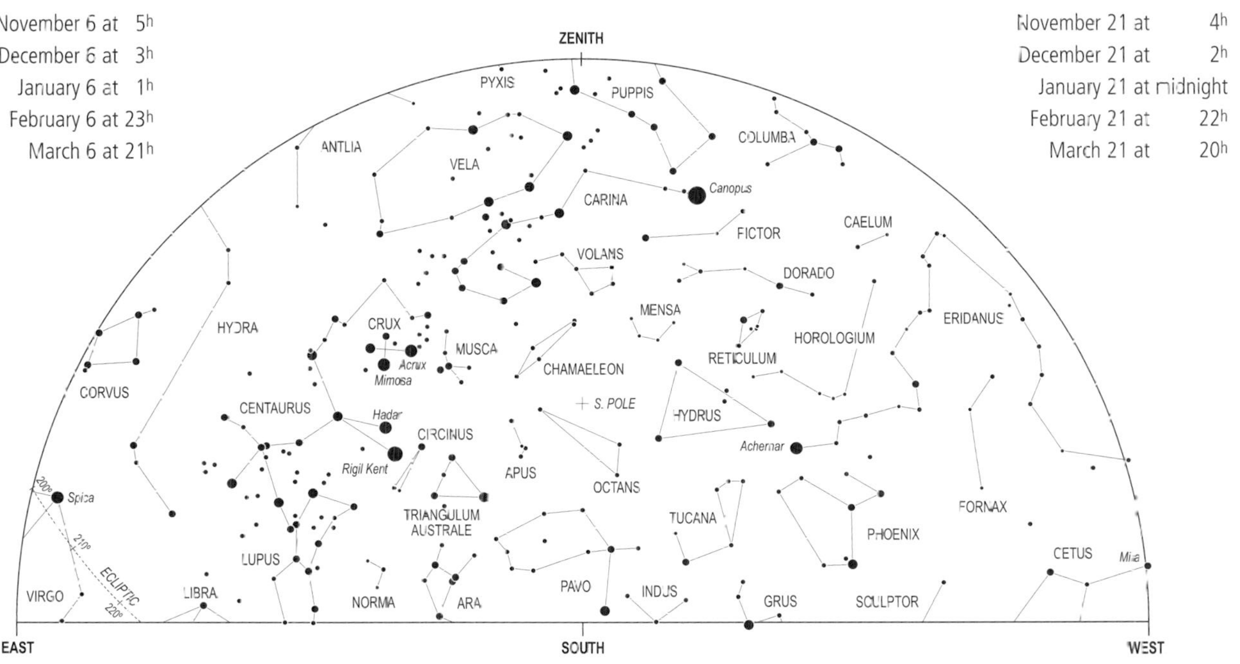

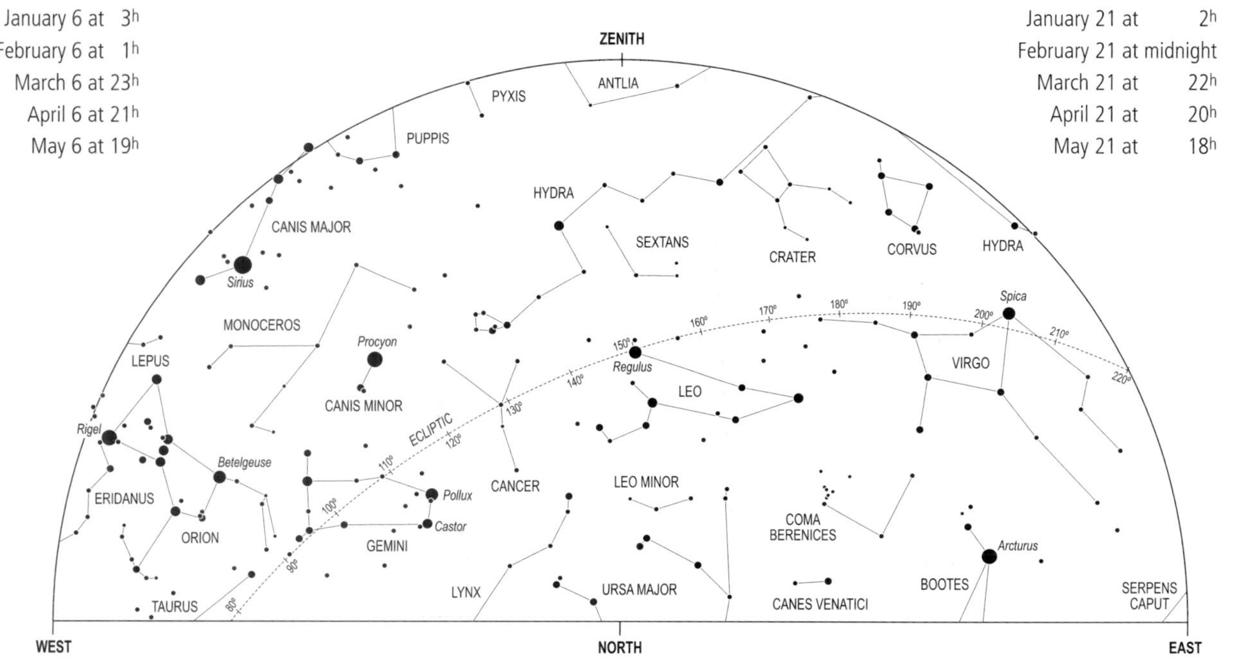
3N
January 6 at 3h
February 6 at 1h
March 6 at 23h
April 6 at 21h
May 6 at 19h
January 21 at 2h
February 21 at midnight
March 21 at 22h
April 21 at 20h
May 21 at 18h
ZENITH
WEST
NORTH
EAST
PYXIS
ANTLIA
PUPPIS
HYDRA
CANIS MAJOR
Sirius
SEXTANS
CRATER
CORVUS
HYDRA
Spica
MONOCEROS
Procyon
LEPUS
CANIS MINOR
Regulus
LEO
VIRGO
Rigel
ECLIPTIC
Betelgeuse
Pollux
Castor
CANCER
LEO MINOR
ERIDANUS
ORION
GEMINI
COMA
BERENICES
Arcturus
LYNX
URSA MAJOR
CANES VENATICI
BOOTES
SERPENS
CAPUT
TAURUS
80°
90°
100°
110°
120°
130°
140°
150°
160°
170°
180°
190°
200°
210°
220°

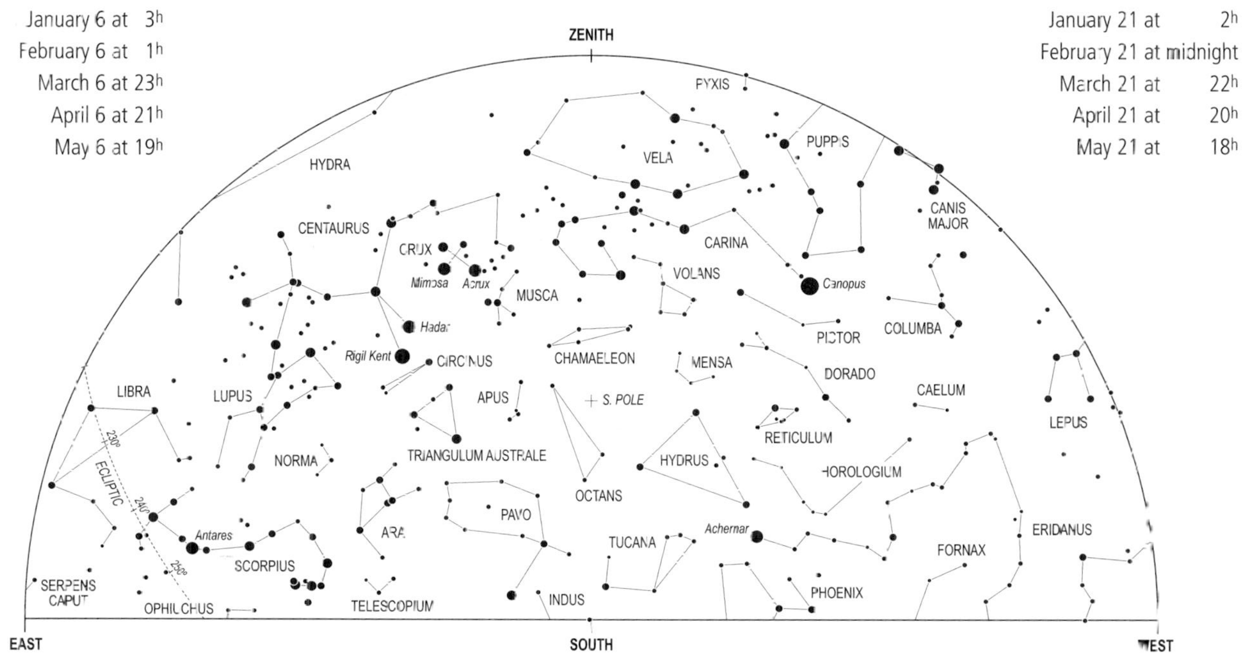
January 6 at 3h
February 6 at 1h
March 6 at 23h
April 6 at 21h
May 6 at 19h
January 21 at 2h
February 21 at midnight
March 21 at 22h
April 21 at 20h
May 21 at 18h
ZENITH
EAST
SOUTH
WEST
HYDRA
CENTAURUS
CRUX
Mimosa
Acrux
MUSCA
Hadar
Rigil Kent
CIRCINUS
LIBRA
LUPUS
NORMA
APUS
TRIANGULUM AUSTRALE
CHAMAELEON
S. POLE
MENSA
OCTANS
HYDRUS
PYXIS
VELA
PUPPIS
CARINA
VOLANS
Canopus
CANIS MAJOR
COLUMBA
PICTOR
DORADO
CAELUM
RETICULUM
HOROLOGIUM
LEPUS
ERIDANUS
FORNAX
PHOENIX
Achernar
TUCANA
PAVO
INDUS
ARA
TELESCOPIUM
SCORPIUS
Antares
OPHIUCHUS
SERPENS CAPUT
ECLIPTIC
230°
240°
250°

3S

4N

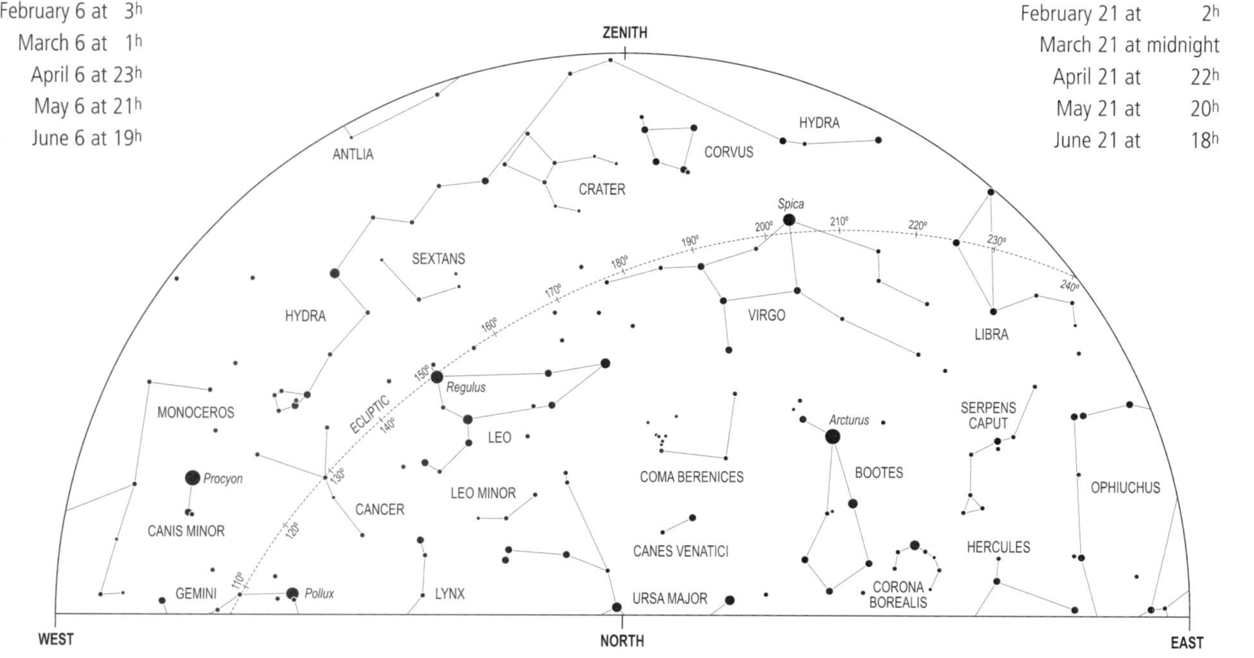
February 6 at 3h
March 6 at 1h
April 6 at 23h
May 6 at 21h
June 6 at 19h
February 21 at 2h
March 21 at midnight
April 21 at 22h
May 21 at 20h
June 21 at 18h
ZENITH
ANTLIA
CRATER
CORVUS
HYDRA
Spica
SEXTANS
HYDRA
VIRGO
LIBRA
ECLIPTIC
Regulus
MONOCEROS
LEO
Arcturus
SERPENS CAPUT
Procyon
CANCER
LEO MINOR
COMA BERENICES
BOOTES
OPHIUCHUS
CANIS MINOR
CANES VENATICI
HERCULES
GEMINI
Pollux
LYNX
URSA MAJOR
CORONA BOREALIS
WEST
NORTH
EAST
110°
120°
130°
140°
150°
160°
170°
180°
190°
200°
210°
220°
230°
240°

4S

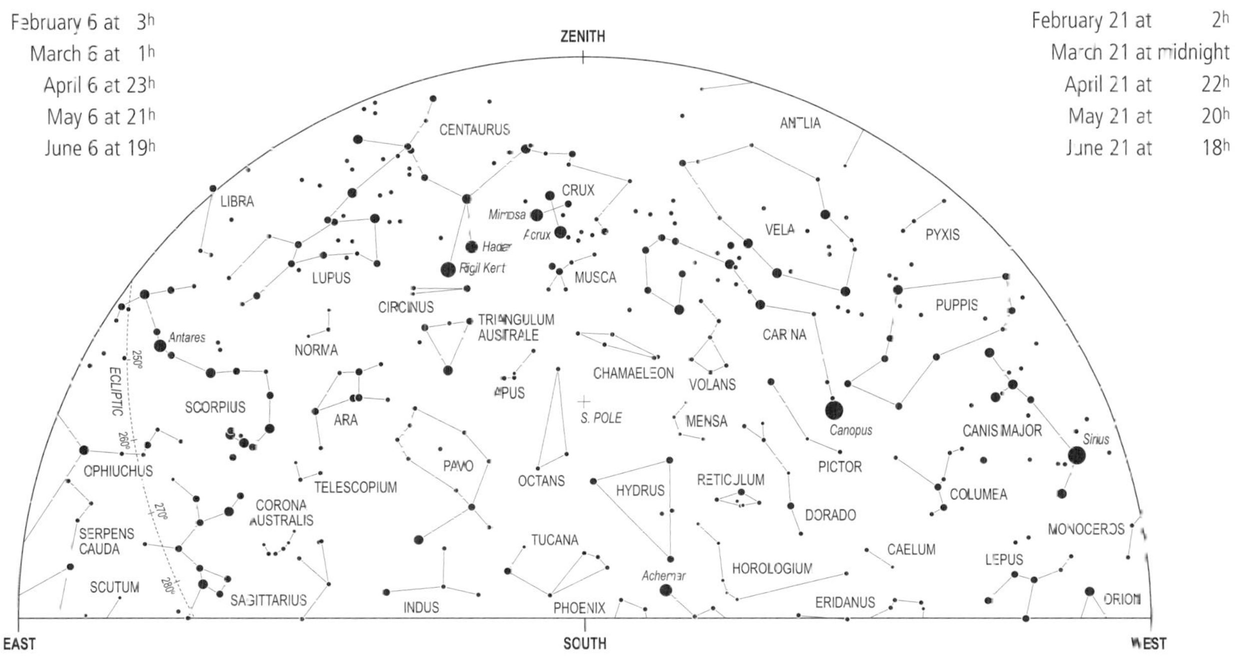
February 21 at 2h
March 21 at midnight
April 21 at 22h
May 21 at 20h
June 21 at 18h
February 6 at 3h
March 6 at 1h
April 6 at 23h
May 6 at 21h
June 6 at 19h
ZENITH
WEST
SOUTH
EAST
ORION
MONOCEROS
LEPUS
Sirius
CANIS MAJOR
COLUMBA
CAELUM
PYXIS
PUPPIS
ANTLIA
VELA
CARINA
Canopus
PICTOR
DORADO
ERIDANUS
HOROLOGIUM
RETICULUM
MENSA
VOLANS
CHAMAELEON
S. POLE
HYDRUS
Achernar
PHOENIX
TUCANA
OCTANS
APUS
MUSCA
CRUX
Acrux
Mimosa
Hadar
Rigil Kent
CENTAURUS
TRIANGULUM AUSTRALE
CIRCINUS
PAVO
INDUS
TELESCOPIUM
ARA
NORMA
LUPUS
LIBRA
SCORPIUS
Antares
CORONA AUSTRALIS
SAGITTARIUS
ECLIPTIC
250°
260°
270°
280°
OPHIUCHUS
SERPENS CAUDA
SCUTUM

5N

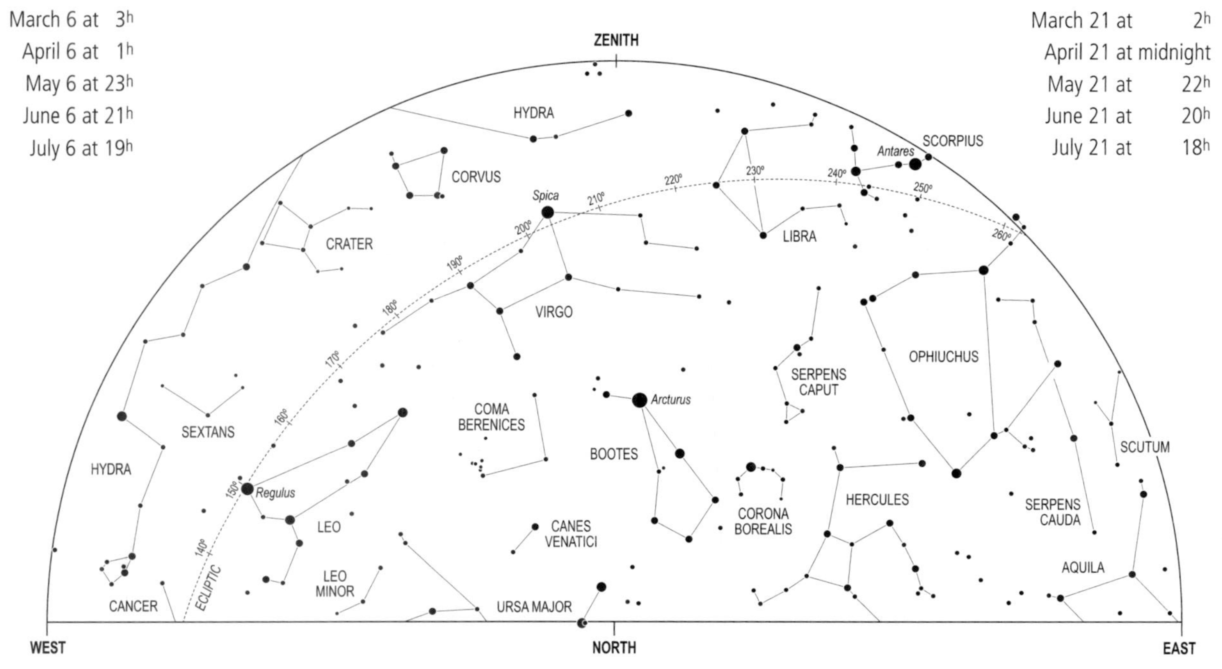
March 6 at 3h
April 6 at 1h
May 6 at 23h
June 6 at 21h
July 6 at 19h
March 21 at 2h
April 21 at midnight
May 21 at 22h
June 21 at 20h
July 21 at 18h
ZENITH
WEST
NORTH
EAST
ECLIPTIC
140°
150°
160°
170°
180°
190°
200°
210°
220°
230°
240°
250°
260°
HYDRA
CORVUS
CRATER
Spica
VIRGO
LIBRA
Antares
SCORPIUS
SEXTANS
HYDRA
Regulus
LEO
LEO MINOR
CANCER
COMA BERENICES
Arcturus
BOOTES
CANES VENATICI
URSA MAJOR
SERPENS CAPUT
OPHIUCHUS
CORONA BOREALIS
HERCULES
SERPENS CAUDA
SCUTUM
AQUILA

5S

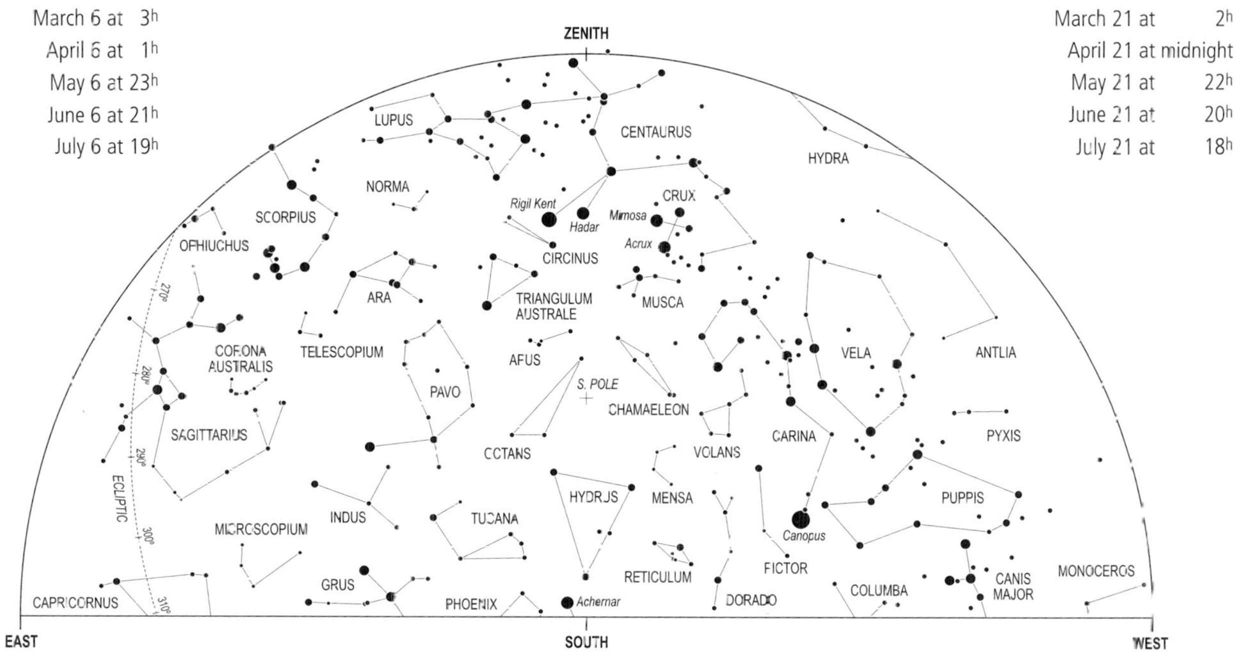
March 6 at 3h
April 6 at 1h
May 6 at 23h
June 6 at 21h
July 6 at 19h
March 21 at 2h
April 21 at midnight
May 21 at 22h
June 21 at 20h
July 21 at 18h
ZENITH
LUPUS
CENTAURUS
HYDRA
NORMA
Rigil Kent
Hadar
Mimosa
CRUX
Acrux
SCORPIUS
OPHIUCHUS
CIRCINUS
ARA
TRIANGULUM AUSTRALE
MUSCA
VELA
ANTLIA
CORONA AUSTRALIS
TELESCOPIUM
APUS
PAVO
S. POLE
CHAMAELEON
SAGITTARIUS
OCTANS
VOLANS
CARINA
PYXIS
HYDRUS
MENSA
PUPPIS
INDUS
TUCANA
Canopus
MICROSCOPIUM
RETICULUM
PICTOR
GRUS
CANIS MAJOR
MONOCEROS
COLUMBA
CAPRICORNUS
PHOENIX
Achernar
DORADO
ECLIPTIC
270°
280°
290°
300°
310°
EAST
SOUTH
WEST

6N

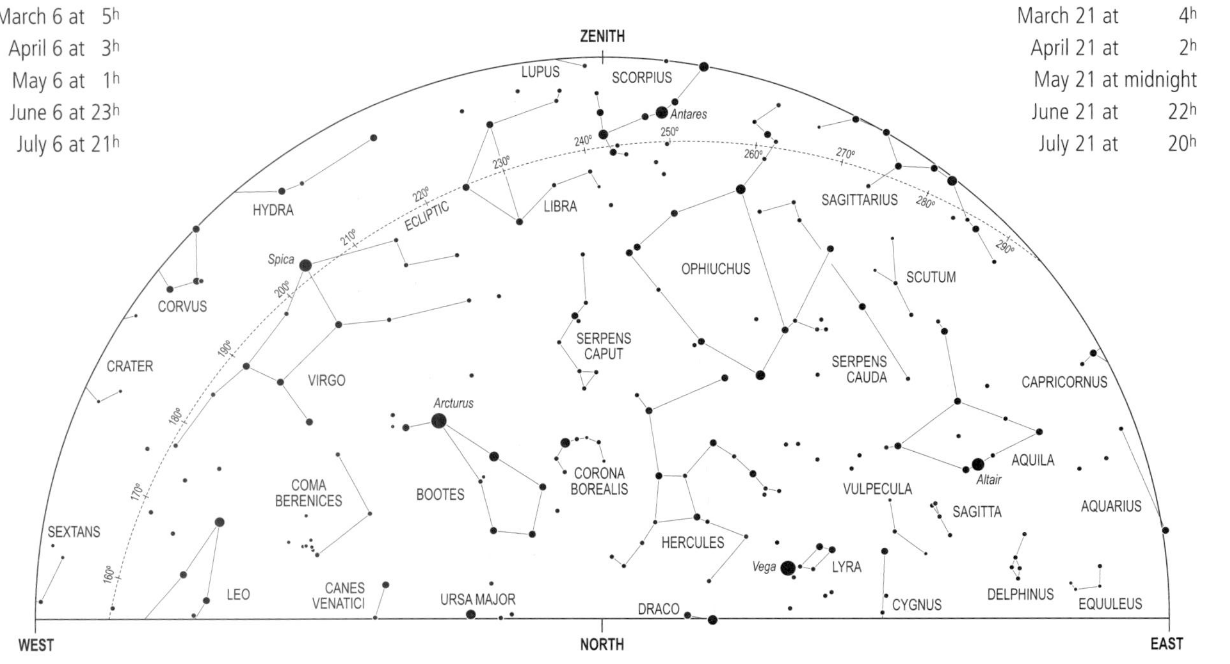
March 6 at 5h
April 6 at 3h
May 6 at 1h
June 6 at 23h
July 6 at 21h
March 21 at 4h
April 21 at 2h
May 21 at midnight
June 21 at 22h
July 21 at 20h
ZENITH
WEST
NORTH
EAST
ECLIPTIC
160°
170°
180°
190°
200°
210°
220°
230°
240°
250°
260°
270°
280°
290°
LUPUS
SCORPIUS
Antares
HYDRA
LIBRA
SAGITTARIUS
Spica
OPHIUCHUS
SCUTUM
CORVUS
SERPENS CAPUT
CRATER
VIRGO
SERPENS CAUDA
CAPRICORNUS
Arcturus
AQUILA
Altair
COMA BERENICES
BOOTES
CORONA BOREALIS
VULPECULA
SAGITTA
AQUARIUS
SEXTANS
HERCULES
Vega
LYRA
LEO
CANES VENATICI
URSA MAJOR
DRACO
CYGNUS
DELPHINUS
EQUULEUS

6S

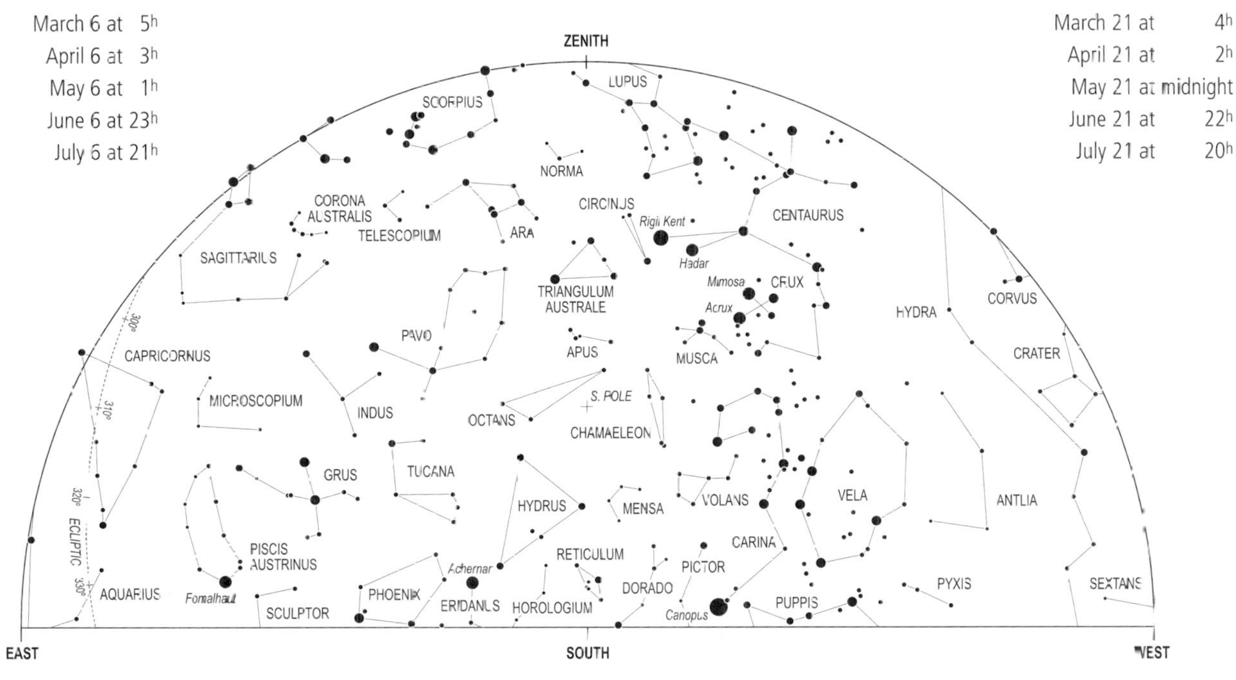
March 6 at 5h
April 6 at 3h
May 6 at 1h
June 6 at 23h
July 6 at 21h
March 21 at 4h
April 21 at 2h
May 21 at midnight
June 21 at 22h
July 21 at 20h
ZENITH
EAST
SOUTH
WEST
SCORPIUS
LUPUS
NORMA
CORONA AUSTRALIS
TELESCOPIUM
ARA
CIRCINUS
Rigil Kent
CENTAURUS
Hadar
SAGITTARIUS
TRIANGULUM AUSTRALE
Mimosa
CRUX
Acrux
HYDRA
CORVUS
PAVO
APUS
MUSCA
CAPRICORNUS
CRATER
MICROSCOPIUM
INDUS
OCTANS
S. POLE
CHAMAELEON
300°
310°
320°
330°
ECLIPTIC
GRUS
TUCANA
HYDRUS
MENSA
VOLANS
VELA
ANTLIA
PISCIS AUSTRINUS
RETICULUM
CARINA
Achernar
PICTOR
AQUARIUS
Fomalhaut
PHOENIX
ERIDANUS
HOROLOGIUM
DORADO
PUPPIS
PYXIS
SEXTANS
SCULPTOR
Canopus

7N

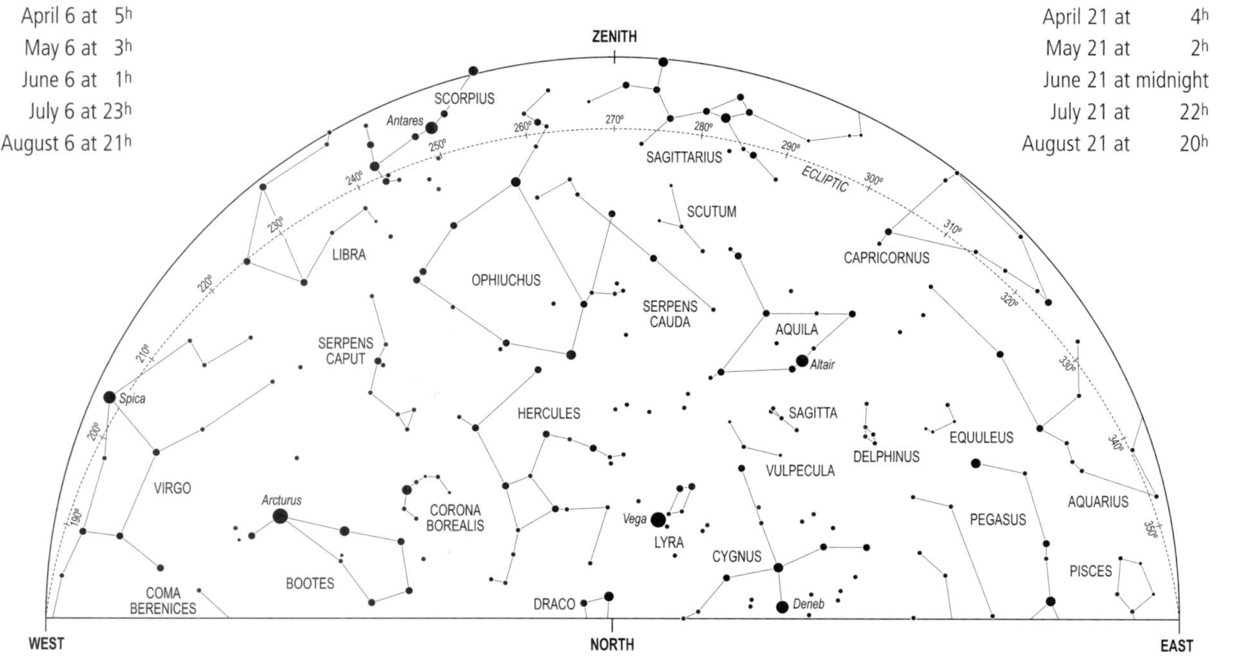
April 6 at 5h
May 6 at 3h
June 6 at 1h
July 6 at 23h
August 6 at 21h
April 21 at 4h
May 21 at 2h
June 21 at midnight
July 21 at 22h
August 21 at 20h
ZENITH
WEST
NORTH
EAST
ECLIPTIC
SCORPIUS
Antares
SAGITTARIUS
SCUTUM
LIBRA
OPHIUCHUS
CAPRICORNUS
SERPENS CAUDA
AQUILA
Altair
SERPENS CAPUT
Spica
HERCULES
SAGITTA
EQUULEUS
DELPHINUS
VULPECULA
VIRGO
Arcturus
CORONA BOREALIS
Vega
LYRA
PEGASUS
AQUARIUS
CYGNUS
PISCES
BOOTES
COMA BERENICES
DRACO
Deneb
190°
200°
210°
220°
230°
240°
250°
260°
270°
280°
290°
300°
310°
320°
330°
340°
350°

7S

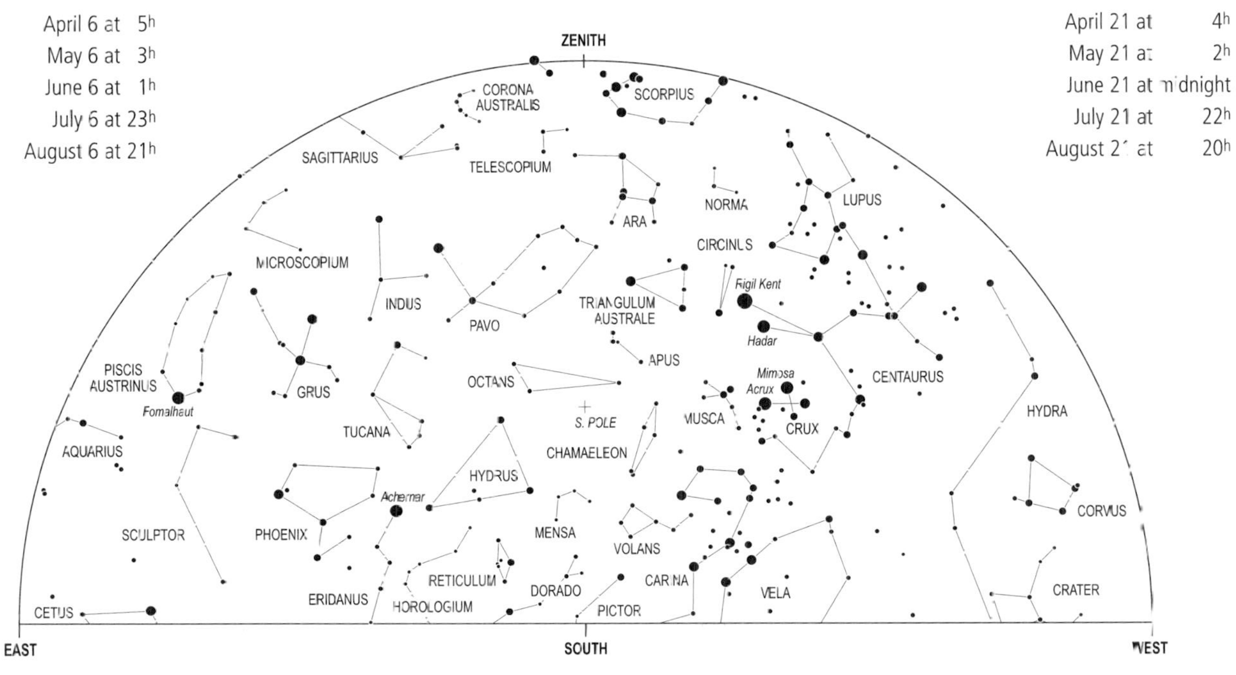
April 6 at 5^{h}
May 6 at 3^{h}
June 6 at 1^{h}
July 6 at 23^{h}
August 6 at 21^{h}
April 21 at 4^{h}
May 21 at 2^{h}
June 21 at midnight
July 21 at 22^{h}
August 21 at 20^{h}
ZENITH
EAST
SOUTH
WEST
CORONA AUSTRALIS
SCORPIUS
SAGITTARIUS
TELESCOPIUM
NORMA
LUPUS
ARA
CIRCINUS
MICROSCOPIUM
INDUS
PAVO
TRIANGULUM AUSTRALE
Rigil Kent
Hadar
APUS
CENTAURUS
PISCIS AUSTRINUS
Fomalhaut
GRUS
OCTANS
Mimosa
Acrux
S. POLE
MUSCA
CRUX
HYDRA
AQUARIUS
TUCANA
CHAMAELEON
HYDRUS
Achernar
CORVUS
SCULPTOR
PHOENIX
MENSA
VOLANS
RETICULUM
DORADO
CARINA
VELA
CRATER
ERIDANUS
HOROLOGIUM
PICTOR
CETUS

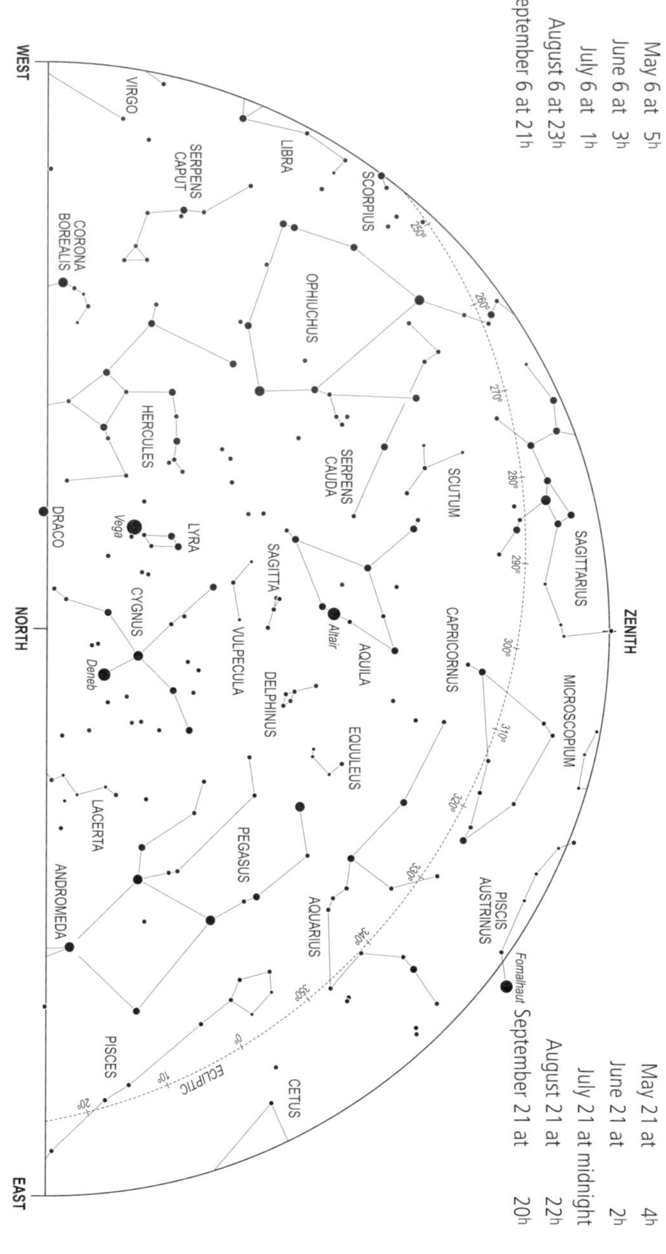

8N
May 6 at 5h
June 6 at 3h
July 6 at 1h
August 6 at 23h
September 6 at 21h
May 21 at 4h
June 21 at 2h
July 21 at midnight
August 21 at 22h
September 21 at 20h
WEST
NORTH
EAST
ZENITH
VIRGO
LIBRA
SERPENS CAPUT
SCORPIUS
CORONA BOREALIS
OPHIUCHUS
HERCULES
SERPENS CAUDA
SCUTUM
SAGITTARIUS
DRACO
Vega
LYRA
SAGITTA
CYGNUS
Deneb
VULPECULA
Altair
AQUILA
CAPRICORNUS
MICROSCOPIUM
DELPHINUS
EQUULEUS
LACERTA
PEGASUS
ANDROMEDA
AQUARIUS
PISCIS AUSTRINUS
Fomalhaut
PISCES
CETUS
ECLIPTIC
250°
260°
270°
280°
290°
300°
310°
320°
330°
340°
350°
0°
10°
20°

8S

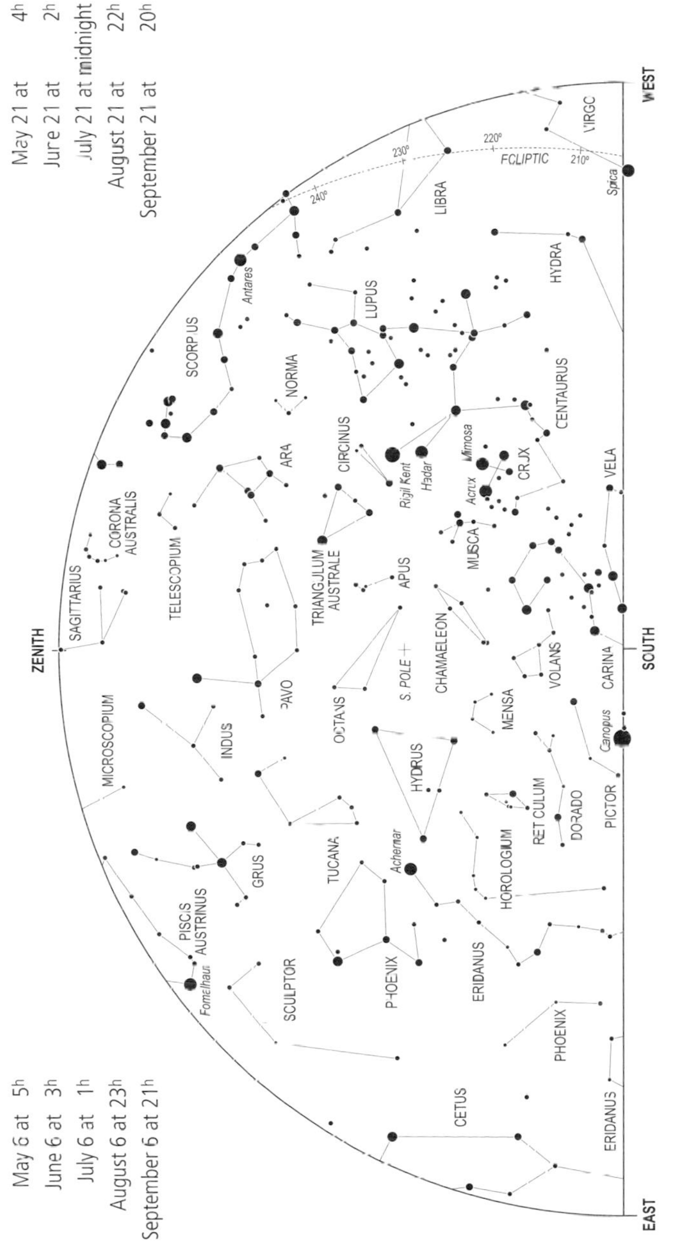

May 21 at 4^h
June 21 at 2^h
July 21 at midnight
August 21 at 22^h
September 21 at 20^h
May 6 at 5^h
June 6 at 3^h
July 6 at 1^h
August 6 at 23^h
September 6 at 21^h
WEST
SOUTH
EAST
ZENITH
ECLIPTIC
210°
220°
230°
240°
VIRGO
Spica
LIBRA
HYDRA
Antares
SCORPIUS
LUPUS
NORMA
CENTAURUS
CIRCINUS
ARA
Rigil Kent
Hadar
Mimosa
Acrux
CRUX
VELA
CORONA AUSTRALIS
SAGITTARIUS
TELESCOPIUM
TRIANGULUM AUSTRALE
APUS
MUSCA
CHAMAELEON
CARINA
VOLANS
S. POLE
PAVO
OCTANS
MENSA
Canopus
MICROSCOPIUM
INDUS
HYDRUS
RETICULUM
DORADO
PICTOR
GRUS
TUCANA
Achernar
HOROLOGIUM
PISCIS AUSTRINUS
Fomalhaut
SCULPTOR
PHOENIX
ERIDANUS
CETUS

9N

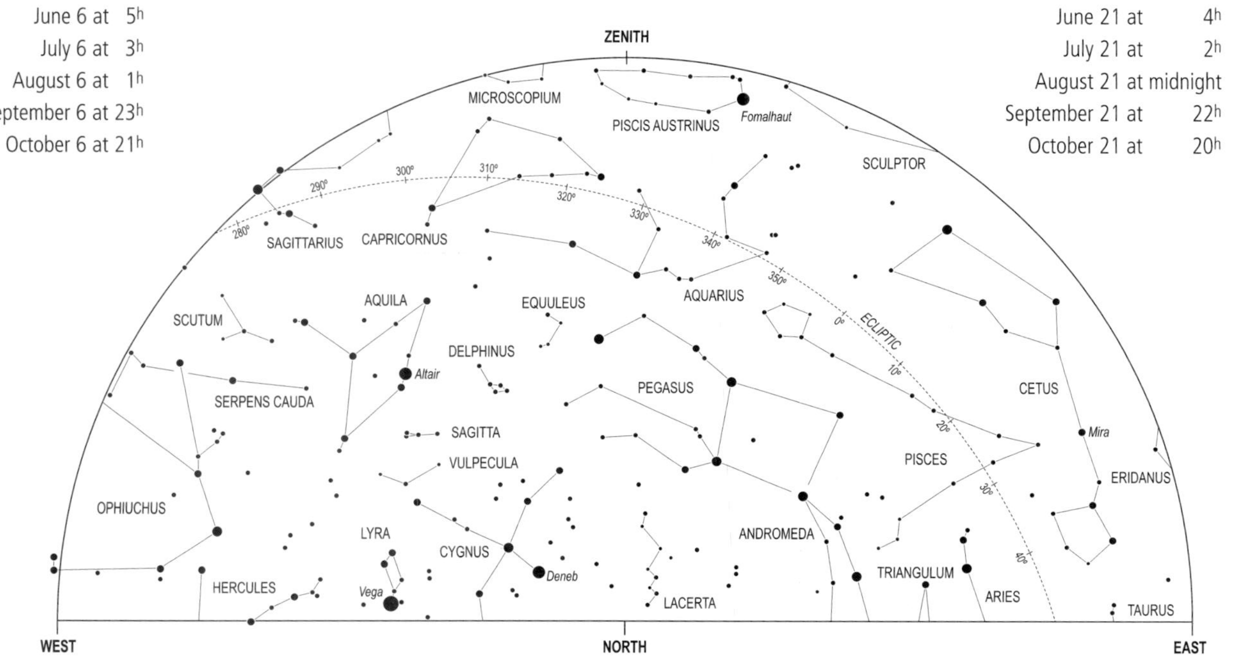
June 6 at 5h
July 6 at 3h
August 6 at 1h
September 6 at 23h
October 6 at 21h
June 21 at 4h
July 21 at 2h
August 21 at midnight
September 21 at 22h
October 21 at 20h
ZENITH
MICROSCOPIUM
PISCIS AUSTRINUS
Fomalhaut
SCULPTOR
SAGITTARIUS
CAPRICORNUS
AQUILA
SCUTUM
EQUULEUS
AQUARIUS
ECLIPTIC
DELPHINUS
Altair
PEGASUS
SERPENS CAUDA
CETUS
SAGITTA
VULPECULA
Mira
PISCES
ERIDANUS
OPHIUCHUS
LYRA
CYGNUS
ANDROMEDA
Deneb
TRIANGULUM
HERCULES
Vega
LACERTA
ARIES
TAURUS
WEST
NORTH
EAST
280°
290°
300°
310°
320°
330°
340°
350°
0°
10°
20°
30°
40°

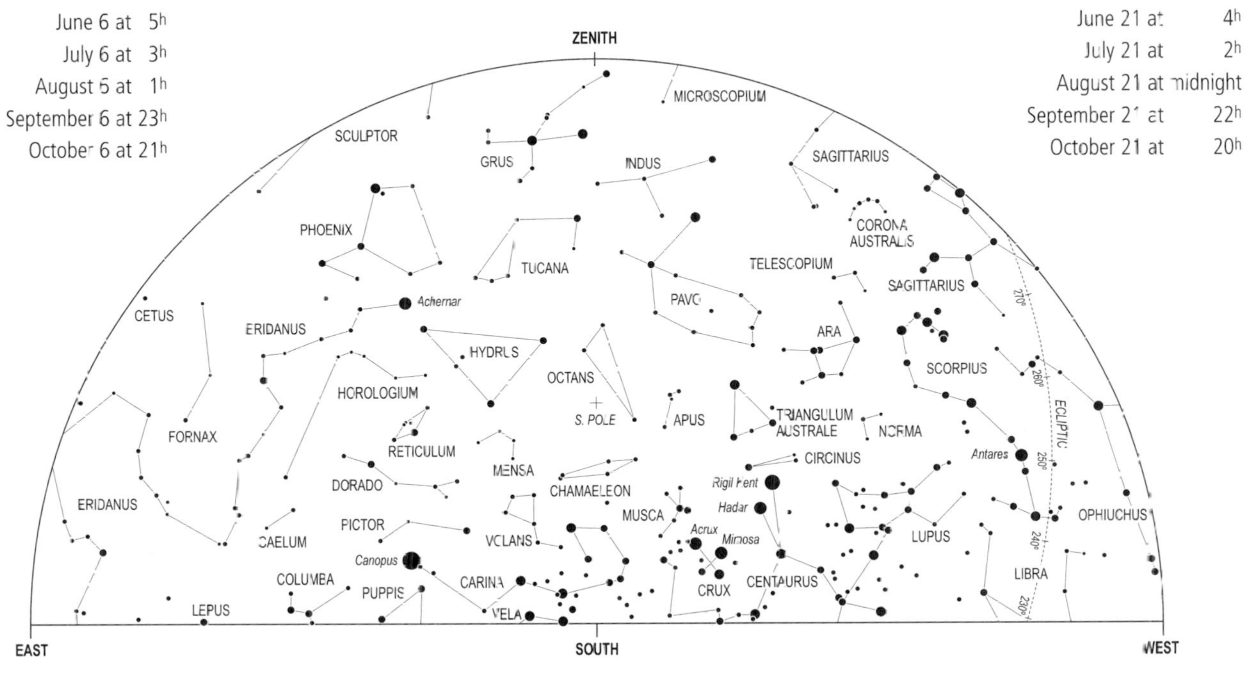
June 6 at 5h
July 6 at 3h
August 5 at 1h
September 6 at 23h
October 6 at 21h
June 21 at 4h
July 21 at 2h
August 21 at midnight
September 21 at 22h
October 21 at 20h
ZENITH
MICROSCOPIUM
SCULPTOR
GRUS
INDUS
SAGITTARIUS
PHOENIX
TUCANA
CORONA AUSTRALIS
TELESCOPIUM
CETUS
Achernar
ERIDANUS
PAVO
ARA
HYDRUS
OCTANS
SCORPIUS
HOROLOGIUM
S. POLE
APUS
TRIANGULUM AUSTRALE
NORMA
ECLIPTIC
FORNAX
RETICULUM
MENSA
CIRCINUS
Antares
DORADO
CHAMAELEON
Rigil Kent
MUSCA
Hadar
OPHIUCHUS
PICTOR
Acrux
Mimosa
LUPUS
CAELUM
VOLANS
LIBRA
Canopus
COLUMBA
CARINA
PUPPIS
CRUX
CENTAURUS
LEPUS
VELA
EAST
SOUTH
WEST

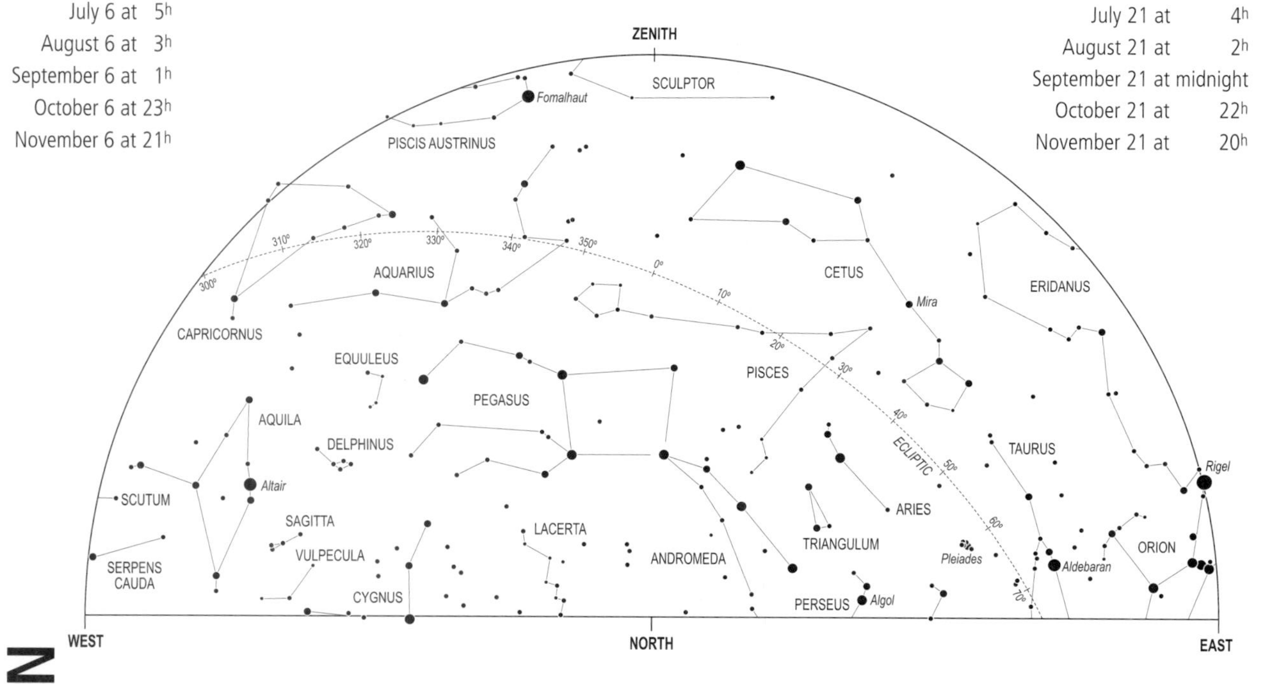
10N
July 6 at 5^h
August 6 at 3^h
September 6 at 1^h
October 6 at 23^h
November 6 at 21^h
July 21 at 4^h
August 21 at 2^h
September 21 at midnight
October 21 at 22^h
November 21 at 20^h
ZENITH
WEST
NORTH
EAST
SCULPTOR
Fomalhaut
PISCIS AUSTRINUS
CETUS
Mira
ERIDANUS
AQUARIUS
CAPRICORNUS
EQUULEUS
PEGASUS
PISCES
AQUILA
DELPHINUS
Altair
SCUTUM
SAGITTA
VULPECULA
SERPENS CAUDA
CYGNUS
LACERTA
ANDROMEDA
TRIANGULUM
ARIES
ECLIPTIC
TAURUS
Rigel
ORION
Pleiades
Aldebaran
PERSEUS
Algol
300°
310°
320°
330°
340°
350°
0°
10°
20°
30°
40°
50°
60°
70°

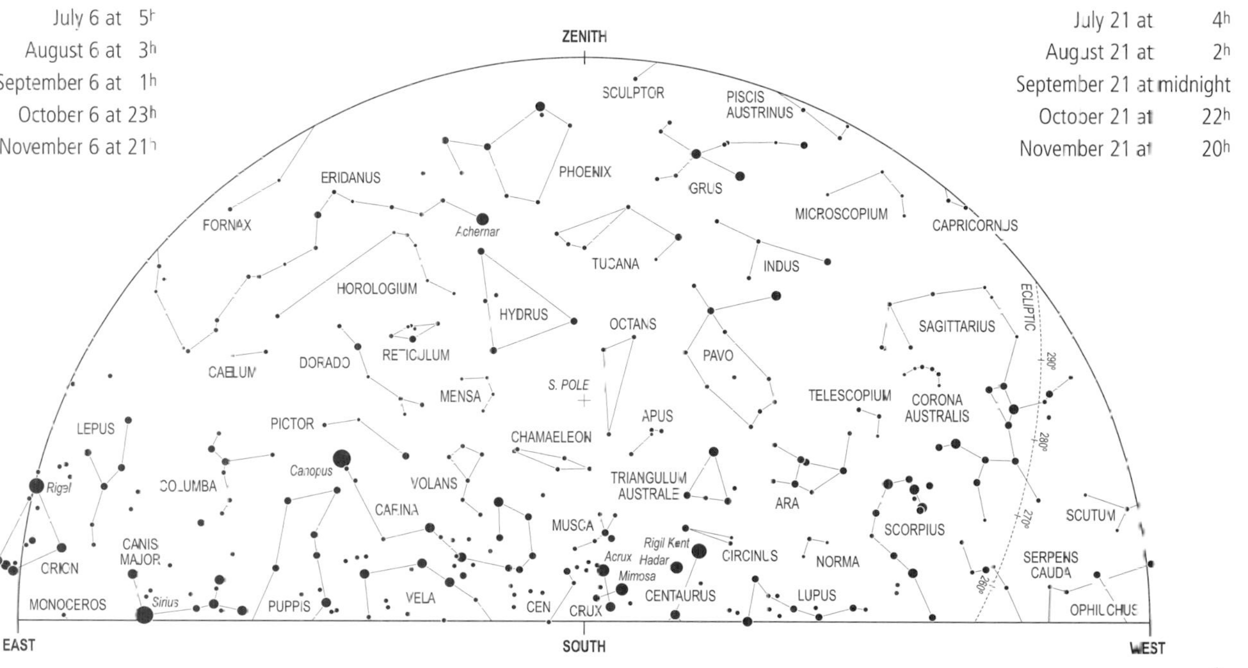

July 6 at 5h
August 6 at 3h
September 6 at 1h
October 6 at 23h
November 6 at 21h
July 21 at 4h
August 21 at 2h
September 21 at midnight
October 21 at 22h
November 21 at 20h
ZENITH
EAST
SOUTH
WEST
SCULPTOR
PISCIS AUSTRINUS
PHOENIX
GRUS
ERIDANUS
FORNAX
Achernar
MICROSCOPIUM
CAPRICORNUS
TUCANA
INDUS
HOROLOGIUM
HYDRUS
OCTANS
ECLIPTIC
SAGITTARIUS
PAVO
290°
280°
270°
260°
CAELUM
DORADO
RETICULUM
MENSA
S. POLE
TELESCOPIUM
CORONA AUSTRALIS
LEPUS
PICTOR
CHAMAELEON
APUS
Rigel
COLUMBA
Canopus
VOLANS
TRIANGULUM AUSTRALE
ARA
CARINA
SCORPIUS
SCUTUM
MUSCA
Rigil Kent
CANIS MAJOR
Acrux
Hadar
CIRCINUS
NORMA
SERPENS CAUDA
ORION
Mimosa
MONOCEROS
Sirius
PUPPIS
VELA
CEN
CRUX
CENTAURUS
LUPUS
OPHIUCHUS

10S

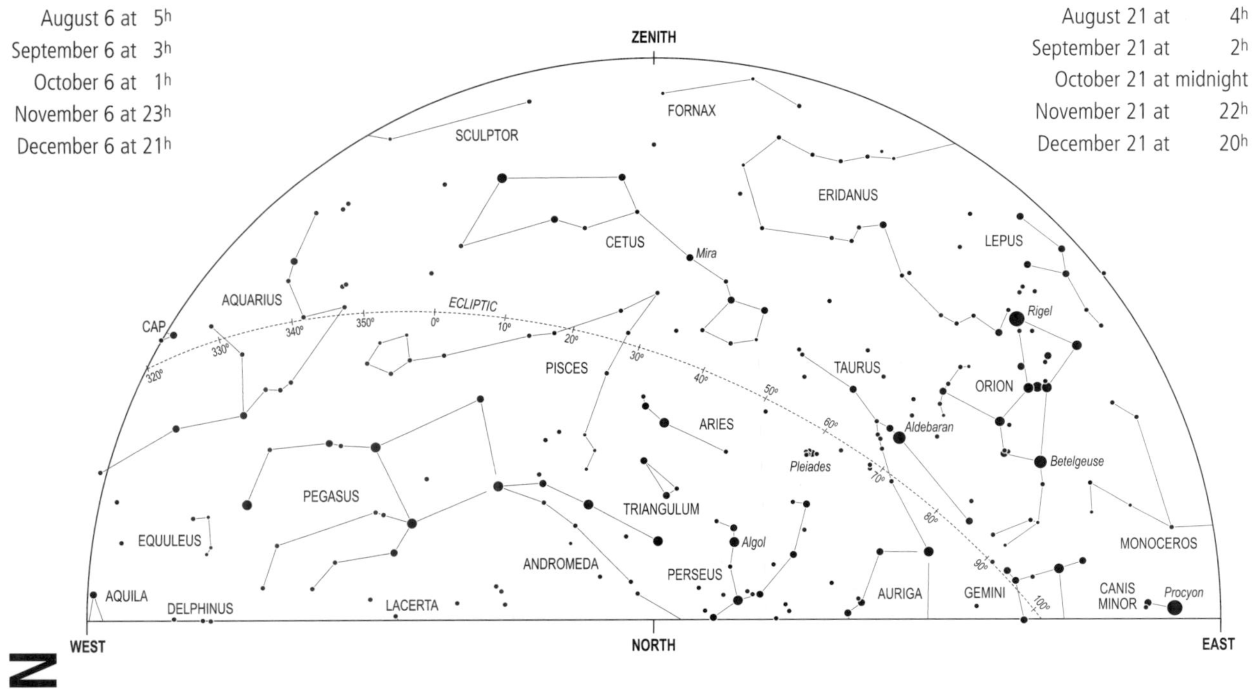
August 6 at 5h
September 6 at 3h
October 6 at 1h
November 6 at 23h
December 6 at 21h
ZENITH
August 21 at 4h
September 21 at 2h
October 21 at midnight
November 21 at 22h
December 21 at 20h
FORNAX
SCULPTOR
ERIDANUS
CETUS
Mira
LEPUS
AQUARIUS
ECLIPTIC
CAP
Rigel
320°
330°
340°
350°
0°
10°
20°
30°
40°
50°
60°
70°
80°
90°
100°
PISCES
TAURUS
ORION
ARIES
Aldebaran
Pleiades
Betelgeuse
PEGASUS
TRIANGULUM
EQUULEUS
Algol
MONOCEROS
ANDROMEDA
PERSEUS
AURIGA
GEMINI
CANIS MINOR
Procyon
AQUILA
DELPHINUS
LACERTA
WEST
NORTH
EAST
11N

August 6 at 5h
September 6 at 3h
October 6 at 1h
November 6 at 23h
December 6 at 21h

August 21 at 4h
September 21 at 2h
October 21 at midnight
November 21 at 22h
December 21 at 20h

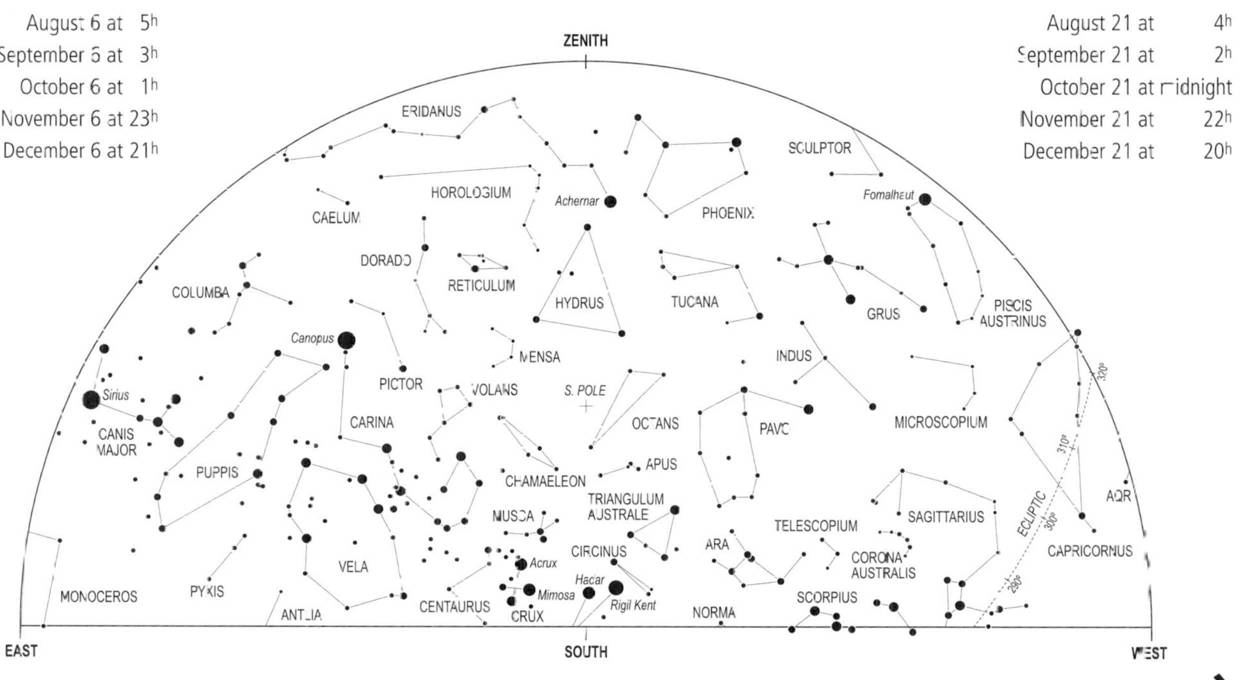

11S

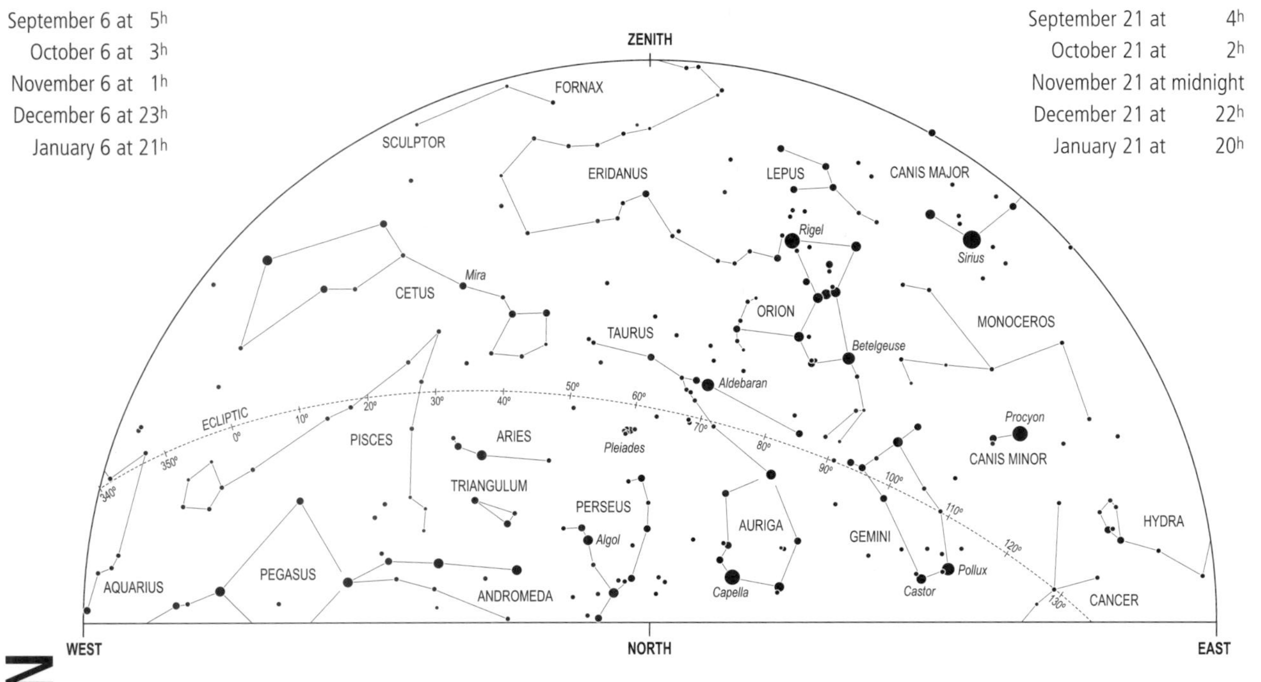
September 6 at 5h
October 6 at 3h
November 6 at 1h
December 6 at 23h
January 6 at 21h
September 21 at 4h
October 21 at 2h
November 21 at midnight
December 21 at 22h
January 21 at 20h
ZENITH
FORNAX
SCULPTOR
ERIDANUS
LEPUS
CANIS MAJOR
Rigel
Sirius
Mira
CETUS
ORION
MONOCEROS
TAURUS
Betelgeuse
Aldebaran
ECLIPTIC
340°
350°
0°
10°
20°
30°
40°
50°
60°
70°
80°
90°
100°
110°
120°
130°
Procyon
PISCES
ARIES
Pleiades
CANIS MINOR
TRIANGULUM
PERSEUS
AURIGA
GEMINI
HYDRA
Algol
AQUARIUS
PEGASUS
ANDROMEDA
Capella
Castor
Pollux
CANCER
WEST
NORTH
EAST
12N

12S

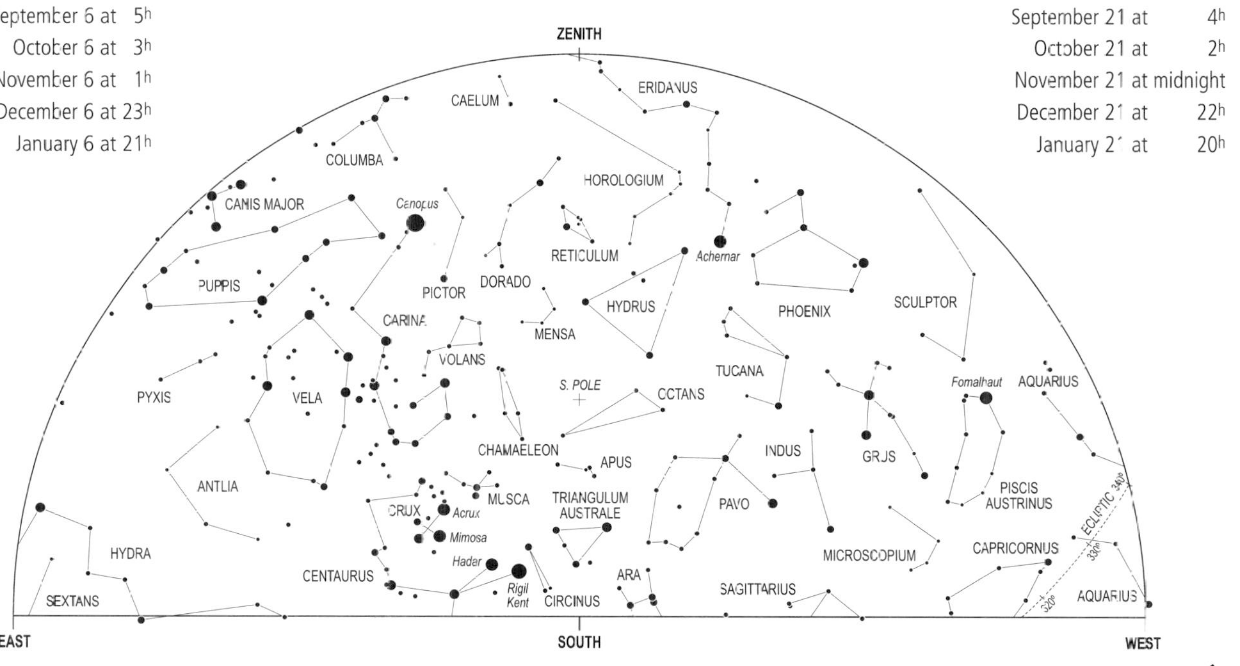
September 6 at 5^h
October 6 at 3^h
November 6 at 1^h
December 6 at 23^h
January 6 at 21^h
September 21 at 4^h
October 21 at 2^h
November 21 at midnight
December 21 at 22^h
January 21 at 20^h
ZENITH
EAST
SOUTH
WEST
CAELUM
ERIDANUS
COLUMBA
HOROLOGIUM
CANIS MAJOR
Canopus
RETICULUM
Achernar
PUPPIS
PICTOR
DORADO
HYDRUS
PHOENIX
SCULPTOR
CARINA
MENSA
VOLANS
TUCANA
PYXIS
VELA
S. POLE
OCTANS
Fomalhaut
AQUARIUS
CHAMAELEON
APUS
INDUS
GRUS
ANTLIA
MUSCA
TRIANGULUM AUSTRALE
PAVO
PISCIS AUSTRINUS
CRUX
Acrux
Mimosa
ECLIPTIC 340°
330°
320°
HYDRA
MICROSCOPIUM
CAPRICORNUS
Hadar
CENTAURUS
ARA
Rigil Kent
SAGITTARIUS
SEXTANS
CIRCINUS
AQUARIUS

The Planets and the Ecliptic

The paths of the planets about the Sun all lie close to the plane of the ecliptic, which is marked for us in the sky by the apparent path of the Sun among the stars, and is shown on the star charts by a broken line. The Moon and naked-eye planets will always be found close to this line, never departing from it by more than about 7°. Thus the planets are most favourably placed for observation when the ecliptic is well displayed, and this means that it should be as high in the sky as possible. This avoids the difficulty of finding a clear horizon, and also overcomes the problem of atmospheric absorption, which greatly reduces the light of the stars. Thus a star at an altitude of 10° suffers a loss of 60 per cent of its light, which corresponds to a whole magnitude; at an altitude of only 4°, the loss may amount to two magnitudes.

The position of the ecliptic in the sky is therefore of great importance, and since it is tilted at about 23.5° to the Equator, it is only at certain times of the day or year that it is displayed to best advantage. It will be realized that the Sun (and therefore the ecliptic) is at its highest in the sky at noon in midsummer, and at its lowest at noon in midwinter. Allowing for the daily motion of the sky, it follows that the ecliptic is highest at midnight in winter, at sunset in the spring, at noon in summer and at sunrise in the autumn. Hence these are the best times to see the planets. Thus, if Venus is an evening object in the western sky after sunset, it will be seen to best advantage if this occurs in the spring, when the ecliptic is high in the sky and slopes down steeply to the horizon. This means that the planet is not only higher in the sky, but also will remain for a much longer period above the horizon. For similar reasons, a morning object will be seen at its best on autumn mornings before sunrise, when the ecliptic is high in the east. The outer planets, which can come to opposition (i.e. opposite the Sun), are best seen when opposition occurs in the winter months, when the ecliptic is high in the sky at midnight.

The seasons are reversed in the Southern Hemisphere, spring beginning at the September Equinox, when the Sun crosses the Equator on its way south, summer beginning at the December Solstice, when the

Sun is highest in the southern sky, and so on. Thus, the times when the ecliptic is highest in the sky, and therefore best placed for observing the planets, may be summarized as follows:

	Midnight	**Sunrise**	**Noon**	**Sunset**
Northern latitudes	December	September	June	March
Southern latitudes	June	March	December	September

In addition to the daily rotation of the celestial sphere from east to west, the planets have a motion of their own among the stars. The apparent movement is generally *direct*, i.e. to the east, in the direction of increasing longitude, but for a certain period (which depends on the distance of the planet) this apparent motion is reversed. With the outer planets this *retrograde* motion occurs about the time of opposition. Owing to the different inclination of the orbits of these planets, the actual effect is to cause the apparent path to form a loop, or sometimes an S shaped curve. The same effect is present in the motion of the inferior planets, Mercury and Venus, but it is not so obvious, since it always occurs at the time of inferior conjunction.

The *inferior planets*, Mercury and Venus, move in smaller orbits than that of the Earth, and so are always seen near the Sun. They are most obvious at the times of greatest angular distance from the Sun (greatest elongation), which may reach 28° for Mercury, and 47° for Venus. They are seen as evening objects in the western sky after sunset (at eastern elongations) or as morning objects in the eastern sky before sunrise (at western elongations). The succession of phenomena, conjunctions and elongations always follows the same order, but the intervals between them are not equal. Thus, if either planet is moving round the far side of its orbit its motion will be to the east, in the same direction in which the Sun appears to be moving. It therefore takes much longer for the planet to overtake the Sun – that is, to come to superior conjunction – than it does when moving round to inferior conjunction, between Sun and Earth. The intervals given in the table at the top of p.60 are average values; they remain fairly constant in the case of Venus, which travels in an almost circular orbit. In the case of Mercury, however, conditions vary widely because of the great eccentricity and inclination of the planet's orbit.

		Mercury	Venus
Inferior Conjunction	to Elongation West	22 days	72 days
Elongation West	to Superior Conjunction	36 days	220 days
Superior Conjunction	to Elongation East	35 days	220 days
Elongation East	to Inferior Conjunction	22 days	72 days

The greatest brilliancy of Venus always occurs about thirty-six days before or after inferior conjunction. This will be about a month after greatest eastern elongation (as an evening object), or a month before greatest western elongation (as a morning object). No such rule can be given for Mercury, because its distances from the Earth and the Sun can vary over a wide range.

Mercury is not likely to be seen unless a clear horizon is available. It is seldom as much as 10° above the horizon in the twilight sky in northern temperate latitudes, but this figure is often exceeded in the Southern Hemisphere. This favourable condition arises because the maximum elongation of 28° can occur only when the planet is at aphelion (furthest from the Sun), and it then lies well south of the Equator. Northern observers must be content with smaller elongations, which may be as little as 18° at perihelion. In general, it may be said that the most favourable times for seeing Mercury as an evening object will be in spring, some days before greatest eastern elongation; in autumn, it may be seen as a morning object some days after greatest western elongation.

Venus is the brightest of the planets and may be seen on occasions in broad daylight. Like Mercury, it is alternately a morning and an evening object, and it will be highest in the sky when it is a morning object in autumn, or an evening object in spring. Venus is to be seen at its best as an evening object in northern latitudes when eastern elongation occurs in June. The planet is then well north of the Sun in the preceding spring months, and is a brilliant object in the evening sky over a long period. In the Southern Hemisphere a November elongation is best. For similar reasons, Venus gives a prolonged display as a morning object in the months following western elongation in October (in northern latitudes) or in June (in the Southern Hemisphere).

The *superior planets*, which travel in orbits larger than that of the Earth, differ from Mercury and Venus in that they can be seen opposite the Sun in the sky. The superior planets are morning objects after conjunction with the Sun, rising earlier each day until they come to opposition. They will then be nearest to the Earth (and therefore at their brightest), and will be on the meridian at midnight, due south in northern latitudes, but due north in the Southern Hemisphere. After opposition they are evening objects, setting earlier each evening until they set in the west with the Sun at the next conjunction. The difference in brightness from one opposition to another is most noticeable in the case of Mars, whose distance from Earth can vary considerably and rapidly. The other superior planets are at such great distances that there is very little change in brightness from one opposition to the next. The effect of altitude is, however, of some importance, for at a December opposition in northern latitudes the planets will be among the stars of Taurus or Gemini, and can then be at an altitude of more than 60° in southern England. At a summer opposition, when Mars is in Sagittarius, it may only rise to about 15° above the southern horizon, and so makes a less impressive appearance. In the Southern Hemisphere the reverse conditions apply, a June opposition being the best, with the planet in Sagittarius at an altitude which can reach 80° above the northern horizon for observers in South Africa.

Mars, whose orbit is appreciably eccentric, comes nearest to the Earth at oppositions at the end of August. It may then be brighter even than Jupiter, but rather low in the sky in Aquarius for northern observers, though very well placed for those in southern latitudes. These favourable oppositions occur every fifteen or seventeen years (e.g. in 1988, 2003 and 2018). In the Northern Hemisphere the planet is probably better seen at oppositions in the autumn or winter months, when it is higher in the sky – such as in 2005 when opposition was in early November. Oppositions of Mars occur at an average interval of 780 days, and during this time the planet makes a complete circuit of the sky.

Jupiter is always a bright planet, and comes to opposition a month later each year, having moved, roughly speaking, from one zodiacal constellation to the next. There is no opposition of Jupiter in 2013.

Saturn moves much more slowly than Jupiter, and may remain in the same constellation for several years. The brightness of Saturn depends on the aspects of its rings, as well as on the distance from Earth and Sun. The Earth passed through the plane of Saturn's rings in 1995 and 1996, when they appeared edge-on; we saw them at maximum opening, and Saturn at its brightest, in 2002. The rings last appeared edge-on in 2009, and they are now opening once again.

Uranus and *Neptune* are both visible with binoculars or a small telescope, but you will need a finder chart to help you locate them (such as those reproduced in this *Yearbook* on pages 136 and 121). *Pluto* (now officially classified as a 'dwarf planet') is hardly likely to attract the attention of observers without adequate telescopes.

Phases of the Moon in 2013

NICK JAMES

New Moon				First Quarter				Full Moon				Last Quarter			
	d	h	m		d	h	m		d	h	m		d	h	m
												Jan	5	03	58
Jan	11	19	44	Jan	18	23	45	Jan	27	04	38	Feb	3	13	56
Feb	10	07	20	Feb	17	20	31	Feb	25	20	26	Mar	4	21	53
Mar	11	19	51	Mar	19	17	27	Mar	27	09	27	Apr	3	04	37
Apr	10	09	35	Apr	18	12	31	Apr	25	19	57	May	2	11	14
May	10	00	28	May	18	04	35	May	25	04	25	May	31	18	58
June	8	15	56	June	16	17	24	June	23	11	32	June	30	04	54
July	8	07	14	July	16	03	18	July	22	18	16	July	29	17	43
Aug	6	21	51	Aug	14	10	56	Aug	21	01	45	Aug	28	09	35
Sept	5	11	36	Sept	12	17	08	Sept	19	11	13	Sept	27	03	56
Oct	5	00	35	Oct	11	23	02	Oct	18	23	38	Oct	26	23	41
Nov	3	12	50	Nov	10	05	57	Nov	17	15	16	Nov	25	19	28
Dec	3	00	22	Dec	9	15	12	Dec	17	09	28	Dec	25	13	48
Jan	1	11	14												

All times are UTC (GMT)

Longitudes of the Sun, Moon and Planets in 2013

NICK JAMES

Date		Sun	Moon	Venus	Mars	Jupiter	Saturn	Uranus	Neptune
		°	°	°	°	°	°	°	°
Jan	6	286	206	266	309	67	220	5	331
	21	301	53	285	320	66	221	5	332
Feb	6	317	259	305	333	66	221	6	332
	21	333	97	324	345	67	222	7	333
Mar	6	346	270	340	355	68	221	7	333
	21	1	104	359	7	70	221	8	334
Apr	6	16	323	18	19	73	220	9	334
	21	31	149	37	30	75	219	10	335
May	6	46	359	56	42	79	218	11	335
	21	60	184	74	52	82	217	11	335
June	6	75	46	94	64	85	216	12	335
	21	90	235	112	75	89	215	12	335
July	6	104	79	130	85	92	215	12	335
	21	118	273	148	95	96	215	13	335
Aug	6	134	124	167	106	99	215	12	335
	21	148	327	185	115	102	216	12	334
Sept	6	164	170	204	126	105	218	12	334
	21	178	18	221	135	107	219	11	333
Oct	6	193	205	238	144	109	221	10	333
	21	208	52	254	153	110	222	10	333
Nov	6	224	257	271	163	111	224	9	333
	21	239	96	284	171	110	226	9	333
Dec	6	254	296	294	179	109	228	9	333
	21	269	128	299	187	108	229	9	333

Moon: Longitude of the ascending node: Jan 1: 234° Dec 31: 214°

Mercury moves so quickly among the stars that it is not possible to indicate its position on the star charts at convenient intervals. The monthly notes should be consulted for the best times at which the planet may be seen.

The positions of the Sun, Moon and planets other than Mercury are given in the table on p. 64. These objects move along paths which remain close to the ecliptic and this list shows the apparent ecliptic longitude for each object on dates which correspond to those of the star charts. This information can be used to plot the position of the desired object on the selected chart.

EXAMPLES

Two planets are seen close together low in the eastern sky at around 4 a.m. BST (3 a.m. GMT) in late July. What are they?

> The northern star chart 10N shows the northern sky on 6 August at 3h. The range of ecliptic longitude visible low in the east ranges from 80° to 110°. Reference to the table on p. 64 for 21 July shows that two planets are in this range: Mars is at longitude 95° and Jupiter is at 96°. The two planets are therefore Mars and Jupiter in Gemini. From the table it can be seen that by 6 August, Mars has overtaken Jupiter and, in fact, the two planets come to within 1° of each other on the morning of 23 July 2013.

The positions of the Sun and Moon can be plotted on the star maps in the same way as the planets. This is straightforward for the Sun since it always lies on the ecliptic and it moves on average at only 1° per day. The Moon is more difficult since it moves rapidly along at an average of 13° per day and it moves up to 5° north or south of the ecliptic during the month. An indication of the Moon's position relative to the ecliptic may be obtained by considering its longitude relative to that of the ascending node. The longitude of the ascending node decreases by around 1.7° per month, as will be seen from the values for the first and last day of the year given on p. 64. If d is the difference in longitude between the Moon and its ascending node, then the Moon is on the ecliptic when $d = 0°$, 180° or 360°. The Moon is 5° north of the ecliptic if $d = 90°$, and the Moon is 5° south of the ecliptic if $d = 270°$.

As an example, the Moon is full at 04h on 25 May. The table shows that the Moon's longitude is 184° at 0h on 21 May. Extrapolating at 13° per day the Moon's longitude at 04h on 25 May is around 238°. At this time the longitude of the node is found by interpolation to be around 225°. Thus $d = 13°$ and the Moon is just north of the ecliptic on the border of Libra and Scorpius. Its position may be plotted on northern star charts 5S, 6S and 7S and southern star charts 5N, 6N and 7N.

Some Events in 2013

Jan	2	*Earth* at Perihelion
	10	Moon at Perigee (360,045 km)
	11	New Moon
	18	*Mercury* in Superior Conjunction
	22	Moon at Apogee (405,310 km)
	27	Full Moon
Feb	7	Moon at Perigee (365,315 km)
	10	New Moon
	16	*Mercury* at Greatest Eastern Elongation (18°)
	19	Moon at Apogee (404,475 km)
	21	*Neptune* in Conjunction with Sun
	25	Full Moon
Mar	4	*Mercury* in Inferior Conjunction
	5	Moon at Perigee (369,955 km)
	11	New Moon
	19	Moon at Apogee (404,260 km)
	20	Equinox (Spring Equinox in Northern Hemisphere)
	27	Full Moon
	28	*Venus* in Superior Conjunction
	28	*Uranus* in Conjunction with Sun
	31	Summer Time Begins in the UK
	31	Moon at Perigee (367,495 km)
	31	*Mercury* at Greatest Western Elongation (28°)
Apr	10	New Moon
	15	Moon at Apogee (404,865 km)
	17	*Mars* in Conjunction with Sun
	25	Full Moon
	25	Partial Eclipse of the Moon
	27	Moon at Perigee (362,265 km)
	28	*Saturn* at Opposition in Libra

May 9–10	Annular Eclipse of the Sun
10	New Moon
11	*Mercury* in Superior Conjunction
13	Moon at Apogee (405,825 km)
25	Full Moon
25	Penumbral Eclipse of the Moon
26	Moon at Perigee (358,375 km)
June 8	New Moon
9	Moon at Apogee (406,485 km)
12	*Mercury* at Greatest Eastern Elongation (24°)
19	*Jupiter* in Conjunction with Sun
21	Solstice (Summer Solstice in Northern Hemisphere)
23	Full Moon
23	Moon at Perigee (356,990 km)
July 2	*Pluto* at Opposition in Sagittarius
5	*Earth* at Aphelion
7	Moon at Apogee (406,490 km)
8	New Moon
9	*Mercury* in Inferior Conjunction
21	Moon at Perigee (358,400 km)
22	Full Moon
30	*Mercury* at Greatest Western Elongation (20°)
Aug 3	Moon at Apogee (405,835 km)
6	New Moon
19	Moon at Perigee (362,265 km)
21	Full Moon
24	*Mercury* in Superior Conjunction
27	*Neptune* at Opposition in Aquarius
30	Moon at Apogee (404,880 km)
Sep 5	New Moon
15	Moon at Perigee (367,385 km)
19	Full Moon
22	Equinox (Autumnal Equinox in Northern Hemisphere)
27	Moon at Apogee (404,310 km)

Oct	3	*Uranus* at Opposition in Pisces
	5	New Moon
	9	*Mercury* at Greatest Eastern Elongation (25°)
	10	Moon at Perigee (369,810 km)
	18	Full Moon
	18	Penumbral Eclipse of the Moon
	25	Moon at Apogee (404,560 km)
	27	Summer Time Ends in the UK
Nov	1	*Venus* at Greatest Eastern Elongation (47°)
	1	*Mercury* in Inferior Conjunction
	3	New Moon
	3	Annular/Total Eclipse of the Sun
	6	Moon at Perigee (365,360 km)
	6	*Saturn* in Conjunction with Sun
	17	Full Moon
	18	*Mercury* at Greatest Western Elongation (19°)
	22	Moon at Apogee (405,445 km)
Dec	3	New Moon
	4	Moon at Perigee (360,065 km)
	10	*Venus* attains greatest brilliancy (mag. -4.7)
	17	Full Moon
	19	Moon at Apogee (406,265 km)
	21	Solstice (Winter Solstice in Northern Hemisphere)
	29	*Mercury* in Superior Conjunction

Monthly Notes 2013

January

New Moon: 11 January *Full Moon:* 27 January

EARTH is at perihelion (nearest to the Sun) on 2 January at a distance of 147 million kilometres (91.3 million miles).

MERCURY passes through superior conjunction, on the far side of the Sun, on 18 January. Consequently, it will be unsuitably placed for observation throughout the month.

VENUS, magnitude -3.9, is rising only about one-and-a-half hours before the Sun from all latitudes at the beginning of January, but its elongation from the Sun continues to decrease steadily during the month. By the end of January, the planet will be unobservable from northern temperate latitudes, although observers in the tropics and more southerly latitudes may still glimpse the planet in the dawn twilight sky slightly to the south of east.

MARS, magnitude +1.2, moves from Capricornus into Aquarius during January. The planet sets just under two hours after the Sun, and is inconveniently low down in the south-western sky at dusk.

JUPITER was at opposition in early December 2012, and continues to be observable for most of the night. It fades from magnitude -2.7 to -2.5 during the month, but remains a brilliant object; its northerly declination means that observers in the Northern Hemisphere will have the best view. The planet is moving retrograde in Taurus, close to the northern arm of the V-shaped Hyades cluster, until the end of the month when it reaches its second stationary point and resumes direct motion once more. Figure 1 shows the path of Jupiter against the background stars during 2013. The four Galilean satellites, which Galileo first saw in January 1610, are readily observable with a small telescope or even a good pair of binoculars provided that they are held rigidly. The waxing gibbous Moon will appear quite close to Jupiter on the night of 21/22 January.

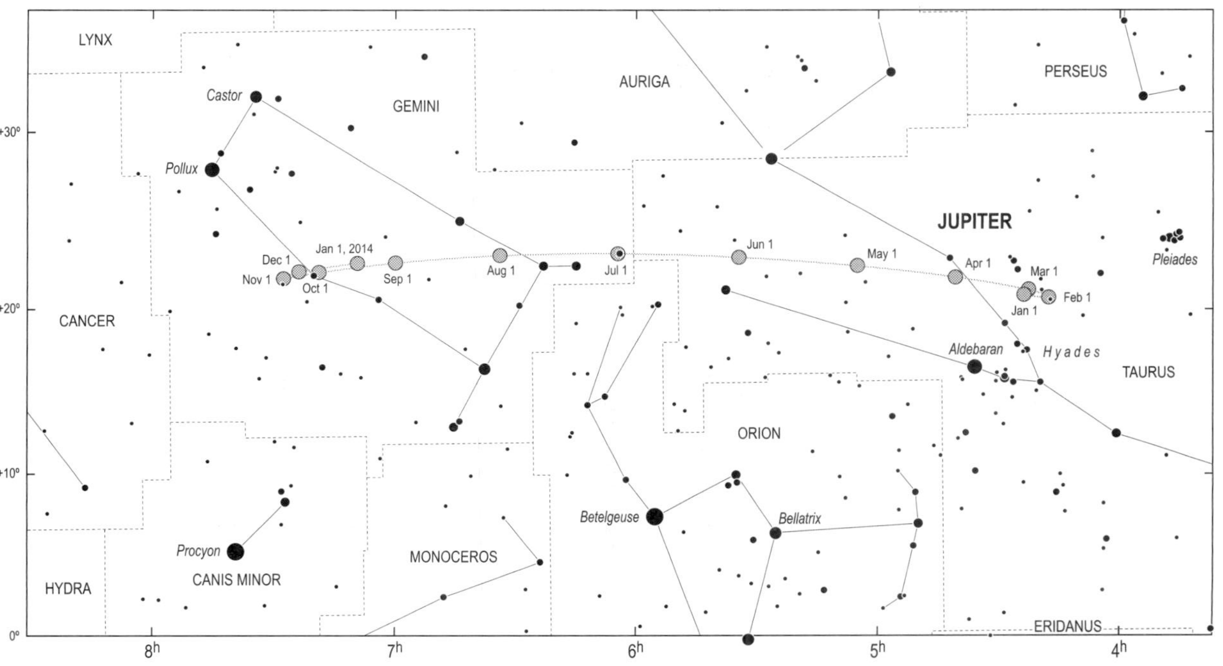

Figure 1. The path of Jupiter as it moves against the background stars of Taurus and Gemini during 2013.

SATURN, magnitude +0.6, is in Libra this month and continues to be visible in the south-eastern sky before dawn. The apparent tilt of the rings, as viewed from Earth, is now almost 19°, so the planet's ring system is beautifully displayed.

A Serendipitous Discovery. Delving back in astronomical history, we find various cases of 'serendipity' – that is, finding something which was not expected and which was not being sought. Perhaps the most remarkable case concerns no less a person than Galileo, the first great telescopic observer, who built his tiny 'optick tube' and used it, in the winter of 1609–10, to make a series of discoveries which overturned many long-cherished theories of the universe. Galileo made an early observation of the four bright satellites of Jupiter – Io, Europa, Ganymede and Callisto – in January 1610. He may not have been the first to see them, but undoubtedly he was the first to study them and their movements in considerable detail, which is why they are usually called the Galilean satellites. Following the discovery, Galileo routinely observed the satellites of Jupiter, and he not only measured their positions relative to the planet but he also carefully drew the observed arrangement of the satellites in his notebooks.

It was by carefully studying these drawings in Galileo's notebooks and comparing them with his own calculations that the astronomer Charles T. Kowal made a most surprising and exciting discovery. There was another object which Galileo saw at least twice, probably three times – in December 1612 and again in January 1613. It was the planet Neptune! Late in 1612 Neptune had moved into the same telescopic field with Jupiter and its satellites. On 27–28 December 1612, Galileo had plotted the position of what he thought was a fixed star on his sketches of Jupiter and its satellites. In fact, it was not a star at all; this was the first known sighting of Neptune (Figure 2a). Shortly thereafter, on 3 January 1613, Neptune was actually occulted by Jupiter (an event which happened again on 19 September 1702). Subsequently, Neptune remained in close proximity to Jupiter throughout January 1613. It is probable that Galileo plotted the position of Neptune again on 5–6 January, because in the second of his sketches that night there is a black spot in exactly the position where Neptune was at the time. Magnification of the spot shows that it was an intentional mark made with ink.

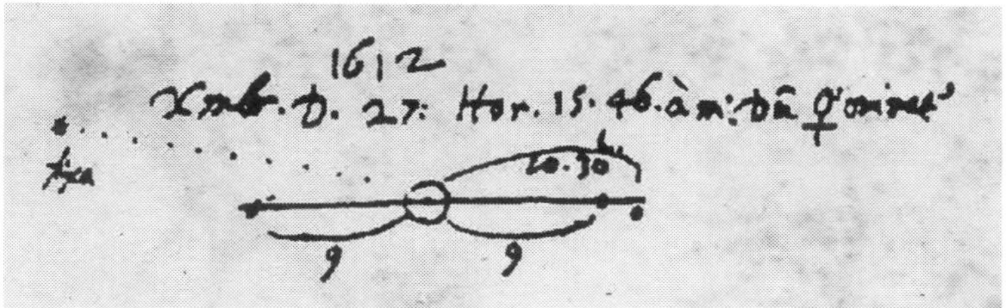

Figure 2a. An extract from Galileo's notebook from the night of 27/28 December 1612. The Galilean satellites and their separations from Jupiter are shown along with the location of a 'fixed star' which turns out to have been Neptune. The top line reads 'December 27: Hour 15.46 after noon: while Venus was rising', and the text describes the configuration of the satellites.

On 28 January 1613, Galileo plotted Neptune's position once again and this observation is perhaps the most interesting. On this occasion Galileo drew two 'stars' near Jupiter. Star 'a' was SAO 119234, but star 'b' was Neptune (Figure 2b). Galileo made a separate drawing showing just these two stars. Charles Kowal notes in his account of the discovery that Galileo wrote a comment in Latin which he translated as: 'After

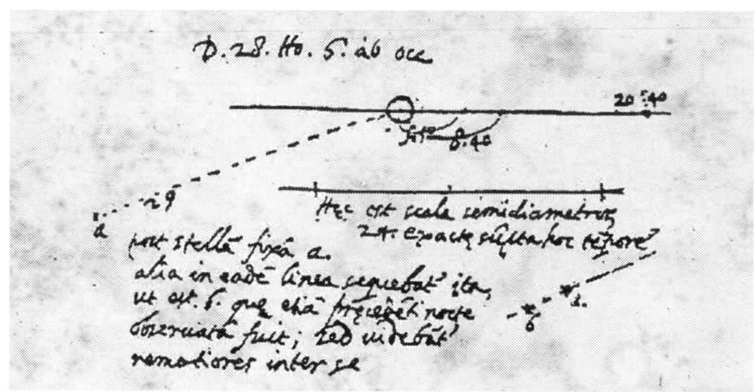

Figure 2b. An extract from Galileo's notebook for 28 January 1613 which shows Jupiter, three Galilean satellites, and two 'stars' near Jupiter. Star 'a' was SAO 119234, but star 'b' was Neptune. Galileo's separate sketch showing just these two 'stars' is lower right. Galileo noted that the two 'stars' seemed further apart than they had been the previous night.

fixed star "a", following in the same line, is star "b" which I saw in the preceding night, but they seemed farther apart from one another.' So, as Kowal notes with great excitement in his narrative, not only had Galileo spotted Neptune, but he had even noticed that it had moved from night to night!

So, four hundred years ago this very month, Galileo recorded the planet Neptune 168 years before the discovery of the seventh planet (Uranus) by William Herschel and well over 233 years before Neptune itself was positively identified by Johann Galle and Heinrich D'Arrest at the Berlin Observatory in September 1846.

Oppositions of the Planets. During 2013 there are oppositions of Saturn, Uranus, Neptune and the dwarf planet Pluto – as indeed there are in almost every year. Jupiter was last at opposition in early December 2012 and the next opposition will be in early January 2014 so, rather unusually, there will be no opposition of the giant planet in 2013. With Mars, the situation is quite different. There has been no opposition since 3 March 2012, and the next will not occur until 8 April 2014, because the synodic period of Mars – 780 days – is much longer than that of any other planet.

The synodic period is the mean interval between successive oppositions. To explain what is meant, let us first consider Pluto, which moves round the Sun at a mean distance of about 5,920 million kilometres (3,680 million miles) – though its orbit is unusually eccentric, and at its closest it can come within the orbit of Neptune. As well as having a large orbit, Pluto is slow-moving. In one year, the time taken for the Earth to go once round the Sun, Pluto covers only a tiny fraction of its orbit – so that the Sun, Earth and Pluto are lined up every 366.7 days. Having been right round the Sun in 365.25 days, the Earth takes only about an extra day and a half to catch up with Pluto.

With the closer planets, more time is required; the synodic periods for Neptune, Uranus, Saturn and Jupiter are respectively 367.5, 369.7, 378.1, and 398.9 days – so that, for instance, oppositions of Jupiter occur on average about 34 days later each year. Coincidentally, with Jupiter last coming to opposition on 3 December 2012, adding 34 days takes the date of the following opposition to 5 January 2014, missing out 2013 altogether! In subsequent years, Jupiter will come to opposition on 6 February 2015, 8 March 2016, and so on.

Mars is a special case. Its mean orbital velocity is comparable with that of the Earth, and the synodic period is therefore a great deal longer. Moreover, the greater eccentricity of Mars's orbit leads to considerable variation in the interval between successive oppositions; although the synodic period is 780 days, the interval may be as short as 764 days or as long as 810 days. Consequently, oppositions of Mars do not occur every year; thus there were oppositions in 2010 and 2012, but there will be none in 2013 – and, following the opposition of 2014, 2015 will be another 'blank year' so far as Mars is concerned.

Of course, Mars will not be invisible during 2013. At the start of the year it is becoming lost in the twilight at dusk as it draws in towards superior conjunction, which it reaches in mid-April. By July, it is just becoming visible in the early morning twilight sky, and by the end of the year it will be visible for several hours in the south-eastern sky before dawn. However, it will be a long way away, and not even large telescopes will show much upon its disk. It must be admitted that observers of Mars must resign themselves to inactivity through most of 2013.

February

New Moon: 10 February *Full Moon*: 25 February

MERCURY becomes visible in the western sky in the evenings for observers in tropical and northern latitudes after the first week of the month; it is not favourably placed for observers in southern temperate latitudes. For Northern Hemisphere observers this is the most favourable evening apparition of the year. Figure 3 shows, for observers in latitude 52° N, the changes in azimuth (true bearing from the north through east, south and west) and altitude of Mercury on successive evenings when the Sun is 6° below the horizon. This is at the end of evening civil twilight and in this latitude and at this time of year occurs about thirty-five minutes after sunset.

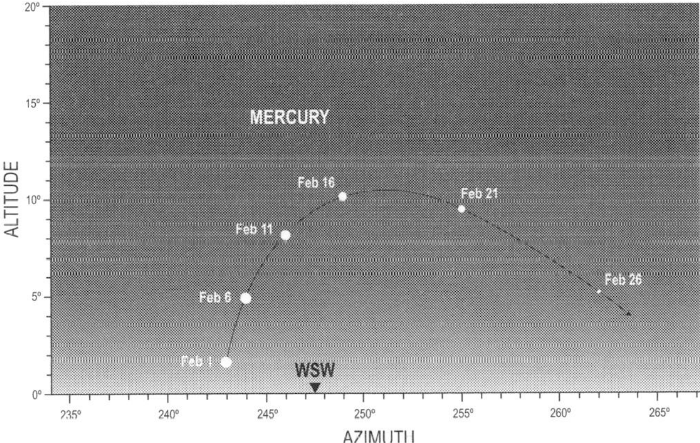

Figure 3. Evening apparition of Mercury from latitude 52°N. The planet reaches greatest eastern elongation on 16 February 2013. It will be at its brightest in early February, before elongation.

The changes in the brightness of the planet are indicated on the diagram by the relative sizes of the white circles marking Mercury's position at five-day intervals: Mercury is at its brightest before it reaches greatest eastern elongation (18°) on 16 February. Between 6 and 25 February, Mercury fades from magnitude -1.1 to +1.8. The diagram gives positions for a time at the end of evening civil twilight on the Greenwich meridian on the stated date. Observers in different longitudes should note that the actual positions of Mercury in azimuth and altitude will differ slightly from those given in the diagram due to the motion of the planet. In late February, Mercury's elongation from the Sun rapidly decreases as the planet draws in towards inferior conjunction early next month.

VENUS, magnitude -3.9, is lost in the morning twilight for observers in northern temperate latitudes as it draws in towards superior conjunction. Observers in the tropics and more southerly latitudes may still glimpse the planet in the dawn twilight sky slightly to the south of east in early February, but even they will lose sight of the planet by month end.

MARS, magnitude +1.2, is in Aquarius and sets less than an hour after the Sun by the end of February. Consequently, it will be unsuitably placed for observation this month.

JUPITER will be seen in the southern sky as soon as darkness falls, and continues to be visible until well into the early morning hours. The planet is moving direct in Taurus and fades very slightly from magnitude -2.5 to -2.3 during February. The waxing first quarter Moon will appear quite close to Jupiter on the early evening of 18 February and the two objects will be a pleasing sight in the sky.

SATURN rises around midnight by the end of February, brightening slightly from magnitude +0.5 to +0.4 during the month. The planet reaches its first stationary point, in Libra, on 21 February and thereafter its motion is retrograde. The waning last quarter Moon will be about 3° south of Saturn in the early morning hours of 3 February.

Asteroid's Near Miss. The tiny near-Earth object 2012 DA14 will pass within about 22,250 kilometres (13,825 miles) of the Earth's surface on 15 February 2013. Although its size is not known with great certainty, it is thought to be about forty metres across. It was discovered by the LaSagra observatory in southern Spain. During its close encounter with

Earth, asteroid 2012 DA14 will actually pass inside the ring of geosynchronous satellites girdling Earth's equator (Figure 4), but its solar orbit can bring it no closer to the Earth's surface than about 20,300 kilometres (12,615 miles), so there is no chance of a collision.

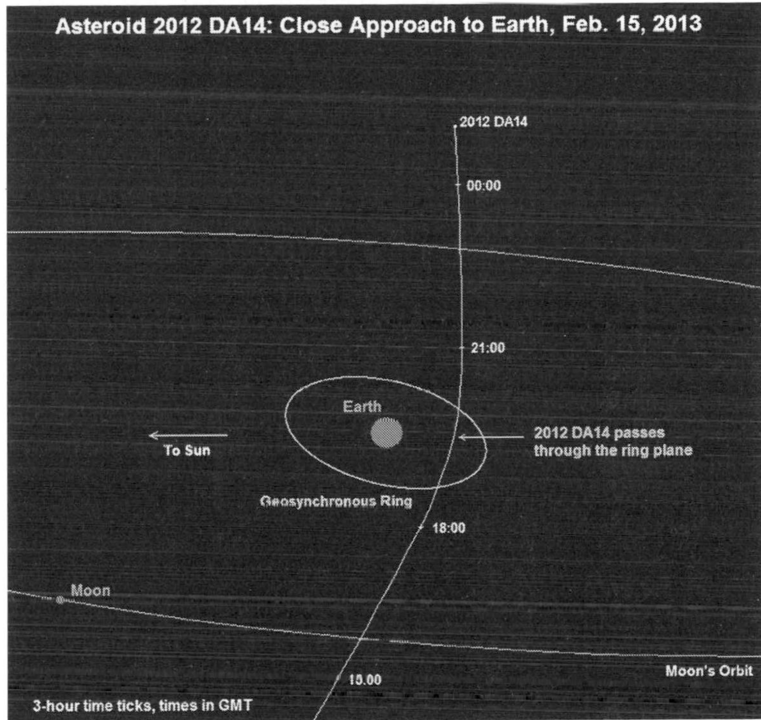

Figure 4. An oblique view showing the path of near-Earth asteroid 2012 DA14 as it passes close to Earth on 15 February 2013, inside the ring of geosynchronous satellites girdling Earth's equator. (Diagram courtesy of Paul Chodas, Jon Giorgini and Don Yeomans, NASA/JPL Near-Earth Object Program Office.)

During its flyby of Earth, the asteroid will move rapidly northwards from the Southern Hemisphere into the Northern, with closest approach occurring at about 19h 26m UT (at a favourable elongation from the Sun) when it will attain eighth magnitude – visible in good binoculars – but it will quickly fade in brightness. Shortly after close approach, the asteroid may pass through Earth's shadow for about eighteen minutes before reappearing from eclipse.

A Remarkable Procession of Meteors. One hundred years ago this month, on 9 February 1913, a great procession of slow-moving fireballs was witnessed by observers across Canada, north-eastern parts of the USA, Bermuda and from several ships at sea, including one off the coast of Brazil. The details of this remarkable event, which appears to be without parallel in the history of meteoritics, may well have been forgotten were it not for the painstaking work of Professor Clarence A. Chant, Director of the David Dunlap Observatory in Toronto and Editor of the *Journal of the Royal Astronomical Society of Canada.* Although not a witness to the extraordinary meteor display himself, Chant was alerted to its occurrence when, shortly after 9 p.m. Eastern Standard Time on the night in question, the Observatory received numerous telephone calls, reporting the display, which had lasted several minutes. The following day, newspapers carried a request from Chant for people to send in reports and he collected these together and undertook a thorough investigation.

The eyewitness accounts that Chant obtained revealed that the fireball procession was first seen west of Regina in Saskatchewan, and then traversed successively Manitoba, Minnesota, Michigan, Ontario, New York, Pennsylvania and New Jersey, just to the south of New York City. It then went out over the sea and was sighted from Bermuda. A painting of the meteor procession, as observed from Toronto, is shown in Figure 5.

Figure 5. An impression of the great fireball procession of 9 February 1913, as seen from Toronto, by Canadian artist Gustav Hahn. (Image courtesy of University of Toronto.)

To gain some insight into the remarkable phenomenon witnessed that night (from Western Ontario), here are a few extracts from the summary of Chant's detailed account of the 9 February event, published in the May–June 1913 issue of the *Journal of the Royal Astronomical Society of Canada*:

> At about 9:05 on the evening in question there suddenly appeared in the north-western sky a fiery red body which quickly grew larger as it came nearer, and which was then seen to be followed by a large tail . . . In the streaming of the tail behind, as well as in the color, both of the head and the tail, it resembled a rocket; but, unlike a rocket, the body showed no indication of dropping to the earth. On the contrary, it moved forward on a perfectly horizontal path . . . continuing in its course, without the least apparent sinking towards the earth, it moved to the south-west where it simply disappeared in the distance. Before the astonishment aroused by this first meteor had subsided, other bodies were seen coming from the north-west, emerging from precisely the same place as the first one. Onward they moved, at the same deliberate pace, in twos or threes or fours, with tails streaming behind . . . They all traversed the same path and were headed for the same point in the south-eastern sky . . . To most observers the outstanding feature of the phenomenon was the slow, majestic motion of the bodies; almost equally remarkable was the perfect formation which they retained . . . Just as the bodies were vanishing, or shortly afterwards, there was heard in many places a distinct rumbling sound, like distant thunder . . . As to the number of bodies, there is great diversity of statement. The usual estimate is from 15 to 20, but some say 60 or 100, while some say there were thousands.

Observers' estimates of the duration of this great display varied enormously; from one-and-a-half minutes all the way up to eight minutes; the average was 3.3 minutes. The two observers who actually timed the event gave durations of three minutes forty seconds and five minutes forty seconds. Apparently, they both saw the whole of the fire ball procession.

Following publication of Chant's account, which aroused intense interest, further eyewitness reports came to light over time. Appeals for observations and various enquiries over a number of years by astronomers such as William F. Denning and William H. Pickering revealed, as has already been said, that the event had been witnessed from ships in the Atlantic Ocean beyond Bermuda and even off the coast of Brazil. The track of the fireball procession thus extended over an astounding distance of about 9,100 kilometres (5,650 miles). Later extensive enquiries conducted by Alexander D. Mebane in the mid-1950s (encouraged by astronomer and meteoriticist Lincoln LaPaz), brought to light previously unknown reports of the event through Canada and the USA and largely confirmed Chant's derived path of the bright meteor procession as published in his 1913 paper.

In the late 1950s and 1960s, planetary scientist John A. O'Keefe re-examined the observations of the great fireball procession, and even ventured to give it a name. Since 9 February was then the feast day of St Cyril of Alexandria, he dubbed the phenomenon the 'Cyrillids'.

O'Keefe's research largely corroborated Chant's original theory (slightly modified by Pickering and others) and ruled out a number of competing ideas. It seems that the bodies had been travelling through space, probably in an orbit about the Sun, and that on approaching the Earth they were captured by its gravity into a satellite-like orbit. The bright meteor procession thus consisted of a large number of natural objects in low Earth orbit which entered the atmosphere at a very low velocity of around eight kilometres per second (five miles per second) and burned up along an extremely long path length. O'Keefe found that a satellite-like orbit could be made to fit the observations over the entire observed track. However, O'Keefe agreed with Mebane in his conclusion that the same fireballs were not seen from every location along the entire track. The procession was caused by multiple objects burning up in the atmosphere, with larger objects fragmenting into greater numbers of smaller ones, all following similar but not identical atmospheric paths, and no single fireball was followed all the way from Saskatchewan to the ocean off the coast of Brazil.

The Power of the Naked Eye. How much can one see without optical aid? Obviously, it depends upon the quality of one's eyesight, but it is interesting to examine some cases which are at the very limit of visibility.

The Pleiades star cluster in Taurus, the Bull is a good test. This cluster is well placed, high in the southern sky early on February evenings from northern temperate latitudes. It is known as the 'Seven

Sisters', and it is true that people with average sight can see seven individual stars under good conditions; but very keen-eyed people can go much further, and the record is said to be held by a nineteenth-century German mathematician and astronomer, Eduard Heis, who could see up to nineteen. (Incidentally, Heis was also the first person to record a count of the Perseid meteor shower in August 1839, giving an hourly rate of 160.)

Then there are the phases of Venus. There seems little doubt that when the planet is in the crescent stage, the phase really can be seen by people with exceptional sight. I (PM) once conducted a rather unkind experiment on television. I showed a photograph of Venus, which at that time was in the crescent stage in the evening sky, and asked viewers to draw the phase – if they could see it with the unaided eye – and send the drawings in to me. What I had done was to show the telescopic view, which is, of course, reversed. Many people sent in sketches reproducing this, so that obviously they were guilty of 'seeing what they expected to see', but three viewers wrote in baffled vein to the effect that they could see the crescent, but the opposite way round to the crescent I had put on the screen. Those were the genuine sightings.

There is also an interesting fact about the Galilean satellites of Jupiter. Jupiter, too, is well placed in February, roughly midway between the Pleiades and the orange-red star Aldebaran (Alpha Tauri). But for the overpowering brilliancy of Jupiter itself, Ganymede (the largest of the planet's four Galilean satellites) at least would be an easy naked-eye object. There are well-authenticated cases that it can be glimpsed, and of these the first goes back to the year 364 BC! One of the earliest known Chinese astronomers was Gan De. Nothing is known about his life, and unfortunately his original works have been long since lost, but portions were reproduced in a book on astrology written by Qutan Xida around AD 720. Gan De is quoted as follows:

> In the year of chan yan . . . Jupiter rose in the morning and went under in the evening together with the Lunar Mansions Xunu, Xu and Wei. It was very large and bright. Apparently there was a small reddish star appended to its side. This is called 'an alliance' (*tong meng*).

Tong meng was a term used to describe a close alliance. And since the year can be checked, there seems little doubt that what Gan De saw was a satellite. It must have been either Ganymede or Callisto, which can move out furthest from Jupiter, but as Callisto is much the fainter of the two, Ganymede is the more probable object. Alternatively, Gan De may have seen two of the Galileans very close together; we cannot be sure, but at least he seems to have the honour of priority (anticipating Galileo by almost 2,000 years), even though he can have had absolutely no idea that he was looking at a satellite of Jupiter rather than at a star.

March

New Moon: 11 March *Full Moon:* 27 March

Equinox: 20 March

Summer Time in the United Kingdom commences on 31 March.

MERCURY is not visible from the latitudes of northern Europe or North America this month. The planet is in inferior conjunction on 4 March, but thereafter moves rapidly out from the Sun to become visible to observers in the tropics and more southerly latitudes. For Southern Hemisphere observers this is the most favourable morning apparition of the year. Figure 6 shows, for observers in latitude 35°S,

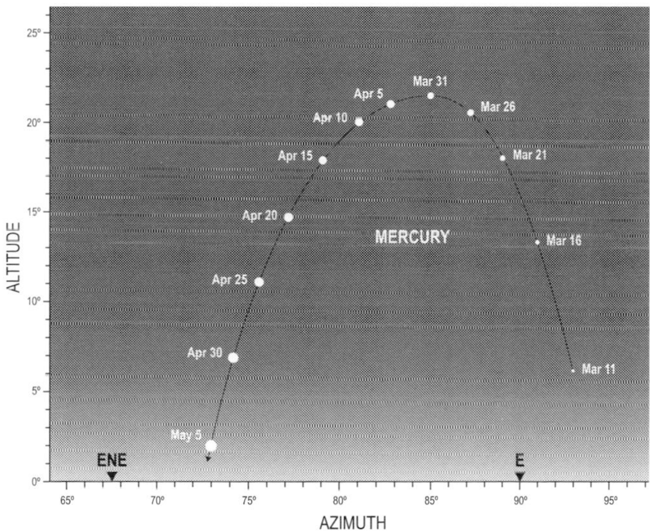

Figure 6. Morning apparition of Mercury from latitude 35°S. The planet reaches greatest western elongation on 31 March 2013. It will be at its brightest in late April, several weeks after elongation.

the changes in azimuth (true bearing from the north through east, south and west) and altitude of Mercury on successive mornings when the Sun is 6° below the horizon. This is at the beginning of morning civil twilight, which in this latitude and at this time of year occurs about twenty-five minutes before sunrise.

During its long period of visibility, which runs from mid-March until the end of April, Mercury brightens from magnitude +1.6 to -1.0. The changes in the brightness of the planet are indicated on the diagram by the relative sizes of the white circles marking Mercury's position at five-day intervals. It should be noted that Mercury is at its brightest over four weeks after it reaches greatest western elongation (28°) on 31 March. The diagram gives positions for a time at the beginning of morning civil twilight on the Greenwich meridian on the stated date. Observers in different longitudes should note that the actual positions of Mercury in azimuth and altitude will differ slightly from those given in the diagram due to the motion of the planet.

VENUS passes through superior conjunction, on the far side of the Sun, on 28 March. Consequently, it will be unsuitably placed for observation throughout the month.

MARS is in conjunction with the Sun in mid-April, and so the planet is unobservable this month.

JUPITER is visible in the southern sky as soon as darkness falls, remaining visible until an hour or so after midnight by the end of the month from northern temperate latitudes, but setting well before midnight from locations in the tropics and further south. The planet is moving direct through Taurus, not far from Aldebaran and the Hyades, and fades very slightly from magnitude -2.3 to -2.1 during March. The waxing crescent Moon will appear quite close to Jupiter on the evening of 17 March and the two objects will make a nice pairing.

SATURN may be seen rising in the eastern sky around midnight at the beginning of March by observers in the latitudes of the British Isles, and a couple of hours earlier by month end; the period of visibility will be even longer for those in the tropics and further south. The planet is moving retrograde in Libra. The waning gibbous Moon will appear quite close to Saturn on the early morning of 2 March and as it rises late evening on 29 March. Saturn is at opposition late next month and it brightens slightly from magnitude +0.4 to +0.3 during March.

A Naked-Eye Comet? As Martin Mobberley describes in the section 'Comets in 2013' elsewhere in this *Yearbook*, there is a good chance that there will be a naked-eye comet on view this month. Mind you, comets are the most erratic members of the Solar System; they may sometimes appear spectacular, but predicting how bright they are going to appear when at their best is notoriously difficult. Nevertheless, it does look as though the comet C/2011 L4 (PANSTARRS) could be the cometary highlight of 2013 and may well be a naked-eye object from mid-February through to mid-April, although it will only be well placed for Northern Hemisphere observers from mid-March onwards. The comet passes just 0.30 AU (44.9 million kilometres) from the Sun at perihelion passage on 11 March and should appear in the western evening twilight sky some days after that. How soon after perihelion the comet appears in the evening sky depends on just how bright it is at that time.

Comet C/2011 L4 (PANSTARRS) will be moving almost directly northwards through the constellations of Pisces and Andromeda, roughly parallel to the eastern side of the Great Square of Pegasus, from mid-March to mid-April. It should be a naked-eye object throughout this time, although fading steadily, and because it will be receding from the Sun it will be travelling 'tail first'. The comet should be displaying both a narrow, straight ion or gas tail and a broader, curving dust tail at this time. For a helpful chart showing its movement relative to the background stars, see the section 'Comets in 2013' on p. 162.

From latitude 52°N, on the evening of 18 March, about 35 minutes after sunset, comet C/2011 L4 (PANSTARRS) will be about 15 degrees high, slightly north of due west in a bright twilight sky. Ten days later, on the evening of 28 March, about 35 minutes after sunset, the comet will be 20° degrees high, about 30° degrees north of west, but it is likely to be a couple of magnitudes fainter by this time, although still a naked-eye object.

Thereafter, the comet races northwards, though continuing to fade, passing a degree or so west of the Andromeda Galaxy (M31) on 5 April. By this time the comet will be visible in a dark sky and although then only a fifth-magnitude object, it should still have a noticeable tail.

One can only hope for the best, because there is always a fair degree of uncertainty where comets are concerned. Interestingly, it was at the same time of the year, in late March and early April of 1997, that observers in northern temperate latitudes enjoyed great views of the bright naked-eye comet, C/1995 O1 Hale-Bopp, the most celebrated comet of recent times. It remained a naked-eye object for many months and was truly beautiful (Figure 7). Marked changes were observed in the tails and a spiral structure within the comet's head.

Figure 7. Comet C/1995 O1 Hale-Bopp imaged by Martin Mobberley on 30 March 1997 using ISO 400 Kodak colour film with a 160mm f/3.3 Takahashi Astrographic reflector. Two types of tail, a bright dusty swath and a fainter, intricate, ion tail are shown, although much of the blue ion tail detail is lost in this black and white reproduction.

'The Old Moon in the New Moon's Arms'. From 13 to 16 March, the crescent Moon will be well placed in the western sky at dusk. Many people believe that the thin crescent Moon, appearing in the evening sky soon after sunset, is the true 'new Moon'. In fact, this is not true. New Moon occurs when the Moon lies between the Earth and the Sun. Its dark side is then turned towards us, and, under normal circumstances (such as on 11 March), we cannot see the Moon at all, though if the alignment is perfect we are treated to the spectacle of a solar eclipse.

During the crescent stage, you will probably see the 'night' side of the Moon shining faintly. This is the appearance which country folk often call 'the Old Moon in the New Moon's arms'. Originally it caused a great deal of bewilderment, and it was even suggested that the Moon itself might be faintly self-luminous. However, Leonardo da Vinci, the great Italian scientist and painter who lived between 1452 and 1519, realized the truth. He wrote:

> The Moon has no light of itself, but so much of it as the Sun sees, it illuminates. Of this illuminated part we see as much as faces us. And its night receives as much brightness as our waters lend it as they reflect on it the image of the Sun, which is mirrored in all those waters that face the Sun and the Moon.

In other words, Leonardo da Vinci was saying that the faint luminous glow is caused by light reflected on to the Moon by the Earth; sometimes it is called 'the da Vinci glow', but nowadays we generally call it 'Earthshine'. The Earth is nearly fully illuminated as viewed from the Moon, when the latter appears as a thin crescent in the evening sky. So sunlight reflected from the Earth on to the night side of the Moon causes it to glow faintly and the entire globe of the Moon is faintly discernible.

Earthshine can be quite bright, and with binoculars or a telescope you can see details in the earth-lit portion. There is one crater, the forty-kilometre (twenty-five-mile) diameter Aristarchus, which is highly reflective by lunar standards, and so is often visible when lit only by earthshine. The great astronomer William Herschel seems to have mistaken the earth-lit Aristarchus for an erupting volcano in 1789, and one can hardly blame him. The earthshine is usually seen whenever the Moon is a slender crescent, though of course a clear sky is needed (mist will drown the faint glow). The fact that the earthshine is not always equally bright has nothing to do with the Moon, but depends upon the state of our own atmosphere. If the Earth is largely cloud-covered, it will reflect more light, and this brightens the earthshine.

It has been said that colour can be seen in the Earth-lit part of the Moon; I (PM) have never seen anything of the kind, but positive reports are quite common. Photographing the earthshine is quite a challenge, because one tends to overexpose the sunlit crescent; but it can be done, and although such photographs are of no real scientific value, it is worth taking a picture of 'the Old Moon in the New Moon's arms' for its beauty alone.

Ursa Major, the Great Bear. Ursa Major is a large constellation. The so-called Plough, an asterism made up of seven stars, is only part of it, and the name is unofficial; it has also been called King Charles's Wain, while Americans know it as the Big Dipper. On the whole, we must admit that the Americans are the most rational. The pattern is nothing like that of a bear or a plough, but it does give a slight impression of a large spoon or ladle (Figure 8).

The seven main stars of the Plough all have proper names. Starting from the end of the plough handle (or the spoon handle), the stars are:

Eta or Alkaid (magnitude 1.9)
Zeta or Mizar (2.1)
Epsilon or Alioth (1.8)
Delta or Megrez (3.3)
Gamma or Phad (2.4)
Beta or Merak (2.4)
Alpha or Dubhe (1.8)

The proper names are Arabic, and some of them have variants; thus Phad has also been called Phekda or Phecda, while Alkaid has the alternative name of Benetnasch.

Ursa Major is extremely easy to find, not because its stars are particularly bright, but because the pattern is so well marked. Broadly speaking it may be seen rather low in the north during winter evenings, high in the north-east during spring evenings, high in the north-west during summer evenings, and descending in the north-west during autumn evenings. It may pass right overhead, as it does around midnight in March. It may, of course, appear at all sorts of angles. Look for it during a late evening in January, and you will see the Bear 'standing on its tail'. But wherever it may be, it stands out at once.

Alioth, Dubhe and Alkaid are all very slightly above the second magnitude, and to all intents and purposes they appear equally bright, but this does not mean that their true luminosities are the same. Alkaid is over a hundred light years away, Dubhe and Alioth less than eighty, so that Alkaid is really the most powerful of the three. Megrez, or Delta Ursae Majoris, is obviously fainter than its six companions, and there is

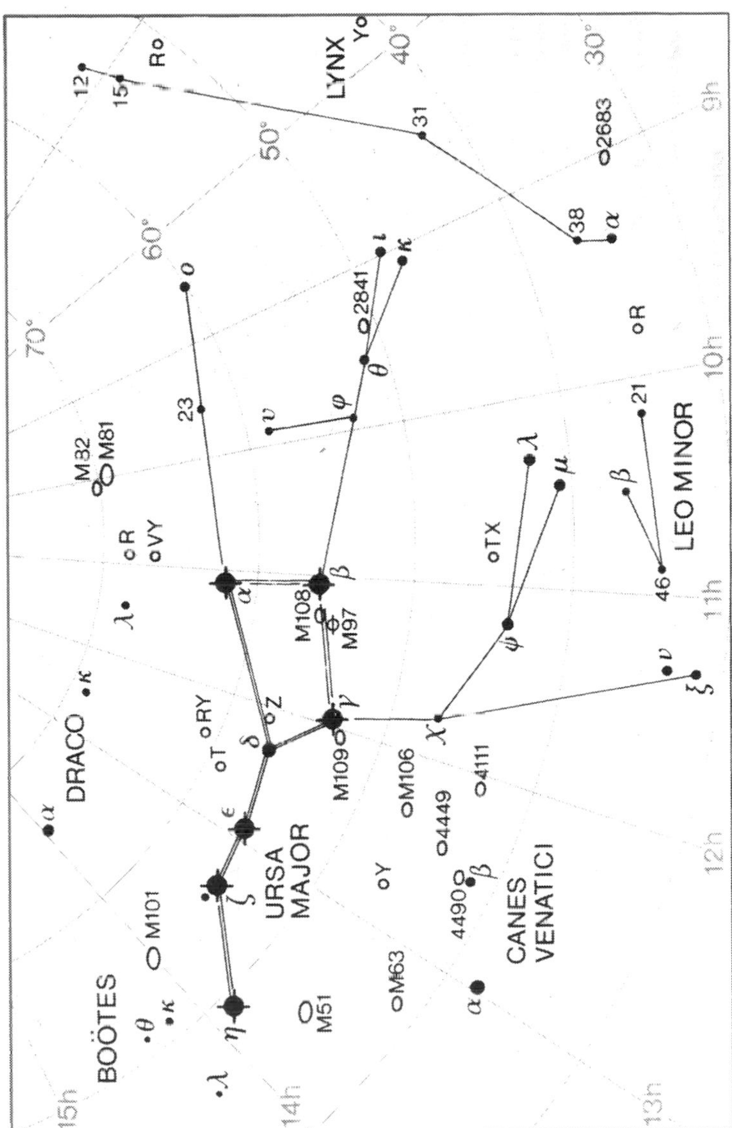

Figure 8. Map showing the principal stars of Ursa Major, the Great Bear, a fairly large constellation which includes the seven stars making up the very well-known and quite distinctive grouping known as the Plough, or Big Dipper.

a minor mystery about it, because the astronomers of more than a thousand years ago stated that it was as bright as the other Ursa Major stars. Either they were wrong, or (less probably) Megrez has actually faded by about a magnitude since then.

With one exception, all of the brightest Ursa Major stars are white (of spectral types A and B). The exception is Dubhe, which is somewhat orange in hue (spectral type K), showing that its surface temperature is lower. The difference is not hard to detect with the naked eye; binoculars, of course, bring it out very clearly.

The most famous member of Ursa Major is Mizar, because it is accompanied by the fourth-magnitude Alcor. The Arab astronomers of a thousand years ago were familiar with it, but, strangely, they regarded it as a test of eyesight, which is certainly not true today; it is easy to detect against any reasonably clear and dark sky, though haze will conceal it.

Apart from Alkaid and Dubhe, the stars of Ursa Major are travelling through space in much the same direction at much the same rate, so they make up what is termed a 'moving cluster'. The individual movements are too slight to be noticed with the naked eye over periods of many lifetimes, but eventually the shape will become distorted; 50,000 years hence, for example, Alkaid will lie 'below' Mizar, and Dubhe will have made off away from Merak.

Undoubtedly, Ursa Major is the most useful signpost in the sky to northern Europeans and North Americans. Moreover, the position of Ursa Major itself will indicate how many of the surrounding groups are on view and how many are not. When Ursa Major is standing upright over the horizon, for instance, there is no point in looking for Arcturus, the leading star of Boötes, the Herdsman, which will be out of sight.

The remainder of Ursa Major covers a wide area, but is not particularly conspicuous. There are two triangles to the south and west of the Plough (marking the front and rear paws of the bear), made up of stars between magnitudes 3 and 4, easy to identify when the constellation is high in the sky, but often obscured by horizon haze when Ursa Major is at its lowest.

April

New Moon: 10 April *Full Moon:* 25 April

MERCURY is on view before dawn, in the east-north-eastern twilight sky, for observers in the tropics and Southern Hemisphere throughout April. From more northerly latitudes the planet will not be visible this month. Figure 6, given with the notes for March, shows, for observers in latitude 35°S, the changes in azimuth (true bearing from the north through east, south and west) and altitude of Mercury on successive mornings when the Sun is six degrees below the horizon, about twenty-five minutes before sunrise in this latitude. The changes in the brightness of the planet are indicated on the diagram by the relative sizes of the white circles marking Mercury's position at five-day intervals. It will be noted that Mercury is at its brightest at the end of April, over four weeks after greatest western elongation; the planet brightens from magnitude +0.3 to -1.0 during the month. Towards the end of April, Mercury's elongation from the Sun rapidly decreases as the planet draws in towards superior conjunction on 11 May. The thin waning crescent Moon will appear fairly close to Mercury in the dawn twilight sky on the morning of 8 April.

VENUS passed through superior conjunction towards the end of March and so is unobservable during April.

MARS is in conjunction with the Sun on 17/18 April and is therefore too close to the Sun to be observable this month.

JUPITER is visible throughout the evening in the south-western sky, but setting before midnight by the end of the month from northern temperate latitudes and even earlier from locations further south. The planet continues to move direct through Taurus; its brightness decreases from magnitude -2.1 to -2.0 during April. The waxing crescent Moon will appear quite close to Jupiter on the evening of 14 April, between the 'horns' of Taurus, the Bull.

SATURN, magnitude +0.3, is at opposition in Libra on 28 April, brightening slightly from magnitude +0.2 to +0.1 during the month. Figure 9 shows the path of Saturn against the background stars during the year. The planet becomes visible in the east-south-eastern sky as soon as darkness falls and is observable all night long. Saturn is a lovely sight in even a small telescope, with the rings displayed at an angle of 18.2° as viewed from the Earth. The full Moon lies a few degrees south of Saturn on 25/26 April. At opposition the planet is 1,320 million kilometres from the Earth.

Goodbye to Orion. The seasonal changeover in the night sky from winter to spring is more or less complete by April evenings. Orion, the symbol of winter, has to all intents and purposes disappeared, though the northernmost part of the constellation still lingers above the western horizon and does not actually set until about midnight. Aldebaran and Sirius are barely visible, and of the Hunter's retinue only Procyon in Canis Minor and Castor and Pollux, the Twins of Gemini, remain prominent – apart from the yellowish Capella, high in the west. The spring groups such as Boötes, Leo and Virgo have come well into view.

Ursa Major is practically overhead, which means that Cepheus and Cassiopeia, on the other side of the Pole Star, are reaching their lowest positions in the north, though from northern Europe and North America they never drop low enough to be difficult to find. Vega and Deneb are rising in the north-east, though they have not yet become prominent.

The best way to locate Vega at this time of the year is to use Capella and Polaris as pointers. Now that Sirius is out of the reckoning, no star in the evening sky apart from the lovely orange Arcturus in Boötes, the Herdsman, is anything like so brilliant. Only three stars (Sirius, Canopus and Alpha Centauri) are brighter than Arcturus, but all these lie to the south of the celestial equator.

The southern aspect is dominated by Leo, the Lion. Between Leo and the Twins is a rather obscure Zodiacal group, Cancer, the Crab, whose sole claim to distinction is that it contains the naked-eye open cluster Praesepe, also known as the 'Beehive'. From Cancer, the long, straggly outline of Hydra, the Watersnake (Figure 10 overleaf), stretches down to the horizon, passing beneath Leo, while perched on the Watersnake's back is a crow, Corvus, marked by a quadrilateral of four moderately bright stars which can be quite conspicuous on a clear April evening.

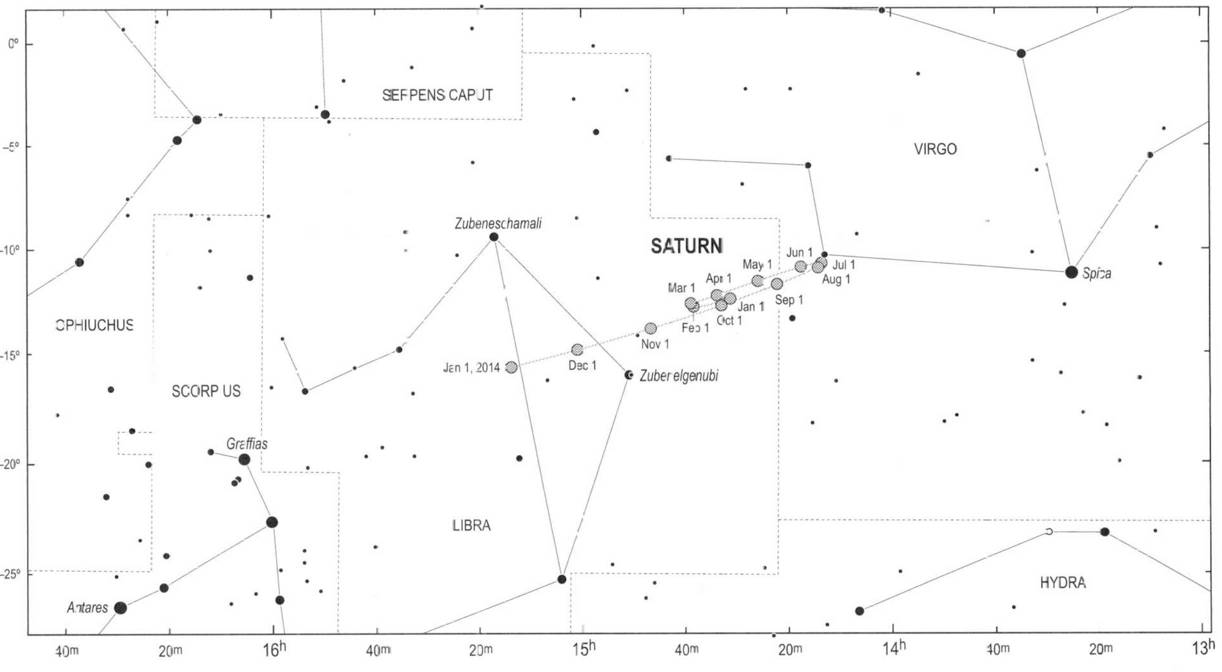

Figure 9. The path of Saturn against the background stars of Virgo and Libra during 2013

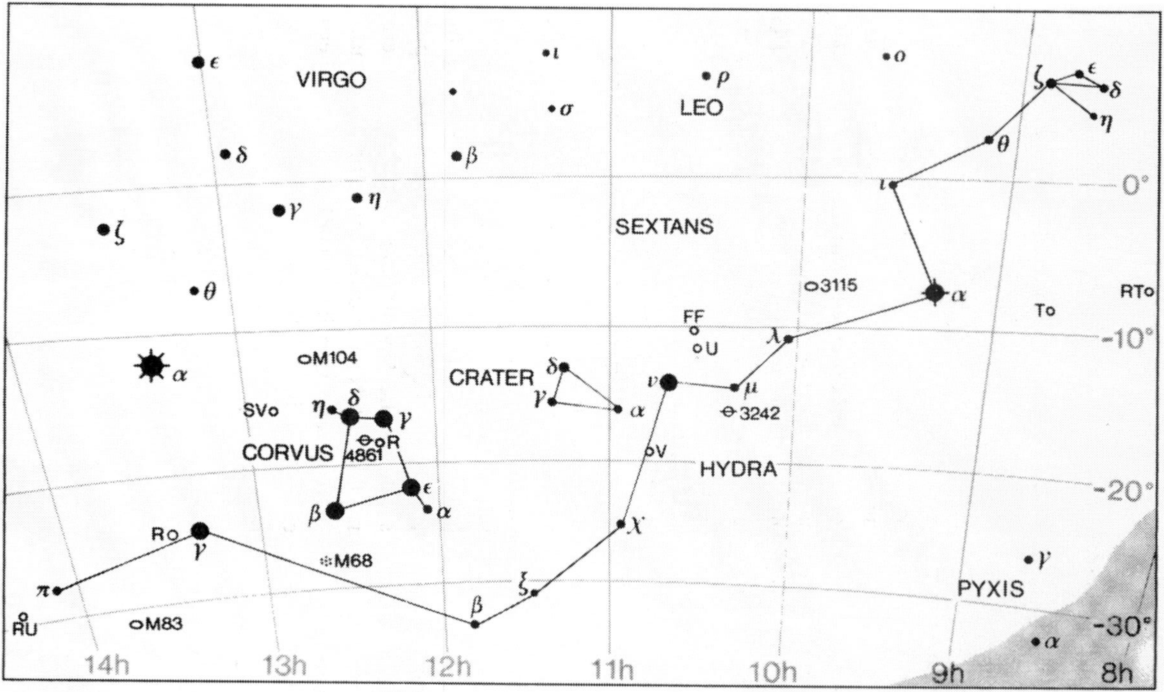

Figure 10. Map showing the long, straggly pattern of Hydra, the Watersnake, which extends from Cancer and passes south of Leo. Perched on the Watersnake's back is the pattern of Corvus, the Crow, marked by a quadrilateral of four moderately bright stars.

A curving line from the handle of the Plough through Arcturus will lead to another first-magnitude star, Spica, the brightest star of Virgo, the Maiden. At this time of the year Arcturus is visible all through the hours of darkness, while Spica does not set until shortly before sunrise.

The Meaning of Magnitude. When the Moon is full, its light drowns out all but the brightest stars, although even at its most brilliant the Moon is vastly inferior to the Sun. Such differences in apparent brightness are described by the apparent magnitude scale. The apparent magnitude of the full Moon is about –12.7, while that of the Sun is nearer –26.7. Apparent magnitude is a measure of the apparent brightness of a star, planet or other object observed from the Earth. The scale works rather in the manner of a golfer's handicap, with the more brilliant performers having the lower magnitudes. Thus magnitude 1 is brighter than 2, magnitude 2 is brighter than 3, and so on. The modern magnitude system, as proposed by Norman Pogson in 1856, defines a typical first-magnitude star as a star that is exactly 100 times as bright as a typical sixth-magnitude star. In other words, a difference of five magnitudes corresponds to a 100-fold difference in apparent brightness. So, a first-magnitude star is about 2.512 times as bright as a second-magnitude star; a second-magnitude star is about 2.512 times as bright as a third-magnitude star, etc.

The four brightest stars in the sky have magnitudes of below zero, so that they have minus values. Pride of place goes to Sirius, at –1.44. Its only rival is Canopus, at –0.72. The other 'minus' stars are Alpha Centauri in the far south of the sky (–0.27) and the Northern Hemisphere star Arcturus (– 0.05). Next in order come three stars near zero magnitude: Vega (+0.03) and Capella (+0.08) in the Northern Hemisphere, and Rigel (+0.12) in the Southern (though Rigel, in Orion, is not so very far south of the celestial equator, and is visible from every inhabited continent).

Conventionally, the first-magnitude stars are taken as those ranging from Sirius down as far as Regulus in Leo (the Lion), whose magnitude is +1.36. The list of first-magnitude stars includes Procyon in the Little Dog, Achernar in Eridanus, Betelgeux in Orion, Beta Centauri, Altair in Aquila, Alpha Crucis and so on. The Pole Star is almost exactly of magnitude 2, though admittedly it is very slightly variable. The faintest stars normally visible with the naked eye on a clear night are of magnitude 6. Binoculars will extend the range down to at least 9, and the world's largest telescopes can record objects down to around magnitude 30. Note that we are talking about 'apparent magnitude', because a star's apparent magnitude is no firm clue to its true intrinsic luminosity. The stars are at very different distances from us. Thus Sirius looks considerably brighter than Deneb, but only because it is so much nearer to us. Sirius has 26 times the luminosity of the Sun, while, according to a recent estimate, Deneb (whose exact distance – and hence luminosity – is not accurately known) could be almost 60,000 times as luminous as the Sun; if it were so close to us as Sirius (8.6 light years) it would cast strong shadows.

Of the planets, Venus can attain magnitude –4.4, Mars –2.8 and Jupiter –2.6, so that all seem much brighter than any of the stars. Even Saturn, with a maximum apparent magnitude of –0.3, can, on occasions when the rings are 'wide-open', outshine all the stars apart from Sirius, Canopus and Alpha Centauri. It is true that apparent magnitude can be a little misleading, but it has been used for many centuries and is not likely to be altered now. Once one becomes used to the system, it is convenient enough.

May

New Moon: 10 May   *Full Moon:* 25 May

MERCURY passes through superior conjunction on 11 May, and so is unsuitably placed for most of the month. However, during the last week of May observers in northern temperate latitudes may glimpse the planet low above the north-western horizon at the end of evening civil twilight (about forty minutes after sunset), although the lengthening twilight at these latitudes will make observation difficult. Mercury's brightness decreases from magnitude -0.9 to -0.3 during the last week of May. Fortunately, from 23–25 May, Mercury will pass about a degree north of the very much brighter Venus, which should be an aid to locating the fainter planet. The planet Jupiter will also lie in the same part of the sky at the end of May, but it is rather fainter than Venus and slightly higher up from the twenty-third to the twenty-fifth of the month. Observers in the tropics and further south should also be able to locate Mercury at the end of May.

VENUS, magnitude -3.9, is beginning to draw slowly out to the east of the Sun, becoming visible to observers in northern temperate latitudes low above the north-western horizon at the end of evening civil twilight during the last week of May. Observers in the tropics may well be able to spot the planet a little earlier in the month, but the planet's northerly declination will make it more difficult for those living further south. Venus and Jupiter will pass within one degree of each other in the bright evening twilight sky on 28 May; Venus will be about six times brighter than Jupiter at this time, and will lie to the north of the fainter planet.

MARS was in conjunction with the Sun in mid-April and remains unobservable this month.

JUPITER, magnitude -1.9, is in conjunction with the Sun in mid-June, and will be lost in the glare of the evening twilight sky by the end of

May. Earlier in May, the planet may be spotted by Northern Hemisphere observers low in the west-north-west about forty minutes to an hour after sunset. The close approach of Jupiter to Venus on 28 May has already been mentioned above; observers should go out about forty minutes after sunset to glimpse the two brightest planets, but they will be very low above the north-western horizon. Those situated in the tropics slightly north of the Equator may fare rather better.

SATURN, just past opposition, is visible in the south-south-east as darkness falls and is observable for most of the night. Its retrograde motion carries it from Libra across the border into neighbouring Virgo on 14 May. It fades slightly from magnitude +0.1 to +0.3 during the month.

The Sooty Star. One of the most remarkable stars in the sky is R Coronae Borealis. It is never brighter than about the sixth magnitude, and is therefore on the fringe of naked-eye visibility, but it is of special interest.

To find it, first locate Corona Borealis, the Northern Crown. This is easy enough. Follow round the curve in the tail of the Great Bear, which is virtually overhead on May evenings in northern temperate latitudes, and identify the brilliant orange star Arcturus in Boötes, the Herdsman. Corona lies not far to the east (Figure 11), and is made up of a semicircle of stars, of which the brightest, Alphekka or Alpha Coronae Borealis, is of the second magnitude – about equal to the Pole Star. Inside the bowl of the Crown, binoculars will usually show two stars. One is of magnitude 6.5, and the other is R Coronae. For a chart showing the location of R Coronae and some suitable comparison stars, see the chart provided on p. 381 in the section 'Some Interesting Variable Stars', elsewhere in this *Yearbook*.

Sometimes, on examining the bowl of the Crown, you will find that R Coronae is missing. At unpredictable intervals it drops in brightness, taking only a few days or a week or two to fall to magnitude 12, or even 15 – well beyond the range of binoculars or small telescopes. After remaining faint for a while, it slowly and jerkily, over successive months, recovers its normal brightness, remaining more or less steady until the onset of the next minimum, which may be after several months or many years.

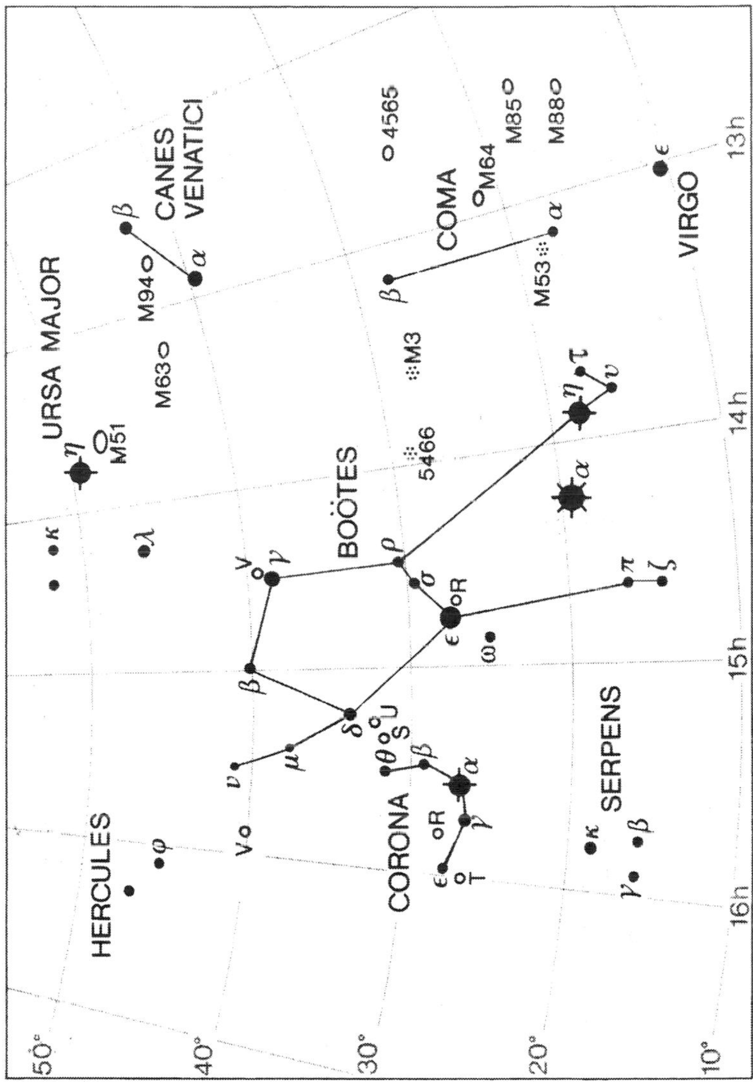

Figure 11. Map showing the location of the small semicircle of stars in Corona Borealis, the Northern Crown, which lies to the east of Boötes, the Herdsman. The location of the unusual variable star R Coronae Borealis is marked by the small circle with the central dot.

Why does R Coronae behave in this peculiar way? Spectroscopic work has shown that it is a yellow supergiant star which contains far less hydrogen than a normal star – so its outer layers are probably largely made up of helium – but the star is especially rich in carbon. Although similar in mass to our Sun, R Coronae is extremely luminous and is losing mass at a prodigious rate. Apparently there are times when carbon accumulates in the star's atmosphere as nothing more or less than soot! In fact, the star retires behind a sooty veil, and much of its light is cut off. Then, when the soot disperses, the veil clears away and the star shines out once more.

R Coronae-type stars are very rare. In fact, there are only eight others which can become brighter than tenth magnitude: S Apodis, XX Camelopardalis, UW Centauri, V854 Centauri, V Coronae Australis, RY Sagittarii, RS Telescopii and SU Tauri. All are luminous and fairly remote, and there seems to be some doubt as to where they fit into the general scheme of stellar evolution. Because their falls to minimum cannot be predicted, amateur astronomers do valuable work by keeping a watch on them, so that professionals, with their spectroscopic equipment, can begin their studies as soon as a fade begins.

So when the sky is clear and Corona Borealis is above the horizon, it is always worth using binoculars to take a quick look inside the bowl of the Crown. If you can see only one star instead of two, you will know that R Coronae has started to put on a performance. But never fear – it will soon be back, and there is no danger that we will permanently lose sight of our sooty star.

The Earliest British Amateur Astronomer? Looking back over the list of British amateur astronomers, there are reasons for suggesting that one of the earliest of them – if not the earliest of all – was the Venerable Bede. Almost everybody has heard his name, though not so many know where he lived or what he did.

Bede was born in the year 673 at Monkwearmouth, in what is now north-east Durham. At that stage in history, of course, there was no single kingdom of England, and there were several independent kingdoms of which Northumbria was the most powerful. There had been wars between Northumbria and the 'middle kingdom' of Mercia, which had been ruled by King Penda, one of the last champions of heathenism in Britain. In 642 King Oswald of Northumbria had been defeated and killed by Penda's forces, but in 655, at the Battle of Winwaed, Oswy, a Christian and the new king of Northumbria, routed Penda's forces and killed Penda himself. By the time of Bede's birth, the conversion of England to Christianity was almost complete, though parts of southern England held out for a few years longer. When Oswy died, Egfrith ascended the throne of Northumbria in 670, and for a while things were peaceful. Certainly Bede did not lead an adventurous life. He became a monk, and spent his days in a monastery at Jarrow, writing voluminously about all sorts of subjects. Evidently he had no liking for travel, and it is possible that he never even visited Hadrian's Wall, though he lived less than fifteen kilometres away from it.

Bede's greatest book was his *Ecclesiastical History of the English People*, completed in 731. In it there are several references to comets. Thus: 'In the month of August 678, in the eighth year of Egfrith's reign, a star known as a comet appeared, which remained visible for three months, rising in the morning and emitting what seemed to be a tall column of bright flame.' Actually, Bede may have the date wrong: according to Chinese sources there was a brilliant comet in September 676, which had a 3-degree tail and moved from Gemini into the region of the Great Bear before it faded away.

Then, in 729, Bede wrote that 'two comets appeared around the Sun, striking terror into all who saw them. One comet rose early and preceded the Sun, while the other followed the setting Sun at evening. They appeared in January, and remained visible for about two weeks, bearing their fiery tails northward as though to set the welkin [the firmament] aflame.' Probably Bede himself saw this comet – it may well have been one rather than two, rising before the Sun in the morning and setting after it in the evening. Bede does not mention Halley's Comet, which appeared during his lifetime; but it was visible when Bede was only eleven years old, so that it is perhaps understandable that he did not record it.

Bede died in 731, busy with his writing almost to the end. His *History* is invaluable; it gives us our only reliable source of information about what happened in England for the first few centuries after the departure of the Romans. Note, too, that when the Jesuit Moon-mapper Riccioli allotted names to lunar craters in 1651, Bede was commemorated by a rather large, though obscure, ring near the bright crater we now call Censorinus. For some reason or other the name was dropped by later astronomers; but I feel that it should be restored, if only because Bede was at least paying some attention to the sky when nobody else in England was likely to be doing so.

June

New Moon: 8 June *Full Moon:* 23 June

Solstice: 21 June

MERCURY is at greatest eastern elongation (24°) on 12 June, so the planet continues to be visible low above the north-western horizon at the end of evening civil twilight in the first three weeks of June, though observers in the latitudes of the British Isles will find it a difficult object to locate in the long evening twilight. During this observing period, the planet fades from magnitude -0.3 to +1.4. Observers in the tropics and in the Southern Hemisphere should find the planet somewhat easier to find, particularly around 20–21 June when it is once again quite close to the brilliant Venus in the evening twilight sky.

VENUS, magnitude -3.9, is becoming easier to locate in the west-north-western sky at dusk as its elongation from the Sun increases, although observers in northern temperate latitudes will be hampered by the bright evening twilight at this time of the year; those in the tropics and further south will fare rather better. As mentioned above, Venus will once again be quite close to Mercury on 20–21 June, a handy guide to locating the very much fainter planet in binoculars.

MARS remains inconveniently placed for observation by those in northern temperate latitudes, but from the tropics and further south it gradually becomes visible as a difficult morning object towards the end of the month, when it may be seen low above the east-north-eastern horizon, about an hour before sunrise. Its magnitude is +1.5.

JUPITER is in conjunction with the Sun on 19 June and consequently will not be visible this month.

SATURN is visible in the southern sky as soon as darkness falls and is observable until well into the early morning hours. The planet continues to move retrograde in Virgo. It fades slightly from magnitude +0.3 to +0.5 during the month.

47 Tucanae. Globular clusters are found in a roughly spherical halo around the edge of our Galaxy, the Milky Way, and it has been said that they provide a sort of outer framework to it. They contain many more stars and their stars are much older than those found in the less compact open or galactic clusters (such as the Pleiades in Taurus) that are found in the galactic disk. As their name implies, globular clusters are spherical in form, and they may contain up to one million stars. The stars within globular clusters are tightly bound by gravity, and this causes their member stars to be highly concentrated towards their centres.

About 150 globular clusters are known around the Milky Way and there are probably some that are as yet undiscovered. It is thought that larger galaxies may have far more globular clusters than the Milky Way. According to some estimates, our nearest large galactic neighbour, the Andromeda Galaxy, may have as many as 500 globular clusters in its vicinity. Some of the giant elliptical galaxies that are found at the hearts of large galaxy clusters may harbour many thousands.

From the latitudes of the British Isles, one of the best-known globulars in our own Galaxy, Messier 13 in the constellation of Hercules, is just visible with the naked eye as a hazy patch; a large telescope will resolve it into stars almost to its centre (Figure 12). This cluster will be found high in the southern sky around midnight in the middle of June; it is located roughly midway between the third-magnitude stars Eta and Zeta Herculis, but slightly nearer to Eta.

However, Messier 13 is far inferior to two globular clusters which are too far south to be seen from northern Europe. One is Omega Centauri, the largest and brightest globular cluster, which may be sighted from locations in southern Europe, south of latitude 40°N. The other is 47 Tucanae in the otherwise unremarkable constellation of Tucana, the Toucan, located in the far southern sky only eighteen degrees from the south celestial pole. In fact, 47 Tucanae lies quite close to the Small Magellanic Cloud (SMC) on the sky, but the SMC is at least ten times more distant than the globular cluster.

Some of the stars in globular clusters vary in brightness. Certain short-period variables known as RR Lyrae stars are particularly common in globular clusters. Indeed about 90 per cent of all variable stars

Figure 12. Messier 13 in the constellation of Hercules is one of the best-known globular clusters to be visible from the latitudes of northern Europe and North America. It contains roughly one million stars and a large telescope will resolve it into stars almost to its centre. This image was obtained by the 2.5-metre telescope of the Sloan Digital Sky Survey (SDSS). (Image courtesy of Robert Lupton, The Sloan Digital Sky Survey.)

in globular clusters are RR Lyrae stars, which explains why they were formerly known as 'cluster-type variables'. The distances to RR Lyrae stars may be determined by carefully measuring their periods of variation, relating this to their true intrinsic brightness, and comparing this with their apparent brightness. This, of course, provides an invaluable clue to the distances of the globular clusters themselves. Indeed, it was in this way, more than ninety years ago, that the American astronomer Harlow Shapley obtained the first reliable estimate of the size of the Galaxy. Interestingly, Shapley used RR Lyrae stars, but thought that they were Cepheid variables. Since RR Lyrae variables are generally fainter than Cepheids, by a factor of ten or so, this led him to overestimate the distance to the clusters, but this error was later corrected. Unfortunately, 47 Tucanae does not contain many short-period variables, and its distance is somewhat uncertain, but recent estimates using other distance determination techniques indicate that it lies between 13,000 and 17,000 light years from us.

Great though this distance may seem, 47 Tucanae is one of the nearest of the globular clusters. The cluster is about 120 light years in diameter and even small telescopes will resolve its outer parts into stars; with large instruments, even the centre is resolvable. It looks as though the stars are so crowded that they are in imminent danger of colliding with each other. Actually this is not so; even in the middle of a globular cluster the stars are still, on average, several light weeks apart, but from a planet in such a system the sky would be ablaze. Many stars would cast shadows, and a large number of them would be red, because globular clusters are rather old systems, and their leading stars have evolved to the red giant stage.

Omega Centauri may be slightly larger and brighter than 47 Tucanae, but seen through a telescope it is probably less imposing, because it more than fills the field when a reasonably high magnification is used. Appearing roughly the size of the full Moon in the sky under ideal conditions, to most observers, 47 Tucanae is the most beautiful of all the globular clusters; it is noted for having a very bright and dense core (Figure 13). European astronomers never cease to regret that it lies so far south in the sky.

Seeing Stars from the Bottom of a Well. There used to be a comic song which began: 'Where do all the flies go in the winter-time?' More scientifically, one is often asked: 'Where do all the stars go in the daytime?' The answer is, of course, that they do not go anywhere. It is just that the bright sunlit sky effectively drowns them out.

The brightest thing in the sky, apart from the Sun and the Moon, is the planet Venus. People with keen eyes can see it well before sunset or after sunrise, and there are a few exceptional people who can pick it out most of the time. I (PM) remember, too, the total solar eclipse of 11 June 1983, which I saw from Tanjung Kodok on the island of Java, thirty years ago this month. Venus was visible at least a quarter of an hour before totality began (I did not see it then myself, because I was too busy setting up my equipment) and it was followed for about half an hour after totality had ended. This, of course, was with the naked eye. Telescopically Venus can always be found in daytime when it is not too near the Sun; indeed, this is the best way to observe Venus – and the same is true for the other inner planet, Mercury.

Figure 13. To most observers living in the Southern Hemisphere, 47 Tucanae (NGC 104) is the most beautiful of all globular clusters. Small telescopes will resolve its outer parts into stars, but with large instruments even the centre is resolvable. 47 Tucanae has a very bright, dense core as shown in this image, acquired using the 4-metre Blanco telescope of the Cerro Tololo Inter-American Observatory. (Image courtesy of NOAO/AURA/NSF.)

If you have a telescope fitted with setting circles so that you can point it accurately, you can locate bright stars easily in the daytime (though always take great care not to search round in the region of the Sun, and in any case always sweep *away* from the Sun so that there is no danger of looking at it by mistake). In twilight, first-magnitude stars are easy enough. But what about the old story that stars can be seen in full daylight if the observer goes down to the bottom of a deep well or mineshaft? I have heard this story many times, but a little thought will show that there is nothing in it. What matters is the *contrast* between the star and the sky background, which is just the same from the bottom of a well as it is from ground level.

If you doubt me, try for yourself. I have done so – from the bottom of Homestake Gold Mine in South Dakota, almost a kilometre-and-a-half below the surface. When I looked straight up the shaft, I could see a circle of blue with no stars. So I fear that another charming old tale has to be disregarded.

Yet what about the airless Moon? The sky there is black all the time. I asked the 'last man on the Moon', Commander Eugene Cernan of *Apollo 17*, whether he saw stars in the daytime. He replied that he had, though only after having 'dark-adapted' by shielding his eyes from the glare of the surrounding sunlit rocks. From a spacecraft, of course, the stars are glorious and unwinking. But I'm afraid that a gold mine or a well will not help!

July

New Moon: 8 July *Full Moon:* 22 July

EARTH is at aphelion (furthest from the Sun) on 5 July at a distance of 152 million kilometres (94.5 million miles).

MERCURY is in inferior conjunction on 9 July, and is thus unsuitably placed for observation for most of the month. However, during the last week of July, it may be glimpsed low above the east-north-eastern horizon at the beginning of morning civil twilight. From northern temperate latitudes the planet will be a difficult object, but rather easier for those in the tropics and the Southern Hemisphere. During this period the planet brightens from magnitude +1.3 to +0.1. Mercury reaches greatest western elongation (20°) on 30 July.

VENUS, magnitude -3.9, is visible in the west-north-western sky at dusk, but is rather low for observers in northern temperate latitudes, setting only one-and-a-quarter hours after the Sun by the end of the month. From equatorial and more southerly latitudes, Venus is more favourably situated, setting over two hours after the Sun by month end from these locations. The phase of the planet decreases from 90 per cent to 83 per cent during July.

MARS, magnitude +1.6, is gradually becoming easier to pick out in the early morning twilight sky, low above the east-north-eastern horizon, about an hour before sunrise. The planet's direct motion carries it from Taurus across the border into neighbouring Gemini during the month. Figure 14 shows the path of Mars against the background stars during 2013. Mars is less than a degree north of the much more brilliant Jupiter on 22 July.

JUPITER, magnitude -1.9, which was in conjunction with the Sun in mid-June, becomes visible as a conspicuous object in the early morning sky during July. The planet is now moving direct in Gemini. On 22 July

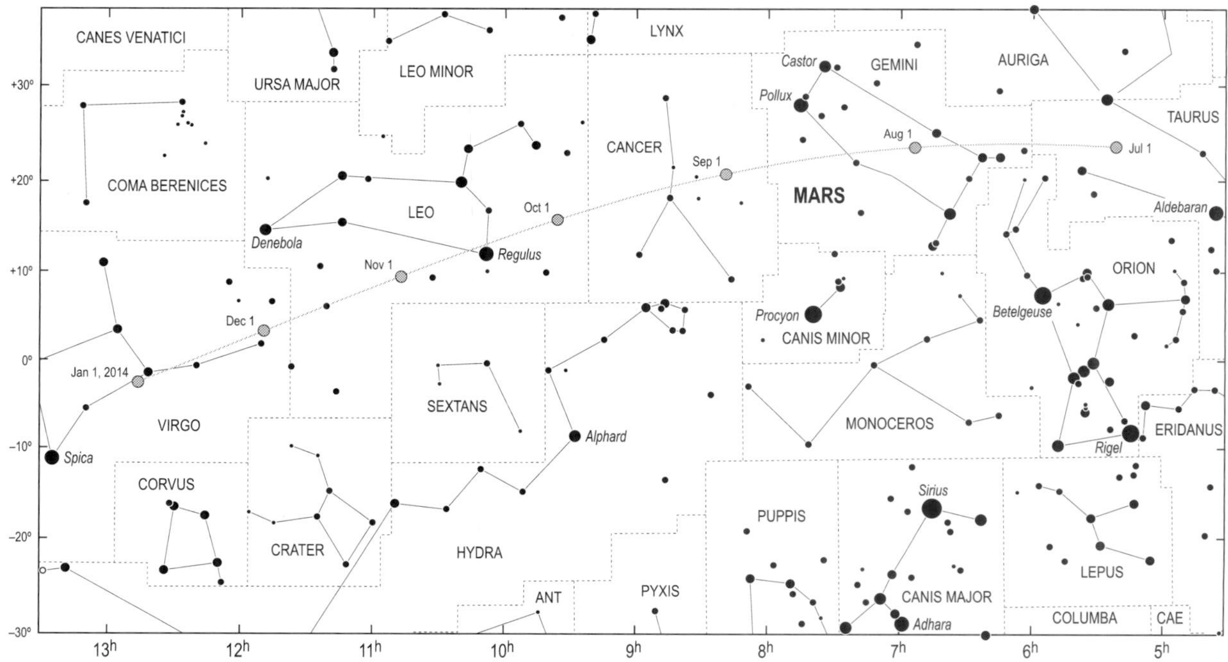

Figure 14. The path of Mars as it moves through the Zodiacal constellations from Taurus to Virgo between July 2013 and the end of the year.

it passes less than a degree south of the much fainter planet Mars, and this will be an aid to locating the Red Planet. Jupiter will be rising more than two hours before the Sun by the end of July.

SATURN starts the month moving retrograde in Virgo, but after reaching its second stationary point on 11 July, it resumes its direct motion once more. The planet fades slightly from magnitude +0.5 to +0.6 during July.

PLUTO reaches opposition on 2 July, in the constellation of Sagittarius, at a distance of 4,705 million kilometres (2,924 million miles). It is visible only with a moderate-sized telescope since its magnitude is +14.

The Star in the Indian Bowl. If you had looked into the sky during July 1054, you would have seen something very unusual – a star so bright that it remained visible even in the middle of the day. We know what it was: a supernova, representing the death-throes of a very massive star which had blown up and sent most of its material hurtling outwards into space. The 1054 supernova was recorded in Asia, though apparently not in Europe. Gradually it faded, and once it dropped below naked-eye visibility it was lost (there were no telescopes in those days!) but we see its remnant today in the form of the expanding cloud of gas known as the Crab Nebula (Figure 15), in Taurus the Bull. At the heart of the gas-cloud is a neutron star or pulsar, the incredibly compact core of the old, exploded star, which is spinning extremely rapidly and sending out pulses of radio waves.

It is always helpful to have more historical records of the supernova itself, and one rather unexpected observation may have been found, originating from New Mexico in the United States. Over half a century ago an Indian burial bowl was discovered there, about 18 centimetres in diameter, and apparently made between the years AD 1000 and 1150. Archaeo-astronomers from the University of Texas at Austin have examined it, and have claimed that on it there is a representation of the supernova of 1054.

The white earthenware bowl shows the stylized black-painted figure of a rabbit curled into a crescent shape to represent the Moon. It was a rabbit which most Native Americans used to symbolize the Moon, because they felt that this was the shape indicated by the light and dark

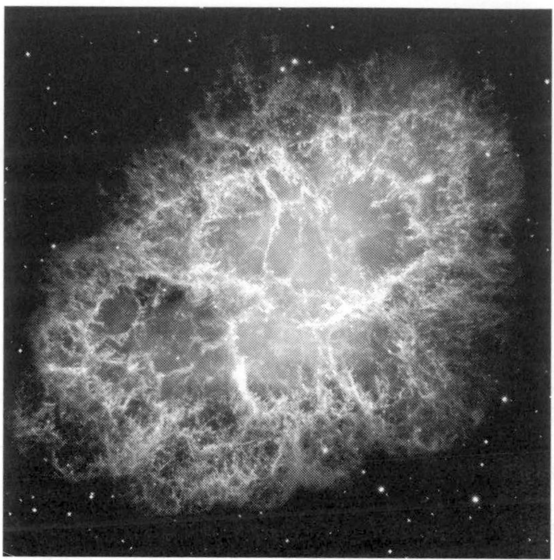

Figure 15. The Crab Nebula, the expanding remnant of a supernova explosion, witnessed not only by Japanese and Chinese astronomers in July 1054, but also by Native Americans. This mosaic image was assembled from twenty-four individual Wide Field and Planetary Camera 2 exposures with NASA's Hubble Space Telescope. The filaments are the tattered remains of the original star and consist mostly of hydrogen. (Image courtesy of NASA, ESA and Allison Loll/Jeff Hester (Arizona State University. Acknowledgement: Davide De Martin (ESA/Hubble).)

patches on the disk. Just as we have a 'Man in the Moon', so they had a 'Rabbit in the Moon'. On the bowl, a small starburst is shown at the tip of one of the rabbit's feet, and this seems to be the position in which the bright new star would have been seen, relative to the crescent Moon, in the eastern sky on the morning of 5 July 1054, when the supernova was at its brightest. Twenty-three rays emanate from the star on the bowl, and the American scientists Ralph Robert Robbins and Russell Westmoreland have suggested that this represents the 23-day period when the supernova was visible in the daytime.

Archaeologists have also found other burial bowls with rabbit images which apparently represent lunar eclipses. The bowls were made by members of the Mimbres tribe from New Mexico, and indicate that considerable attention was paid to the Moon. Unfortunately, the Mimbres left no written records, and they are now extinct, so we cannot ask them what they think about it all, but there is at least a chance that they have left us a record of a colossal stellar explosion which lit up our skies nearly a thousand years ago.

The Hunt for Phobus 1. It is a curious fact that although Russian space scientists have had great success with their missions to Venus, they have had little luck at all with Mars, which should be a much easier target. This month we mark the twenty-fifth anniversary of the launch of two probes on a mission which promised to be exceptionally interesting – but which ended in failure. The craft were called Phobos 1 and Phobos 2 (Fobos 1 and Fobos 2 in Russian).

They were launched in July 1988, not mainly to study Mars itself, but to concentrate upon Phobos, the larger of the two tiny satellites – an irregularly shaped, dark object measuring 27 x 22 x 18 kilometres (Figure 16). The Phobos project was a truly international venture, with scientists from fourteen countries joining in; participants represented Austria, Bulgaria, Hungary, Ireland, East and West Germany, Poland,

Figure 16. Phobos, the larger of Mars's two tiny moons, imaged by the High Resolution Stereo Camera (HRSC) on board ESA's Mars Express spacecraft on 28 July 2008 from a distance of about 350 km. (Image courtesy of ESA/DLR/FU Berlin (G. Neukum).)

Finland, France, Czechoslovakia, Switzerland, Sweden, USA and institutions of the European Space Agency.

The launches themselves went well: Phobos 1 started its journey on 7 July 1988 and Phobos 2 five days later. According to the flight plan, they were scheduled to reach Mars in January 1989, making their Phobos encounters the following April. But then, on 29 August 1988, came a disastrous error by one of the Soviet ground controllers. An incorrect command was sent to Phobos 1, which caused the spacecraft to shut down its attitude control system. By the time controllers had realized the error, the spacecraft's solar panels had swung out of alignment with the Sun, the batteries were drained and the electrical and electronic systems lost all power. Phobos 1 went out of contact.

There was a glimmer of hope that if the 'dead' spacecraft tumbled in flight then its solar panels might eventually point towards the Sun, providing just enough power for vital onboard systems to revive operations. Unfortunately, the most stable natural attitude of the spacecraft was to keep the backside of its solar panels to the Sun! After repeated attempts to re-establish communications during September and October, the mission was finally declared over on 3 November 1988 – but not before European astronomers had received an unusual request from the Russians.

It was thought that there might be a slim chance of restoring the situation if only the Phobos 1 craft could be located. So, on 21 September 1988, the Russians sent a request to the La Silla Observatory in Chile, one of the main international sites of the European Southern Observatory (ESO). Conditions there are as good as they are anywhere in the world, and the Soviet request was straightforward: would La Silla try to photograph Phobos 1? If it were possible to obtain a sequence of images, of course only as faint points of light, knowledge of its exact position and even rotational status might enable the Russians to re-activate it.

La Silla promptly agreed. The first available night, 22 September, was lost due to snow, and then followed a period of bright moonlight, but on the nights of 1/2, 2/3 and 3/4 October, a careful search was made, using the latest electronic equipment on the 1.5-metre Danish telescope. Fifteen ten-minute exposures were made. On 3/4 October, the largest telescope then operating at La Silla, the 3.6-metre reflector, was also used for the hunt by two visiting astronomers, D. Hatzidimitriou and C. A. Collins. On one exposure, the 3.6-metre telescope was set to make a direct exposure of the field in which Phobos 1 was expected to be. On another, the telescope was set to 'follow the spacecraft' in its predicted motion, so that the stars would be drawn out into trails. In each case the limiting magnitude was below 25. Alas, there was no sign of Phobos 1! The images were intensively studied at La Silla, by H.-U. Nørgaard-Nielsen and H. Pedersen, and if there had been a trace of the elusive spacecraft, it could not have been overlooked; but it simply was not there.

La Silla did not give up, and on 9/10 October, two more visiting astronomers, G. Suchail and Y. Mellier, made a final effort. They took four frames, extending over a period of thirty-six minutes, but again Phobos 1 declined to appear. That, sadly, was as much as could be done. Either Phobos 1 was outside the expected field of view (did one of its on-board rockets fire after contact was lost?) or else it was positioned so unluckily that the sunlight reflected off its surface in our direction was too faint to be recorded, even with the most sophisticated equipment.

Phobos 1 was never found, but the whole episode underlined, yet again, the way in which astronomers and space scientists are ready to help each other. If anyone could have located the missing probe, it would have been the observers at La Silla.

Incidentally, Phobos 2 did remain in contact until it reached Mars, entering orbit about the Red Planet on 29 January 1989, but it 'went silent' on 27 March before the Phobos lander experiments could begin, doubtless because of an on-board computer fault. (For further details of Phobos 2, see p. 244 of the article by David M. Harland in the *2012 Yearbook*.)

More recently, the Russians tried again with the launch of their ambitious Phobos-Grunt mission, which promised to actually return samples of the moon Phobos to Earth. The craft launched successfully on its Zenit rocket from the Baikonur Cosmodrome on 8 November 2011 and successfully entered an elliptical orbit around the Earth. Sadly, the two firings of the probe's cruise stage, which would have first raised it to a higher orbit and then sent it on path to Mars, never took place. Only limited radio contact was ever made with the craft and it eventually re-entered the atmosphere and burned up over the Pacific Ocean.

So the Russians have still to achieve real success with the Red Planet.

August

New Moon: 6 August *Full Moon:* 21 August

MERCURY continues to be visible as an early morning object during the first week of August, rather low above the east-north-eastern horizon at the beginning of morning civil twilight. At this time the planet will be a rather easier object from equatorial and more southerly locations than it is from the latitudes of the British Isles. During this period its brightness increases from magnitude 0.0 to -0.7. The planet passes through superior conjunction on 24 August, and so will be unobservable for the rest of the month.

VENUS is visible in the western sky at dusk, but continues to be inconveniently low for observers in northern temperate latitudes, setting just over an hour after the Sun. From equatorial and more southerly latitudes, Venus is much more favourably situated, setting over three hours after the Sun by month end from locations in the Southern Hemisphere. The phase of the planet decreases from 83 per cent to 74 per cent during August. Venus's eastwards motion carries it from Leo into Virgo and it brightens very slightly from magnitude -3.9 to -4.0 during the month.

MARS, magnitude +1.6, is visible in the eastern sky about an hour-and-a-half before sunrise; slightly less from the Southern Hemisphere. The planet's direct motion carries it from Gemini into Cancer during the month. The waning crescent Moon will be a few degrees south of Mars on the early morning of 4 August and will make a nice grouping with both the Red Planet and the much brighter Jupiter which will be in the same part of the sky.

JUPITER is a lovely object in the early morning sky, moving direct amongst the stars of Gemini. The planet's brightness increases very slightly from magnitude -2.0 to -2.1 during August. The waning crescent Moon will make a nice grouping with Jupiter and Mars early on

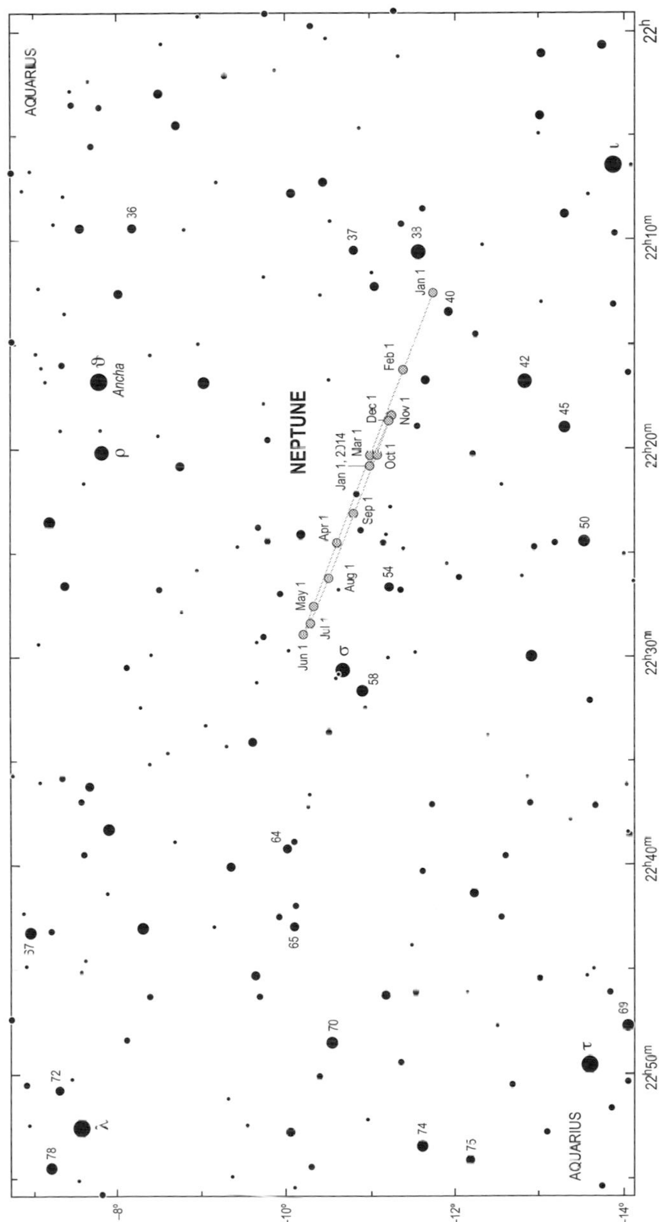

Figure 17. The path of Neptune against the stars of Aquarius during 2013.

4 August, and the crescent Moon will again appear close to Jupiter on the last day of the month.

SATURN is visible as an evening object in the west-south-western sky in Virgo throughout August, but by the end of the month it sets only two hours after the Sun from the latitudes of the British Isles; rather later from further south. Its brightness decreases slightly from magnitude +0.6 to +0.7 during the month. The waxing crescent Moon will pass three degrees south of Saturn on 12/13 August.

NEPTUNE is at opposition on 27 August, in the constellation of Aquarius. It is not visible with the naked-eye since its magnitude is +7.8. At opposition Neptune is 4,334 million kilometres (2,693 million miles) from the Earth. Figure 17 shows the path of Neptune against the background stars during the year.

The Celestial Harp. As seen from Britain, the brilliant blue star Vega is almost overhead during summer evenings – a position occupied during winter evenings by the yellowish Capella. Vega is a star of spectral type A0, fifty-two times as luminous as the Sun and twenty-five light years away; it was one of the first stars to be discovered with a large 'infrared excess' due to a circumstellar disk of warm dust, perhaps a planetary system in the making. Vega's apparent magnitude is +0.03, so that it is surpassed by only four stars: Sirius, Canopus and Alpha Centauri in the Southern Hemisphere, and Arcturus in the Northern. Vega was the northern pole star around 12,000 BC and will be so again in about AD 13,700, when it will lie less than four degrees from the North Celestial Pole.

Lyra, the Lyre or Harp, is a small constellation (Figure 18), covering less than 300 square degrees of the sky, but it contains a wealth of interesting objects. One is the eclipsing binary Beta Lyrae (Sheliak), which has a magnitude range of 3.3 to 4.4; the period is 12.9 days, and there are alternate deep and shallow mimima, so that the two components are not nearly so unequal as in Algol (Beta Persei), for example. Gamma Lyrae, magnitude 3.2, is a convenient comparison star. Another variable star in the same constellation is R Lyrae, a red giant star classed as 'semi-regular' with a rough period of about forty-six days and a magnitude range of 3.9 to 5.0, although there are somewhat erratic wanderings in its maximum and minimum brightness.

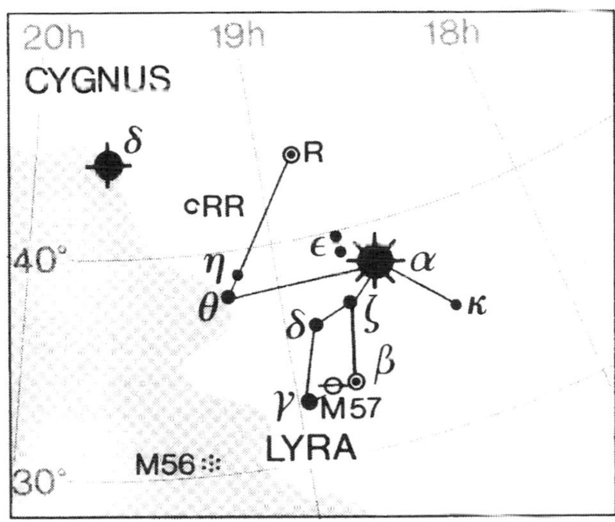

Figure 18. Map showing the principal stars of Lyra, the Lyre or Harp, a small constellation whose principal star, Vega (Alpha Lyrae), lies almost overhead on summer evenings in northern temperate latitudes.

Even though R Lyrae is visible with the naked eye, it is not easy to estimate accurately, because there is a dearth of suitable comparison stars near it. The closest are Eta and Theta Lyrae (each mag. 4.4); there is also 16 Lyrae, but this is of magnitude 5.1, and is rather too faint to be useful except when R Lyrae is near minimum. Most observers will prefer to use binoculars for following its fluctuations, but a wide field is needed as otherwise all the comparison stars will be out of the field of view. Like all M-type stars R Lyrae is decidedly red. This colour cannot be readily seen with the naked eye, but it is very obvious in binoculars, and contrasts with Eta and 16 Lyrae (white) and Theta (yellowish).

Semi-regular stars of small range, such as R Lyrae, are not usually intensively studied by professional astronomers, and amateurs can make themselves very useful. It would be misleading to call R Lyrae a suitable subject for observation by complete beginners, but it is worth attempting to draw up a light-curve, particularly as it is so far north that from Europe and North America it is visible for so much of the year. However, do not be surprised if your first attempts at a light-curve yield some very strange-looking results!

Epsilon Lyrae, close to Vega, is a naked-eye double with almost equal components (magnitudes 4.7 and 5.1); the separation is 3.5 arc minutes. A small telescope will show that each component is again double, so that we have a quadruple system, the separation of each close pair being about 2.4 arc seconds. Another star lies between the two pairs, but this is not connected with the system, and is much further away. Zeta Lyrae, which marks the north-western corner of the little parallelogram that helps makes the shape of the harp, is also double (magnitudes 4.3 and 5.9), wide enough to be 'split' with almost any telescope; the separation is almost 44 arc seconds.

Delta Lyrae, the star at the north-eastern corner of the little parallelogram of stars, is a widely spaced naked-eye 'double', but appearances can be deceptive because the two stars are not genuinely associated (as once thought), but lie along the same line of sight. The brighter, and slightly closer, eastern component, Delta-2 (mag. 4.5), is of spectral type M, and its obvious redness makes a good contrast with

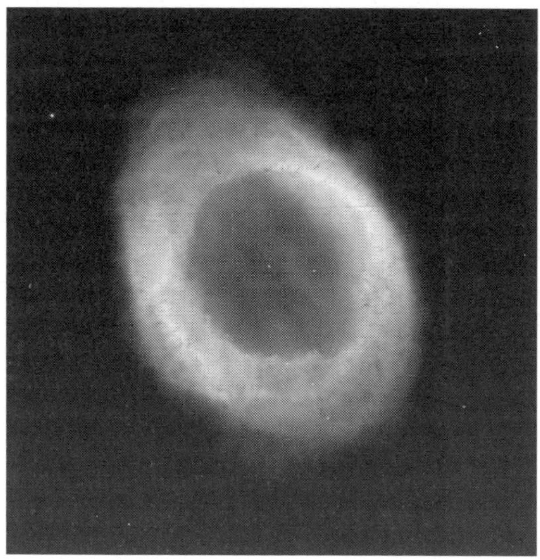

Figure 19. Messier 57, the Ring Nebula in Lyra, imaged using the Hubble Space Telescope. This planetary nebula's simple appearance is thought to be due to perspective – our view from planet Earth looking straight into a barrel-shaped cloud of gas shrugged off by the dying central star. The Ring Nebula is about one light year across and 2,000 light years away. (Image courtesy of H. Bond et al., Hubble Heritage Team (STScI / AURA).)

its fainter, slightly more distant, western companion, Delta-1 (mag. 5.5), which is spectral type B and is white. Binoculars will give good views of this pair.

Between Beta and Gamma Lyrae, the two stars marking the southern end of the little parallelogram, lies the famous Ring Nebula, M57 – the most celebrated of all planetary nebulae (Figure 19). It can be seen with a small telescope, and gives the impression of a tiny, luminous bicycle tyre; the central star is decidedly elusive. Binoculars will also show the eighth-magnitude globular cluster M56, which appears as a faint patch of light.

Lyra is indeed a rich constellation. Mythologically, it represents the harp or lyre which Apollo gave to the great musician Orpheus.

The Mysterious Nova. Among great astronomical observers, Edward Emerson Barnard has an honoured place. He was superbly skilful, and he had the advantage of being able to use very large telescopes, notably the great 36-inch refractor at the Lick Observatory in California. It was with this telescope that he discovered Amalthea, the fifth satellite of Jupiter, on 9 September 1892. (Incidentally, this was the last planetary satellite to be found visually. All later discoveries have been photographic, beginning with W. H. Pickering's identification of Phoebe, the unusual retrograde satellite of Saturn, in 1898.)

But another observation by Barnard remains very much a mystery. It was made on 13 August 1892 at ten minutes to five in the morning Standard Pacific Time, which corresponded to fifty minutes after noon Greenwich Mean Time. The observation was made in broad daylight, half an hour before sunrise; Barnard was looking at the planet Venus, then in its crescent stage. He wrote as follows:

> I saw a star in the field with the planet. The star was estimated to be of at least the 7th magnitude. The position was so low that it was necessary to stand upon the high railing of a tall observing chair. It was not possible to make any measures, as I had to hold on to the telescope with both hands to keep from falling. The star was estimated to be one minute of arc south of Venus . . . There seems to be no considerable star near this place, and the object does not agree with any BD star. [BD stands for the Bonner Durchmusterung, the best star catalogue available at the time.]

Barnard gave the right ascension of the object as 6h 50m 21s, declination N.17° 13'.6. He added that the position excluded the possibility of it being an intra-Mercurial planet but did not preclude the possibility of it being a planet interior to Venus, though such an explanation was improbable.

The object was never seen again, by Barnard or anyone else, and this is the sole record of it. So let us try to decide what it could have been. The existence of unknown planets interior to Venus or to Mercury may be discounted immediately and we can rule out any satellite of Venus; no such satellite exists – and if there had been one, with a magnitude of 7, it could not possibly have been overlooked even before the age of space-probes.

A mistake in the date of the record, then, or perhaps a 'ghost' image of Venus produced in the optics of the telescope? This will not do either. Barnard's records for that night are typically precise, and he was much too experienced and accurate an observer to have been tricked by an optical defect.

We are also able to eliminate all asteroids or minor planets, none of which could have been as bright as magnitude 7 in broad daylight. There seem to be only two possibilities left. The object could have been the nucleus of a comet – but then it would hardly have been so starlike. All in all, it seems that Barnard most likely saw a nova, or new star.

This is plausible. Venus at that time lay in the rich constellation of Gemini, where various novae have appeared; and at that time of the year Gemini was above the horizon only during the hours of daylight, so that in the ordinary way a nova would have been missed. We may assume that by the time Gemini emerged from the morning twilight, the nova had faded away.

Efforts have been made to check all the stars in the area to see whether any of them shows the characteristics of an ex-nova, but with no success. So the mystery remains, and we are unlikely to solve it now.

September

New Moon: 5 September *Full Moon:* 19 September

Equinox: 22 September

MERCURY passed through superior conjunction towards the end of August and remains unsuitably placed for observation from northern temperate latitudes throughout the month. Further south, however, Mercury becomes an evening object after the first week of September, low above the western horizon at the end of evening civil twilight, and remains observable until the third week of October. Indeed, for observers in the tropics and the Southern Hemisphere this is the most favourable evening apparition of the year. Figure 22, given with the notes for October, shows, for observers in latitude 35°S, the changes in azimuth (true bearing from the north through east, south and west) and altitude of Mercury on successive evenings when the Sun is 6° below the horizon, about 25 minutes after sunset in this latitude. The changes in the brightness of the planet are indicated on the diagram by the relative sizes of the white circles marking Mercury's position at five-day intervals Mercury will be at its brightest early in the apparition; it fades from magnitude -0.6 to 0.0 between 8 September and the end of the month.

VENUS is visible in the west-south-western sky at dusk, but remains inconveniently low for observers at the latitudes of the British Isles, setting just over an hour after the Sun. From equatorial and more southerly climes, Venus is much better placed, setting nearly four hours after the Sun by the end of September from locations in the Southern Hemisphere. The planet's eastwards motion carries it from Virgo into Libra and it brightens from magnitude - 4.0 to -4.2 during the month. The phase of Venus decreases from 74 per cent to 63 per cent during September. The waxing crescent Moon will make a pleasing pairing with Venus in the evening sky on 8 September and Venus passes south of the much fainter Saturn between the eighteenth and twentieth of the month.

MARS is visible in the eastern sky for a couple of hours before sunrise to Northern Hemisphere observers, but is rather less favourably situated for those living further south. The planet moves from Cancer into Leo during the month. The magnitude of the planet is +1.6.

JUPITER is a brilliant object in Gemini, brightening from magnitude -2.0 to -2.2 during September. From northern temperate latitudes the planet rises shortly before midnight by the end of the month, but not until three hours later from the Southern Hemisphere. The waning last-quarter Moon will appear a few degrees south of Jupiter in the morning sky on 28 September.

SATURN, magnitude +0.7, is visible in the west-south-western sky as darkness falls. From northern temperate latitudes, Saturn will be rather low in the twilight at dusk by month end, although a somewhat easier object for observers further south. The planet's motion carries it from Virgo across the border into neighbouring Libra at the beginning of September.

The Loveliest Double Star. It is an interesting fact that a surprisingly large number of stars are either double or multiple. It has even been suggested that single stars such as our Sun are in the minority. Some of the doubles are strikingly beautiful; and perhaps the loveliest of them all is the third-magnitude star Albireo, known officially as Beta Cygni.

Albireo is fairly easy to locate, because it is one of the five stars making up the so-called 'Northern Cross' of Cygnus, the Swan, which is well-placed in the mid-evening during September. The brightest member of the Cross is Deneb, which is of first magnitude and is, incidentally, a real cosmic searchlight at least 60,000 times as luminous as the Sun. Deneb marks the tail of the Swan, and stands at the top of the Cross as it rises on its side early on summer evenings from northern temperate latitudes. Deneb also makes the north-eastern corner of the famous 'Summer Triangle' asterism, along with the two bright stars Vega in Lyra (the Lyre) and Altair in Aquila (the Eagle).

Although designated Beta (the second letter of the Greek alphabet), Albireo is actually fifth in order of brightness in Cygnus after Deneb (Alpha Cygni), Sadr (Gamma), Gienah (Epsilon) and Delta. The Cross is quite evident; and Albireo marks the head or beak of the Swan, but it somewhat spoils the symmetry of the pattern, both because it is fainter than the other four main stars and further away from the centre of the Cross. It lies rather above an imaginary line joining Vega and Altair (Figure 20) and Cygnus is well to the north of the celestial equator, so that it is well seen for a large part of the year. (In fact, Deneb never actually sets as seen from the latitude of London.)

With the naked eye there is nothing remarkable about Albireo. But if you have a telescope, even under low power, you will see at once that it is a magnificent double star whose components (magnitudes 3.1 and 5.1) have contrasting golden-yellow and greenish-blue colours. Even powerful binoculars can separate the pair; the apparent distance between them is 34.3 seconds of arc, which, for a double star, is very wide. At their distance of 380 light years, the separation of the two main components is about 620,000 million kilometres, and if the pair is gravitationally bound to one another the orbital period must be over 75,000 years.

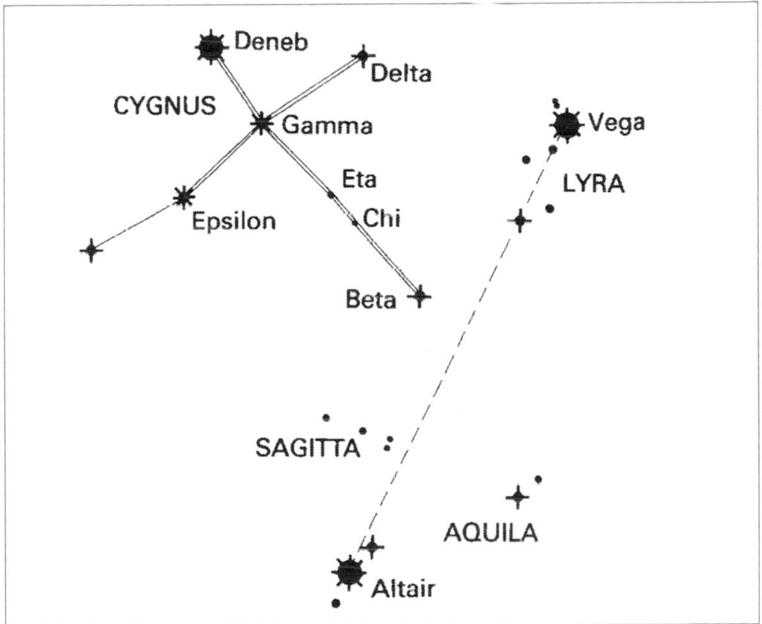

Figure 20. Map showing the principal stars in the area of the Summer Triangle and the location of the beautiful double star Albireo (Beta Cygni), which lies slightly above an imaginary line joining Altair and Vega.

Both components are fairly luminous stars. The golden-yellow primary (Albireo A) is a K3-class giant star of about 5 solar masses with a luminosity about 950 times and a diameter about 50 times that of our Sun. The greenish-blue secondary (Albireo B) is a B8-class dwarf star of 3.3 solar masses with a luminosity about 190 times that of the Sun. It is a rapidly rotating star, spinning in less than 0.6 days and losing gas such that it is probably surrounded by a disk of such material.

It turns out that the golden-yellow star (Albireo A) is itself double so the whole system is actually a triple! Albireo A's companion is a hotter but dimmer (magnitude 5.2) B9-class dwarf star of 3.2 solar masses with a luminosity about 100 times that of the Sun (somewhat similar to Albireo B). On average the components of Albireo A are about 7,000 million kilometres apart, taking 214 years to orbit each other on a highly eccentric path; with a separation of just 0.4 arc seconds, the two components of Albireo A are not readily separable in a telescope.

Imagine the scene from a planet orbiting Albireo B. The two components of Albireo A would appear as brilliant golden-yellow and turquoise-blue points about 0.7 degrees (a little over a full Moon diameter) apart, each shining many times brighter than the Full Moon; the shadow effects would be magnificent.

The Baroness and the Supernova. Mid-evening in early September the Great Square of Pegasus is well-placed in the south-eastern sky. Running from Alpheratz (the top left-hand star in the Square) towards Perseus is a line of three second-magnitude stars, more or less equally spaced, marking the constellation of Andromeda – in mythology the beautiful daughter of Queen Cassiopeia and King Cepheus. The constellation contains Messier 31 (M31), the Andromeda Galaxy, at one time known as the Great Nebula in Andromeda. It is visible to the naked eye on a clear night, not far from the star Nu Andromedae, and was well known in ancient times; it was recorded by the Arabian astronomer Al-Sufi as long ago as the year 964. The first telescopic view of it seems to have been obtained by Simon Marius in 1612; he described it as 'like the flame of a candle seen through horn'. We now know it to be an independent spiral galaxy larger than our own, lying at a distance of about two-and-a-half million light years (Figure 21).

On the evening of 22 August 1885 a Hungarian lady, the Baroness de Podmaniczky, was giving a house-party. One of her guests was an astronomer, Dr de Kövesligethy, and since the baroness herself was

Figure 21. The Andromeda Galaxy, Messier 31 (NGC 224) and its small companions Messier 32 (NGC 221), lower centre, and NGC 205 (sometimes designated Messier 110), to the upper right. The image was made by combining three separate frames acquired using the Burrell Schmidt telescope of the Warner and Swasey Observatory of Case Western Reserve University at Kitt Peak National Observatory. (Image courtesy of Bill Schoening, Vanessa Harvey/REU program/NOAO/AURA/NSF.)

mildly interested in the stars, she owned a 3½-inch refractor. Taking it on to the lawn, she and some of her guests 'looked around'. One object in view was M31, the Great Nebula in Andromeda. Gazing through the telescope at M31, the baroness remarked that she could see a star in the midst of the nebula. Dr de Kövesligethy agreed, but believed that the appearance was due to the presence of the Moon, which drowned the fainter part of the nebula; and neither thought much more about it – until they heard that the star had been seen elsewhere, and was certainly unusual. In fact, Professor Ludovic Gully, at Rouen in France, had noted the star on 17 August, but had put it down to a fault in the telescope which he was testing at the time. The fact that the star was located within the relatively bright nuclear bulge of the galaxy made detection difficult.

But on 20 August, the star was seen by Dr Hartwig, who was on the staff of the Dorpat (Tartu) Observatory in Estonia (at that time controlled by Russia). Hartwig did not fall into the same trap. He realized at once that the star was new, but the Director of the Dorpat Observatory refused to allow him to announce it until the discovery had been confirmed in a moonless sky. Hartwig did write to Kiel, the centre of astronomical information, but apparently his letter never arrived. (Postal services in 1885 were evidently just as unreliable as they are today!)

Max Wolf, in Heidelberg, saw the star on 25 and 27 August, but he, too, put it down to the effect of moonlight. Later, the Irish astronomer Isaac Ward claimed to have seen the star from Belfast on 19 August, but gave its magnitude as 9½ whereas the other observations make it very much brighter. It appears that the star peaked at about magnitude 6. As soon as the Moon had disappeared, Hartwig re-observed the star, and this time there was no room for doubt.

We now know that the star – officially listed as S Andromedae (or SN1885A) – was a supernova, and the first ever noted outside the Milky Way. A supernova is a colossal stellar outburst; a formerly faint star blazes up until it is shining at least a billion times as brilliantly as the Sun. S Andromedae declined quite rapidly in brightness during early September 1885 and had faded to magnitude 16 by February 1886. Certainly it will never be seen again, and it remains the only supernova to have been observed in the Andromeda Galaxy to date, though there have been plenty of ordinary novae.

Back in 1885, the nature of S Andromedae was not appreciated, and some eminent astronomers, such as Étienne Léopold Trouvelot, who worked at the Meudon Observatory near Paris, believed it to be an ordinary nova in the foreground. It is a great pity that it appeared before modern methods of observation could be used. The faint remnant of the supernova was eventually located, just 16 arc seconds from the nucleus of M31, by Robert A. Fesen and co-workers on 9/10 November 1988 using the 4-metre Mayall telescope at the Kitt Peak National Observatory. Further observations of the remnant were made with the Hubble Space Telescope by Robert Fesen and his colleagues in August 1995 and February 1999.

In all events, S Andromedae is unique in one respect: it remains, and probably always will remain, the only supernova to have been independently discovered by a Hungarian baroness!

October

New Moon: 5 October *Full Moon:* 18 October

Summer Time in the United Kingdom ends on 27 October

MERCURY is at greatest eastern elongation (25°) on 9 October, and is visible to the south of due west in the evening twilight sky for observers in equatorial and southern latitudes for the first three weeks of October. Unfortunately, the planet is not suitably placed for observation this month from the latitudes of the British Isles, but for Southern Hemisphere observers this is the most favourable evening apparition of the year. Figure 22 shows the changes in azimuth and

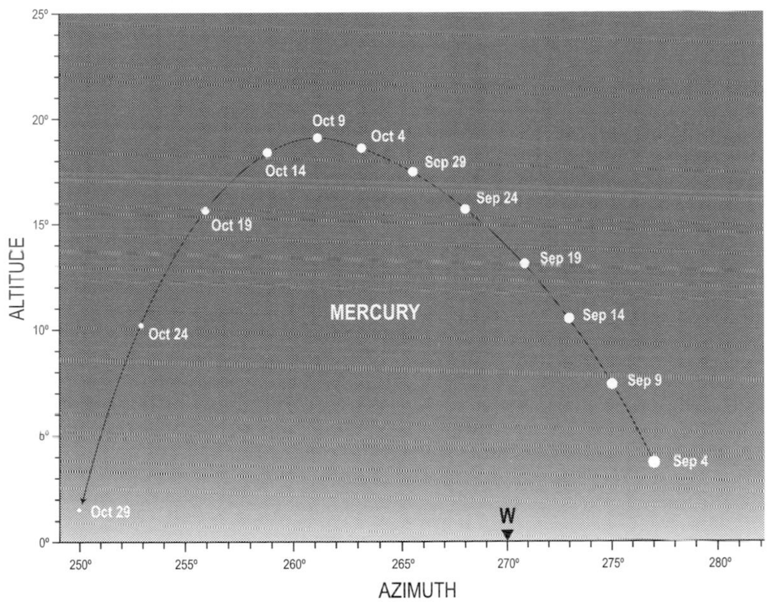

Figure 22. Evening apparition of Mercury from latitude 35°S. The planet reaches greatest eastern elongation on 9 October 2013. It was at its brightest in September, before elongation.

altitude of Mercury at the end of evening civil twilight, about twenty-five minutes after sunset, for observers in latitude 35°S. The changes in the brightness of the planet are indicated by the relative sizes of the white circles marking Mercury's position at five-day intervals: Mercury is at its brightest before it reaches greatest eastern elongation, with the planet fading from magnitude 0.0 to +1.3 between 1 and 24 October. The diagram gives positions for a time at the end of evening civil twilight on the Greenwich meridian on the stated date. Observers in different longitudes should note that the actual positions of Mercury in azimuth and altitude will differ slightly from those given in the diagram due to the motion of the planet. Mercury (magnitude +0.1) passes a few degrees south of Saturn (magnitude +0.7) from 8–11 October.

VENUS is a stunning object in the early evening sky from the tropics and Southern Hemisphere, from where it sets nearly four hours after the Sun by the end of October. Unfortunately, from northern temperate latitudes, the planet remains inconveniently low, being visible for less than half that time. Venus moves from Libra into neighbouring Scorpius and on into Ophiuchus during October, brightening from magnitude -4.2 to -4.4, while the phase decreases from 63 per cent to 50 per cent (half phase or dichotomy). The waxing crescent Moon passes several degrees north of Venus in the evening sky on 8 October.

MARS, magnitude + 1.5, is an early morning object in Leo during October and now visible for several hours before dawn. The waning crescent Moon passes several degrees south of Mars during the early morning hours of 1 and 30 October.

JUPITER is a brilliant object in Gemini, brightening from magnitude -2.2 to -2.4 during October, and rising in the mid evening from northern temperate latitudes, although a couple of hours later from locations much further south. The waning gibbous Moon will appear a few degrees south of Jupiter as it rises on 26 October.

SATURN, magnitude +0.7, is in Libra. From northern temperate latitudes the planet may be glimpsed low down in the west-south-western sky as darkness falls at the beginning of October, but it will soon be lost in the twilight. For a while, Saturn may still be seen rather low down in the western sky after sunset by those living in the tropics and the Southern Hemisphere, but even they will lose sight of it well before month end.

URANUS is at opposition on 3 October in the constellation of Pisces. (The planet passes briefly into Cetus in March.) Uranus is barely visible to the naked-eye as its magnitude is +5.7, but it is easily located in binoculars. Figure 23 shows the path of Uranus against the background stars during the year. At opposition Uranus is 2,848 million kilometres (1,770 million miles) from the Earth.

Transferred Stars. In 1603, the German astronomer Johann Bayer published a star catalogue. It was quite a good one, but is memorable chiefly because Bayer introduced the system of allotting Greek letters to the stars in their various constellations. The first letter of the Greek alphabet is Alpha; therefore, the brightest star in (say) Boötes, the Herdsman, should be Alpha Boötis, the brightest star in Lyra (the Lyre) Alpha Lyrae, and so on through to Omega, the last of the Greek letters. The system was a convenient one, and is still used. Only in the cases of the brightest stars are the individual names commonly accepted; thus Alpha Boötis is better known as Arcturus and Alpha Lyrae as Vega. In some cases the letters are out of order. In Orion, Beta (Rigel) is brighter than Alpha (Betelgeux), and there are many other cases. There are also instances of stars being transferred from one constellation to another.

Originally the constellations were defined somewhat informally by the shapes made by their star patterns, but, as the pace of celestial discoveries quickened in the early twentieth century, astronomers realized it would be helpful to have an official set of constellations, and to agree where one constellation ends and the next begins. Accordingly, on behalf of Commission 3 (Astronomical Notations) of the International Astronomical Union (IAU), Eugène Delporte proposed the eighty-eight 'modern' constellations and their boundaries in 1930. This system will probably not be altered, but it is interesting to look back and see which stars have been given free transfers.

The first, and most illogical, case is that of Alpheratz (Figure 24a), which is of the second magnitude and is one of the four stars making up the Great Square of Pegasus, which is well placed during October

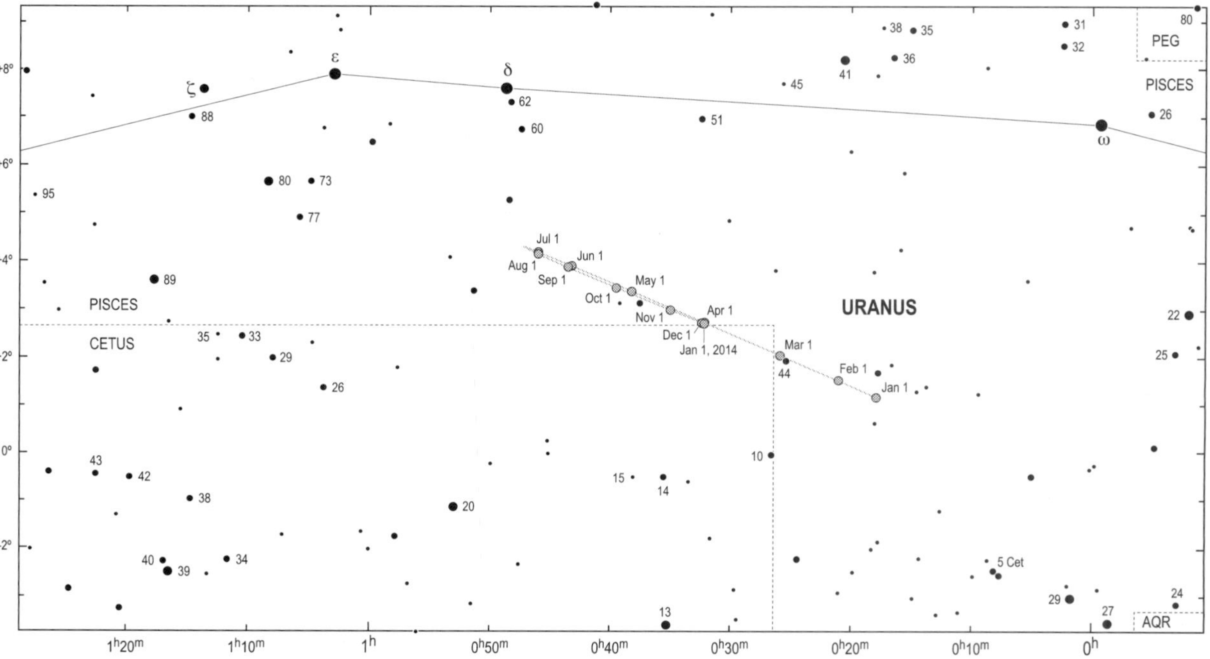

Figure 23. The path of Uranus against the background stars of Pisces during 2013.

evenings. Bayer naturally included it in Pegasus, as Delta Pegasi, but the IAU decree shifted it into the adjacent constellation of Andromeda, and it became Alpha Andromedae (in fact, Bayer had given both designations).

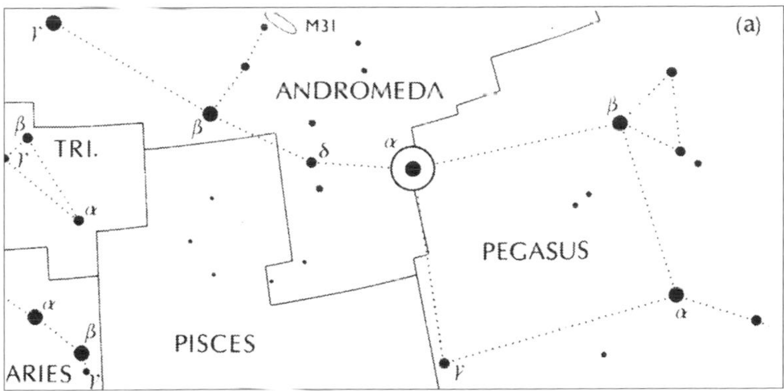

Figure 24a. Map showing the location of the star Alpheratz (circled here), which has been transferred from Pegasus into neighbouring Andromeda (as Alpha Andromedae), even though it marks the north-eastern corner of the 'Great Square'.

The other notable case is that of Al Nath, also of the second magnitude (Figure 24b). It was originally included in Auriga, the Charioteer, as Gamma Aurigae, and this was reasonable enough because Auriga is a well-marked constellation; its leader is, of course, the brilliant yellowish star Capella. For observers in Europe and North America, Auriga will be seen rising in the north-east in the mid-evening during October. But Al Nath has now been transferred to Taurus, the Bull, as Beta Tauri. Taurus itself is one of the Zodiacal constellations, and includes not only the bright orange Aldebaran, but also the two most famous of all open star clusters, the Pleiades and the Hyades. Taurus is frankly rather shapeless, but the transfer of Al Nath is somewhat less irrational than that of Alpheratz. Al Nath marks the tip of the northernmost of the Bull's horns; the star Zeta Tauri marks the tip of the southernmost horn.

Libra, the Scales or Balance, is also in the Zodiac, but is faint and rather indistinct. The third-magnitude star now known as Sigma Librae was originally given the Bayer designation Gamma Scorpii – in spite of the fact that it is well within the constellation Libra and far from the

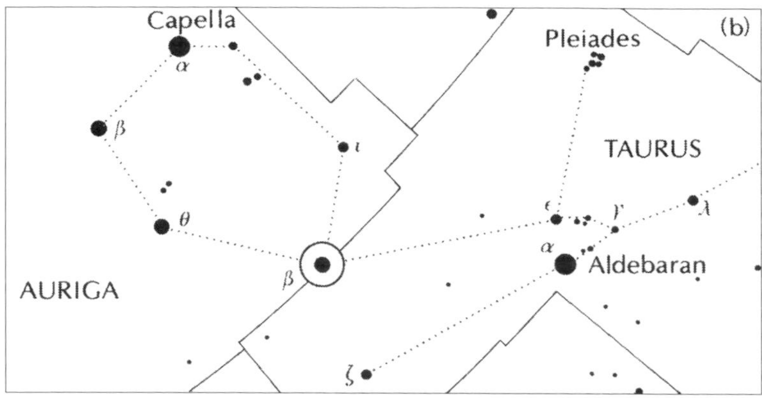

Figure 24b. Map showing the location of the star Al Nath (circled here), which has been transferred from Auriga into neighbouring Taurus (as Beta Tauri), where it now marks the tip of the northernmost horn of the Bull.

boundary with Scorpius. Sigma Librae was also known historically as Zubenalgubi (the south claw). In fact, Libra itself was once Chelae Scorpionis, the Scorpion's Claws.

And Canis Major, the Great Dog, has lost the star lettered 3 by the Reverend John Flamsteed, which has now become Delta Columbae in the neighbouring constellation of the Dove. Finally, three stars in the little constellation of the Shield – Beta, Delta and Epsilon Scuti – were formerly lettered 6, 2 and 3 Aquilae.

It cannot be said that the overall pattern of the eighty-eight constellations is convenient, and if we had followed, say, the Chinese system, everything would be very different. However, we have become so accustomed to them that there is no reason for any new changes.

The Far Side of the Moon. It is now well over fifty years since the first images of the Moon's far side were obtained. They were sent back in October 1959 by the Russian spacecraft Lunik (or Luna) 3, which had been sent on a 'round trip'. Previously only two vehicles had been successfully dispatched to the Moon: Lunik 1, which flew past on 4 January 1959, and Lunik 2, which impacted the surface on 13 September 1959, though without sending back any data.

Lunik 3 passed about 6,200 km above the south pole of the Moon on 6 October 1959, and continued on over the far side. The following day, photography of the far side began and continued for about forty minutes while the probe was between 63,500 km and 66,700 km from the Moon. About thirty pictures were taken, covering nearly three-quarters of the lunar far side. Lunik 3 passed beyond the Moon, and then swung back over its north pole and on towards the Earth. Early attempts to retrieve the images were unsuccessful, but as the probe neared the Earth, eighteen photographs were transmitted by 18 October. All contact with Lunik 3 was lost four days later. The photographs were of quite poor quality by modern standards, and covered roughly one-third of the surface invisible from Earth, but they were good enough to show that the Moon's far side is just as cratered and barren as the known side. One feature was taken to be a huge range of peaks, and was named the Soviet Mountain range; however, it proved to be nothing more than a bright ray, and the name was tactfully deleted from later maps. On 20 July 1965, another Soviet probe, Zond 3, transmitted twenty-three pictures of the lunar far side, taken at distances between 9,960 km and 11,570 km, with a much higher resolution than those of Lunik 3.

The Moon's orbital period is the same as its axial rotation period: 27.32 days. There is no mystery about this so-called 'synchronous rotation'. Tidal action over the ages has been responsible, and the effect is to keep the same face of the Moon turned towards us all the time. The lunar orbit is not circular, so that its speed of orbital motion varies, and the Moon seems to rock very slowly from side to side; this is termed libration in longitude. All in all, it is possible to examine a total of 59 per cent of the Moon's total surface area from the Earth, though of course never more than 50 per cent at any one time.

The 'edges' or libration zones are always very foreshortened, and are difficult to map from Earth, but various points were well established; none of the major seas (maria) crossed the limb to the averted areas, and it was suspected that the far side might have many fewer of the large maria (Latin for 'seas', since early astronomers incorrectly thought that these plains were areas of water) than the side we have always known. This did indeed prove to be the case; only about one per cent of the lunar far side is covered by maria compared with over thirty per cent of the near side. There is just one major sea on the far side, the Mare Orientale, which extends over the limb, but only a small part of its border can be seen from Earth. The far side is also very heavily cratered. The near side has been somewhat shielded from impacts by the body of the Earth, because the synchronous rotation has kept the lunar far side exposed to the steady stream of impactors coming from outer space.

By now the whole of the Moon has been mapped in great detail, most recently by the American spacecraft Lunar Reconnaissance Orbiter (Figure 25), but it is true that only twenty-seven men have had direct views of the far side – the crews of *Apollos 8* and *10* to *17*.

Figure 25. This global mosaic of over 15,000 images, acquired between November 2009 and February 2011 by the Lunar Reconnaissance Orbiter's wide-angle camera, shows the lunar far side. The surface of the far side looks quite different from the more familiar near side which is covered with smooth, dark lunar maria. The probable explanation is that the far-side crust is thicker, making it harder for molten material from the interior to flow to the surface and form the smooth maria. (Image courtesy of NASA/GSFC/Arizona State Univ./Lunar Reconnaissance Orbiter.)

November

New Moon: **3 November** *Full Moon:* **17 November**

MERCURY passes through inferior conjunction on 1 November and rapidly moves out to the west of the Sun, reaching greatest western elongation (19°) on 18 November. The planet is consequently visible from the tropics and northern temperate latitudes as an early morning object from the second week of November until the beginning of December. The planet is inconveniently low from the Southern Hemisphere. For observers in the latitudes of the British Isles this will be the most favourable morning apparition of the year, with the planet being visible in the south-eastern twilight sky before dawn. Figure 26

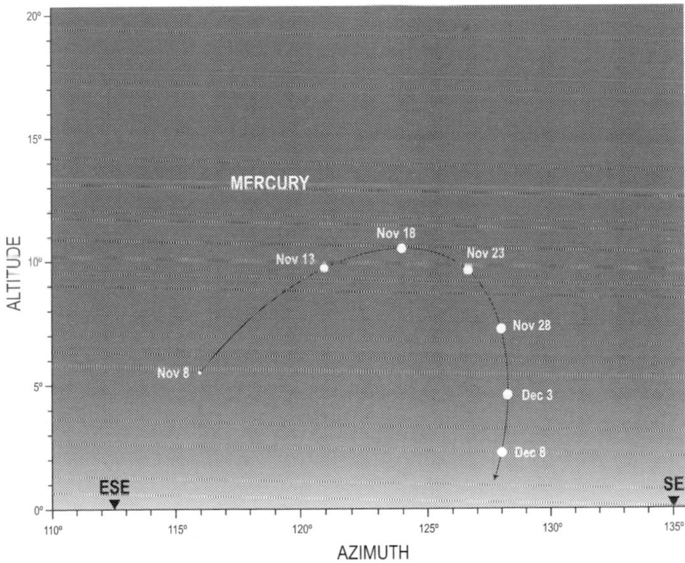

Figure 26. Morning apparition of Mercury from latitude 52°N. The planet reaches greatest western elongation on 18 November 2013. It will be at its brightest in late November, after elongation.

shows the changes in azimuth and altitude of Mercury on successive mornings at the start of morning civil twilight, about forty minutes before sunrise, for observers in latitude 52°N. The changes in the brightness of the planet are indicated by the relative sizes of the white circles marking Mercury's position at five-day intervals: Mercury is at its brightest after it reaches greatest western elongation, with the planet brightening from magnitude +1.4 to -0.7 between 8 and 30 November. The diagram gives positions for a time at the end of evening civil twilight on the Greenwich meridian on the stated date. Observers in different longitudes should note that the actual positions of Mercury in azimuth and altitude will differ slightly from those given in the diagram due to the motion of the planet. Mercury (magnitude -0.7) will pass less than half a degree south of Saturn (magnitude +0.7) early on 26 November.

VENUS reaches greatest eastern elongation (47°) on 1 November, brightening from magnitude -4.4 to -4.6 during the month and moving from Ophiuchus into Sagittarius. The phase shrinks from 50 per cent to 31 per cent during November. Venus is a quite beautiful and brilliant object in the early evening sky from the tropics and Southern Hemisphere, from where it still sets well over three hours after the Sun at the end of November. From northern temperate latitudes, the situation improves during November and the planet becomes a lovely sight low in the south-western sky at dusk by the end of the month. The waxing crescent Moon passes over seven degrees north of Venus early on 6 November.

MARS is an early morning object during November, still visible for several hours before dawn. The planet moves from Leo into neighbouring Virgo during the month and brightens from magnitude +1.5 to +1.2. The waning crescent Moon passes south of Mars on 27 November.

JUPITER is a brilliant object, rising in the early evening from the latitudes of the British Isles – a little later from locations further south – and visible all night long. It brightens from magnitude -2.4 to -2.6 during the month. The planet is in Gemini, reaching its first stationary point on 7 November; thereafter its motion is retrograde.

SATURN is in conjunction with the Sun on 6 November and will be inconveniently placed for observation from northern temperate latitudes this month. However, it may be glimpsed low down in the east-south-eastern twilight sky shortly before dawn by those living in the tropics and the Southern Hemisphere at the very end of November.

A Most Unusual Solar Eclipse. The eclipse of the Sun on 3 November will be seen along a narrow track starting 1000 kilometres off the east coast of the USA at sunrise, travelling across the Atlantic Ocean, curving a couple of hundred kilometres below the south-western coast of West Africa and making landfall in Gabon. The track then crosses Central Africa and ends at sunset close to the Ethiopia/Somalia border. (For a diagram showing the path of this eclipse see the section 'Eclipses in 2013' by Martin Mobberley elsewhere in this *Yearbook.*) Observers in the far south of Europe will see a small partial eclipse but those further north, including in the UK, will miss out.

This eclipse is an example of one of the rarest types of solar eclipse – an annular-total eclipse, also called a 'hybrid' eclipse. In the hundred years from 2001 to 2100, there are only seven such hybrid eclipses. On these rare occasions, the Sun and Moon appear almost *exactly* the same size in the sky. Normally, an annular eclipse will be visible from both ends of the track, while a total eclipse will be seen around the mid-point of the eclipse path, where the cone of the Moon's umbral shadow just reaches the Earth's surface (Figure 27).

The maximum duration of totality on 3 November will be brief, lasting one minute forty seconds at 12:46 UT when the track will be just 58 kilometres wide at a point 300 kilometres south-west of the Liberian coastline, with the Moon a mere 1.6 per cent larger than the Sun. At the beginning, in the Atlantic south of Bermuda, the Moon will be 0.1 per cent smaller than the Sun (annular eclipse) and at the end, near the Ethiopia/Somalia border, the Sun and Moon will be virtually identical in size, producing a total eclipse with a duration of just a split second.

The Equatorial Sky. This month's most unusual hybrid solar eclipse passes across many parts of Central Africa, from Gabon on the West African coast, through Congo, DR Congo, northern parts of Uganda and Kenya, and the far south of Ethiopia to the Ethiopia/Somalia border. For most of this journey, the eclipse path lies very close to the Equator, moving only four degrees north of the Equator by the far eastern end of the track. Anyone travelling to such parts of the world to

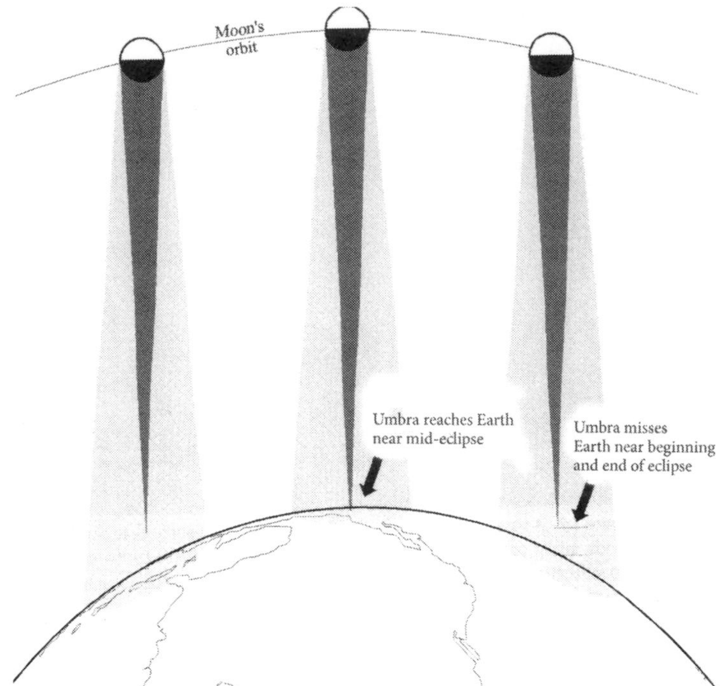

Figure 27. In a rare hybrid or annular-total solar eclipse the narrow cone of the Moon's umbral shadow just reaches the Earth's surface near the middle of the eclipse track (producing a short total eclipse), but because of the curvature of the Earth it usually does not reach the surface towards both ends of the track (producing only an annular eclipse). The hybrid eclipse of 3 November 2013 is even more unusual because it starts annular and ends total, never reverting to annularity.However, the 3 November solar eclipse will be annular only for a few kilometres at its start with a duration of less than one second, and for the remainder it will be a very narrow and short-lasting total eclipse. So this eclipse is even more unusual because it starts annular and ends total, never reverting to annularity.

view the eclipse will obtain excellent views of the equatorial sky during the night-time, and will surely find it quite fascinating.

From the United Kingdom, we miss out on some of the most interesting objects in the sky. The Southern Cross never rises; neither do Canopus, Achernar, the Large and Small Magellanic Clouds or the eccentric variable Eta Carinae. From Southern Hemisphere countries such as Australia, New Zealand and South Africa the situation is better; observers there lose only the northernmost parts of the sky, which are far less brilliant than those towards the South Celestial Pole. The only real losses are the Pole Star, the 'W' of Cassiopeia, and most or all of the Great Bear. Of course, the view is different. From England we see Orion with Betelgeux (or Betelegeuse) at the top left and Rigel at the lower right; the three stars of the Hunter's Belt point upward to Aldebaran, and downward to Sirius. From Australia, Rigel is higher than Betelgeux, and the Belt points downward to Aldebaran and upward to Sirius.

From Central Africa, however, or anywhere very near the Equator, the two celestial poles lie on opposite horizons, with the celestial equator passing overhead and one can see the entire span of the heavens at one time or another. From the far north of Kenya, Polaris is just above the horizon and the South Pole just below, but that is a minor modification. To an observer exactly on the Equator, it would in theory be possible to see the two pole stars (Polaris and Sigma Octantis) at the same time, though I am not sure that this has ever been achieved; Sigma Octantis is so faint that it would be very hard to see so low in the sky. The declination of Polaris is 89° 15' 51" N, and of Sigma Octantis 88° 57' 23" S, so that of the two pole stars Polaris is slightly closer to its polar point.

Around midnight in early November the equatorial observer will see Orion lying on his side in the eastern sky, with Rigel higher than Betelgeux; the Belt points downward and to the right towards Sirius, the Dog Star. Tracking round to the south-east, at the same elevation as Sirius brings us to Canopus (Alpha Carinae). Slightly higher up to the south is Achernar (Alpha Eridani), and slightly higher still, in the south-west, is Fomalhaut (Alpha Piscis Austrini). To ease identification, Canopus, Achernar and Fomalhaut all lie in a straight line.

Obviously, the planets are on view whenever they are sufficiently far from the Sun. In early November 2013, for example, Venus will be beautifully placed in the western sky after sunset, Jupiter will rise just north of east around midnight, and Mars comes up in the early morning hours.

Certainly, the equatorial sky is of special interest. For example, around midnight in early April one can look south and see the Southern Cross (with Alpha and Beta Centauri) and then turn to look north and see the saucepan-shaped pattern of the Plough in the sky at the same time, which may seem a little curious. A little earlier in the evening, one can see Orion sinking down into the western sky, lying in a rather undignified manner on his side. In many respects the equatorial observer is very well placed, because there is virtually nothing which is out of view. This being so, why have no major international observatories been set up at the Equator? The answer is that everything depends upon local conditions, not only meteorological but also political.

December

New Moon: 3 December *Full Moon:* 17 December

Solstice: 21 December

MERCURY passes through superior conjunction on 29 December, and is therefore too close to the Sun for observation after the first week of December. During this short time, the planet may be glimpsed from the tropics and northern temperate latitudes as an early morning object rather low in the south-eastern sky before dawn at the start of morning civil twilight. Mercury is inconveniently low from the Southern Hemisphere. The magnitude of the planet will be -0.7 for this period.

VENUS attains its greatest brilliancy (magnitude -4.7) on 10 December. The planet is a spectacular object in the early evening sky from the tropics and Southern Hemisphere, from where it is still setting about two-and-a-half hours after the Sun in mid-December. At this time, even from northern temperate latitudes, the planet is a lovely sight in the south-western sky at dusk. However, in the second half of December, the planet rapidly draws in towards the Sun and the period of visibility shortens, the planet fading slightly to magnitude -4.4 and its crescent phase shrinking to just 4 per cent by month end. Venus will pass through inferior conjunction on 11 January 2014. The waxing crescent Moon passes nearly seven degrees north of Venus on 5 December.

MARS remains an early morning object moving direct in Virgo, still rising an hour or so after midnight by the end of the month. The planet brightens from magnitude +1.2 to +0.8 during December.

JUPITER is now approaching opposition and is a brilliant object visible all night long. The planet is moving retrograde in Gemini, its brightness increasing very slightly from magnitude -2.6 to -2.7 during December.

SATURN was in conjunction with the Sun in early November and is visible in the south-eastern sky before dawn during December. By the end of the month the planet will be rising four hours before the Sun from the latitudes of the British Isles and only slightly less from locations further south. The planet is in Libra, magnitude +0.6.

The Moon Illusion. When the Moon is nearly full, anyone can see the famous light and dark markings on its face; some people can make out the so-called 'Man in the Moon', though I (PM) admit that I have never been really successful. The dark, roughly circular patches are the waterless lunar seas, which were once great impact basins that filled with molten lava and then solidified; the bright areas are the uplands. Any telescope, or, for that matter, binoculars, will show the Moon's mountains and craters. The lunar landscape is of immense beauty as well as interest, and the fact that men have landed there has not in any way lessened the romance of the Moon.

There is something else which has intrigued many people for many years. If the sky is clear, go out at dusk on 17 December and watch the full Moon rising in the east-north-eastern sky. It is often said that when the full Moon is just rising it looks very large; as it climbs in the sky it shrinks, and when high up it seems comparatively small. Ask the average person whether the full Moon looks largest when it is low down, and he or she is almost certain to answer 'yes'.

Actually, this is not true. The low-down full Moon is no larger than the high-up full Moon. If you want to prove this, you can do so easily. Select a small pebble (or something equivalent) and hold it out in front of you until it just covers the rising full Moon. Repeat the experiment when the Moon is high up, and you will find that there is no difference. Yet it is undeniable that the low-down full Moon *seems* to be large. This is the celebrated Moon Illusion, and it has been known for centuries.

One man who commented upon it was Ptolemy, last of the great astronomers of the Greek school, who lived around AD 150. His explanation was that the low Moon is seen against a background of 'filled space' – trees, houses and so on – while when the Moon is high there is nothing with which to compare it; it is seen against 'empty space'. Hence the Moon Illusion. Many years ago, in one of my BBC television *Sky at Night* programmes, we tried an experiment. With Professor Richard Gregory, we built an apparatus in which we could see the image of the real Moon and compare it with an artificial moon. It was worked by mirrors, and we tested it one evening on the beach at Selsey, much to the amusement of various onlookers. We found that the observer still judged the low Moon to be bigger than the high Moon, though in each case the comparison object was an artificial moon which did not change. We also tried various other tricks. One member of our team covered up one eye for several hours beforehand; the illusion was still present. I even observed while standing on my head, balancing myself precariously against a groyne and creating the general impression that I was insane.

I am never sure whether Ptolemy's explanation is the full answer, but in any case there is no doubt that the effect is an illusion and nothing more. If you do not believe me, experiment for yourself!

Zeta Phoenicis. To many people, one of the most confusing regions in the entire night sky is that of the Southern Birds. There are four bird constellations: Grus (the Crane), Phoenix (the Phoenix), Pavo (the Peacock) and Tucana (the Toucan). All are invisible from the latitudes of northern Europe and North America, so that to see them you have to travel south. Grus is easily the most distinctive of the four. It lies not far from the bright, rather isolated Fomalhaut in the Southern Fish (which does rise briefly over the London horizon), and contains a curved line of stars, giving a vague impression of a long-necked bird in flight. Of its two main stars, Alpha (Alnair or Al Nair) is blue-white and the other (Beta) a warm orange-red.

Phoenix is much less distinctive (Figure 28). It has one second-magnitude star, Alpha Phoenicis (Ankaa), and nothing else above magnitude 3. However, one of its stars, not dignified by a proper name and known merely as Zeta Phoenicis, is of considerable interest. Unfortunately for those living in northern temperate latitudes, you have to be south of latitude 34 degrees North before it can be glimpsed. It lies fairly near the brilliant star Achernar in Eridanus (the River), which is the best way to locate it.

Normally, Zeta Phoenicis is of magnitude 3.9. Every 1.6697664 days (1 day 16 hours 4 minutes and 28 seconds), Zeta Phoenicis fades. It drops by half a magnitude down to magnitude 4.4, so that with any appreciable moonlight around you will be hard pressed to see it with the unaided eye. It does not stay faint; it soon slowly brightens up again, after which there is no other significant fade for another 1.6697664 days. These periodic fades are fairly obvious with the naked

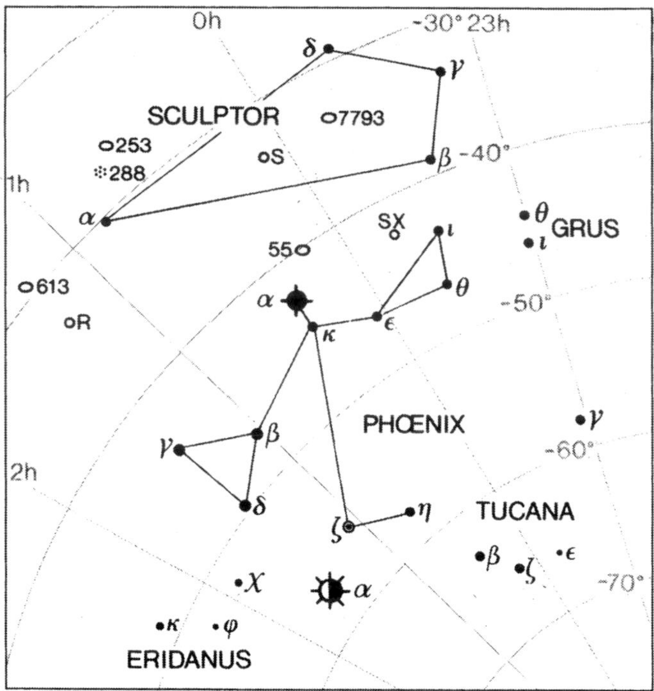

Figure 28. Map showing the principal stars of Phoenix, including the interesting variable star Zeta Phoenicis, indicated here by a small circle with central dot just above the bright star Achernar (Alpha Eridani), which is lower centre on the map.

eye. There is however a much smaller dip in brightness of just 0.18 magnitudes centred right in the middle of the more significant fades.

Although Zeta Phoenicis fluctuates in brightness, it is not a true variable star. It is made up of two B-type dwarf stars, one about five times brighter than the other, and so close together that they cannot be seen separately even with a powerful telescope. When the smaller, slightly cooler, fainter and less massive member of the pair passes in front of its larger, hotter, brighter and more massive companion, it blocks out about 60 per cent of its light causing a fade of about half a magnitude in apparent brightness; this is called the 'primary' eclipse and is the deeper of the two. Since the smaller star passes more or less centrally in front of the larger star, only a ring of light from the larger star is left surrounding the smaller star. It is rather like an annular eclipse. In between each pair of primary eclipses (which are 1.6697664 days apart) is a lesser 'secondary' eclipse, which happens when the larger star completely hides its smaller companion, producing a 'total eclipse' that lasts about an hour. Each of the two types of eclipse lasts for about five hours from start to finish.

Zeta Phoenicis is not unique. There are many more of these so-called 'eclipsing binaries' in the sky, the most famous, and the first-discovered member of the class, being Algol (nicknamed the winking Demon Star) in Perseus. Zeta Phoenicis is not particularly exceptional; it is rather less than 300 light-years distant, and the total luminosity of the system is about 350 times that of the Sun.

All the same, Zeta Phoenicis is worth looking at. If you know when a fade is due, you will be able to check it by observing every ten minutes or so and checking the magnitude against the neighbouring stars which do not vary. Certainly its behaviour would seem strange if the reason for it were not known.

Eclipses in 2013

MARTIN MOBBERLEY

During 2013 there will be five eclipses: a partial eclipse of the Moon, an annular eclipse of the Sun, two penumbral eclipses of the Moon and a hybrid eclipse of the Sun.

1. *A partial eclipse of the Moon* on 25 April will be visible from Western Australia, southern and eastern Africa, the Indian Ocean, the Middle East and the eastern Mediterranean as well as parts of Central Asia and western Russia. However, while this is, technically, a partial lunar eclipse, in practice it is more like a penumbral lunar eclipse because the only part of the lunar disk entering the umbra will be the extreme northern tip, which just clips the umbral shadow between 19h 54m and 20h 21m UT. From the UK the easternmost counties will see this umbral phase when the Moon is rising at an altitude of only five to eight degrees above the horizon, although the low altitude of the lunar

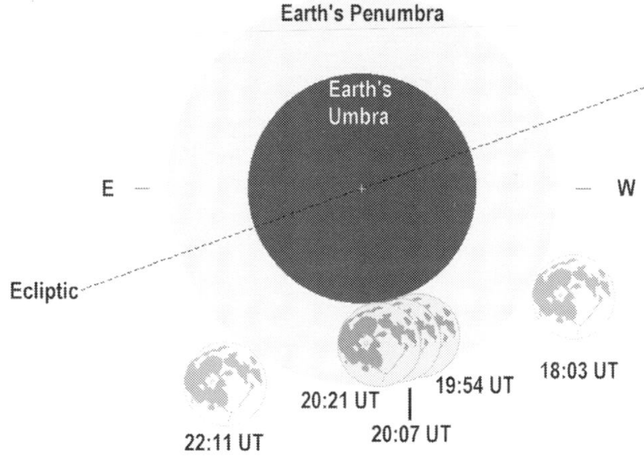

Figure 1. The path of the Moon through the Earth's penumbra and umbra on 25 April. (Diagram courtesy of NASA/Fred Espenak.)

disk at this time often dims the Full Moon enough to make the umbra and penumbra more obvious.

2. *An annular eclipse of the Sun* on 10 May will be visible from parts of Australia and the South Pacific Ocean. The track of annularity is quite similar to that traced by the track of totality in the region six months earlier in November 2012, as it starts at sunrise in Western Australia and tracks across the Northern Territory and northern Queensland before leaving the Australian coast. The track just clips the south-eastern edge of Papua New Guinea and crosses the Solomon Islands before entering the South Pacific Ocean. The eclipse ends at sunset a few thousand kilometres west of the coast of northern Peru. The point of greatest eclipse occurs at the latitude of 2 degrees 12.8 minutes north and the longitude of 175 degrees, 28.3 minutes east where the track width is 172.6 km. The eclipse duration at this point will be six minutes three seconds. At the position where the annular eclipse leaves the Australian coast near Yarraden the Sun will be at an altitude of twenty-eight degrees and the annular phase will last four minutes and forty-three seconds. At the start of the track, just after sunrise, the whole of

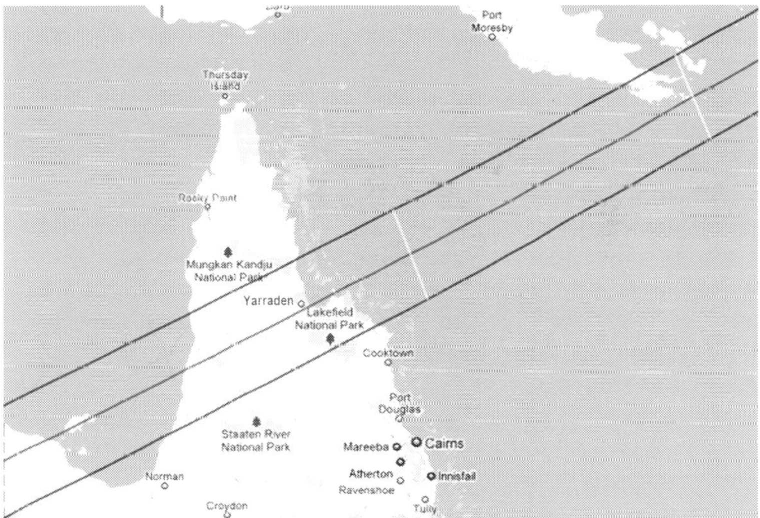

Figure 2. The path of the annular eclipse of 10 May over northern Australia and Papua New Guinea. (Diagram courtesy of NASA.)

the Australia, Papua New Guinea and Indonesia region will experience a partial solar eclipse. From Tasmania, 25 per cent of the Sun will be covered but south-eastern parts of New Zealand will lie just outside the partial zone.

3. *A penumbral eclipse of the Moon* on 25 May will, in theory at least, be visible from western Africa, the Atlantic Ocean and much of the Americas. However, this is the most borderline eclipse imaginable, with the southern limb of the Moon just grazing the penumbral shadow between 03h 53m and 04h 26m UT. In practice, the effect on the appearance of the Moon will be absolutely negligible.

4. *A penumbral eclipse of the Moon* on 18 October will be visible from Africa, the Middle East, the Atlantic Ocean, Europe, Scandinavia, western Russia, Brazil and Newfoundland. Seventy-six per cent of the Moon will be immersed within the penumbra at 23h 51m UT. By the nature of penumbral lunar eclipses the Moon will never appear dark. Its northern limb will never even enter the penumbra and so will be fully illuminated, but its southern limb should appear noticeably dimmer at maximum eclipse.

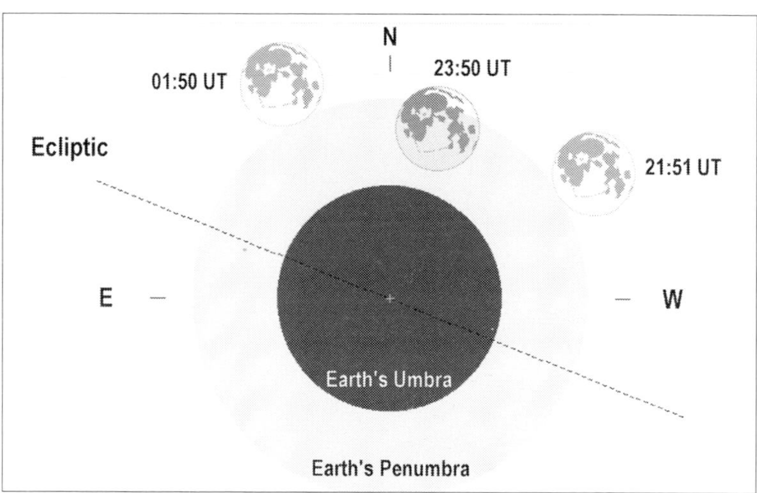

Figure 3. The path of the Moon through the Earth's penumbra on 18 October. (Diagram courtesy of NASA/Fred Espenak.)

5. *A hybrid eclipse of the Sun* on 3 November will be visible along a narrow track starting 1,000 km off the eastern seaboard of the US at sunrise, travelling across the Atlantic Ocean, curving 250 km below the African coastlines of Liberia and Ivory Coast and making landfall at Port Gentil, Gabon. The track then crosses the African continent and finally ends at sunset close to the Ethiopia/Somalia border. A hybrid eclipse is one where the Sun and Moon are so close in size that the eclipse is annular (Moon apparently smaller) at the start and/or end of the track and total (Moon apparently larger) in the middle of the track. The radius of the Earth is enough to raise the observer so that the Moon, four hundred times nearer, increases its angular size enough to blot out the Sun. However, the total phase of a hybrid eclipse is often quite short in duration and the track is relatively narrow compared to a conventional total solar eclipse. In this case the maximum period of totality lasts for one minute forty seconds at 12h 46m UT and the track is 58 km wide at a point 300 km south-west of the Liberian coastline, where the Moon will appear just 1.6 per cent larger than the Sun. In contrast, at the start (sunrise) point the Moon is 0.1 per cent smaller than the Sun in this case (annular) and at the end, near the Ethiopia/Somalia border, the Sun and Moon are virtually identical in

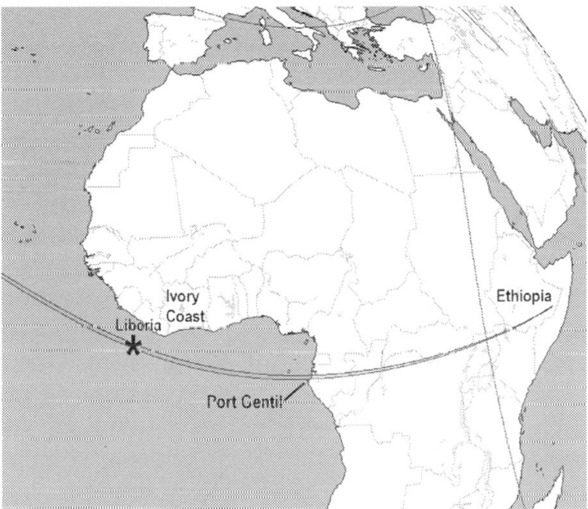

Figure 4. The track of the hybrid solar eclipse of 3 November across the eastern Atlantic Ocean and Africa. (Diagram by the author using WinEclipse by the late Heinz Scsibrany.)

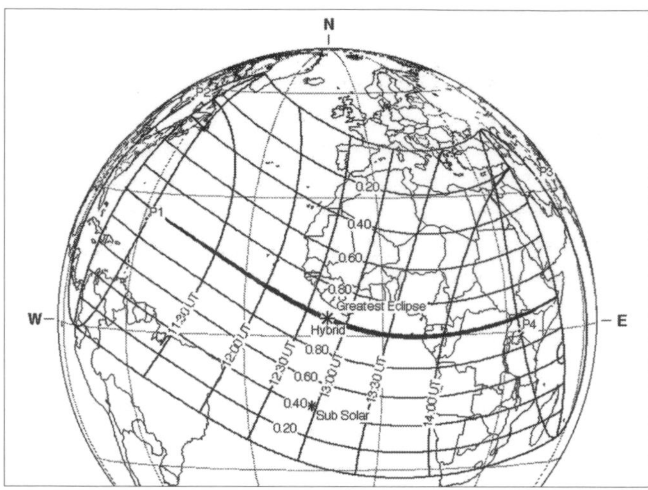

Figure 5. The full extent of the partial solar eclipse of 3 November. Much of the Atlantic hemisphere will see a partial eclipse but the southern tip of Africa and the British Isles will miss out. (Diagram courtesy of NASA/Fred Espenak.)

size, so the total eclipse duration there is just a split second. Even deep valleys on the lunar limb can therefore take such a borderline eclipse from being total to annular if just one brilliant bead of light still remains visible.

For much of the Atlantic hemisphere of the Earth a partial solar eclipse will be seen. Observers in northern South America may see up to 40 per cent of the Sun eclipsed after sunrise and the whole of Africa, excluding the southern tip, will see a partial eclipse. European observers in Spain, southern France, southern Italy and Greece will just see a small partial eclipse but observers further north, including those in the UK, will miss out.

Occultations in 2013

NICK JAMES

The Moon makes one circuit around the Earth in just over twenty-seven days and as it moves across the sky it can temporarily hide, or occult, objects that are further away, such as planets or stars. The Moon's orbit is inclined to the ecliptic by around 5.1° and its path with respect to the background stars is defined by the longitude at which it crosses the ecliptic passing from south to north. This is known as the longitude of the ascending node. After passing the node, the Moon moves eastwards relative to the stars reaching 5.1° north of the ecliptic after a week. Two weeks after passing the ascending node it crosses the ecliptic moving south and then it reaches 5.1° south of the ecliptic after three weeks. Finally it arrives back at the ascending node a week later and the cycle begins again.

The apparent diameter of the Moon depends on its distance from the Earth but at its closest it appears almost 0.6° across. In addition, the apparent position of the Moon on the sky at any given time shifts depending on where you are on the surface of the Earth. This effect, called parallax, can move the apparent position of the Moon by just over 1°. The combined effect of parallax and the apparent diameter of the Moon means that if an object passes within 1.3° of the apparent centre of the Moon as seen from the centre of the Earth it will be occulted from somewhere on the surface of our planet. For the occultation to be visible the Moon would have to be some distance from the Sun in the sky and, depending on the object being occulted, it would have to be twilight or dark.

For various reasons, mainly the Earth's equatorial bulge, the nodes of the Moon's orbit move westwards at a rate of around 19° per year, taking 18.6 years to do a full circuit. This means that, whilst the Moon follows approximately the same path from month to month, this path gradually shifts with time. Over the full 18.6-year period all of the stars that lie within 6.4° of the ecliptic will be occulted.

Only four first-magnitude stars lie within 6.4° of the ecliptic. These are Aldebaran (5.4°), Regulus (0.5°), Spica (2.1°) and Antares (4.6°). As the nodes precess through the 18.6-year cycle there will be a monthly series of occultations of each star followed by a period when the star is not occulted. In 2013 there will be fourteen occultations of first-magnitude stars, all of them Spica. Twelve of these occur at a solar elongation greater than 30° and so should be visible from somewhere on Earth.

In 2013 there will be ten occultations of bright planets, one each of Venus and Mars, two each of Jupiter and Saturn and four of Mercury. Four of these events take place at a solar elongation of greater than 30°.

The following table shows events potentially visible from somewhere on the Earth when the solar elongation exceeds 30°. More detailed predictions for your location can often be found in magazines or the *Handbook of the British Astronomical Association*.

Object	Time of Minimum Distance (UT)		Minimum Distance	Elongation	Best Visibility
			°	°	
Spica	05 Jan 2013	19:54	0.6	−82	Australia
Jupiter	22 Jan 2013	03:06	0.5	124	South America
Spica	02 Feb 2013	01:39	0.3	−109	Southern Africa
Jupiter	18 Feb 2013	11:44	0.9	97	Australia
Spica	01 Mar 2013	07:10	0.1	−137	South America
Spica	28 Mar 2013	14:51	0.0	−164	Australia
Spica	25 Apr 2013	00:27	0.0	169	Southern Africa
Spica	22 May 2013	10:46	0.1	142	Southern Pacific
Spica	18 Jun 2013	20:22	0.1	116	Southern Africa
Spica	19 July 2013	03:49	0.3	90	Central America
Spica	12 Aug 2013	09:20	0.6	64	Japan
Spica	08 Sept 2013	14:51	0.7	38	Middle East
Venus	08 Sept 2013	20:51	0.4	41	South America
Spica	29 Nov 2013	17:30	0.8	−44	Alaska
Spica	27 Dec 2013	03:06	1.0	−72	Russia
Saturn	29 Dec 2013	01:10	0.9	−47	Southern Indian Ocean

Comets in 2013

MARTIN MOBBERLEY

Forty-six short-period comets should reach perihelion in 2013 although one of these, 83D/Russell, has been declared defunct and so has a D/prefix. All these returning comets orbit the Sun with periods of between three and eighteen years and many are too faint for amateur visual observation, even with a large telescope. Bright, or spectacular, comets have much longer orbital periods and, apart from a few notable exceptions like 1P/Halley, 109P/Swift-Tuttle and 153P/Ikeya-Zhang, the best performers usually have orbital periods of many thousands of years and are often discovered less than a year before they come within amateur range. For this reason it is important to regularly check the best comet websites for news of bright comets that may be discovered well after this yearbook is published. Some recommended sites are:

British Astronomical Association Comet Section: www.ast.cam.ac.uk/~jds/

Seiichi Yoshida's bright comet page: www.aerith.net/comet/weekly/current.html/

CBAT/MPC comets site: www.minorplanetcenter.net/iau/Ephemerides/Comets/

Yahoo Comet Images group: http://tech.groups.yahoo.com/group/Comet-Images/

Yahoo Comet Mailing list: http://tech.groups.yahoo.com/group/comets-ml/

The CBAT/MPC web page above also gives accurate ephemerides of the comets' positions in right ascension and declination.

Only four or five periodic comets are expected to reach a magnitude of thirteen or brighter during 2013 but they should all be observable visually with amateur telescopes, or with amateur CCD imaging systems, in a reasonably dark sky. In addition, the long-period comet 2011 L4 (PANSTARRS) reaches perihelion at around 0.3 AU (Astronomical Units) in March 2013 and looks like the most promising comet at the time of writing. One comet which reached perihelion in late 2012, P/1994 X1 (McNaught-Russell), should still be observable in 2013, and 2011 F1 (LINEAR) should be in visual reach of Southern Hemisphere amateur telescopes.

The infamous comet 29P/Schwassmann-Wachmann 1, now near aphelion, will usually be too faint for visual observation, even in large amateur telescopes, but is renowned for going into outburst several times per year when it can reach magnitude 11. It should perhaps be explained that the distances of comets from the Sun and the Earth are often quoted in Astronomical Units (AU) where 1 AU is the average Earth-Sun distance of 149.6 million km or 93 million miles.

The best cometary prospects for 2013 are listed below in the order they reach perihelion.

Comet	Period (years)	Perihelion	Peak Magnitude
29P/Schwassmann-Wachmann 1	14.7	2004 July 3	11 when in outburst
P/1994 X1 (McNaught-Russell)	18.3	2012 Dec 4	12 in Dec 2012
2011 F1 (LINEAR)	Long	2013 Jan 8	10 in Jan 2013
2011 L4 (PANSTARRS)	Long	2013 Mar 11	Naked eye (see text)
102P/Shoemaker	7.2	2013 Aug 13	12 in Sept 2013
2P/Encke	3.3	2013 Nov 21	5 in Nov 2013
154P/Brewington	10.8	2013 Dec 12	10 in Oct 2013

WHAT TO EXPECT

As the year 2013 starts, amateur astronomers equipped with large telescopes and CCDs may still be able to image comet P/1994 X1 (McNaught-Russell). This comet was predicted to arrive at perihelion on 4 December 2012. P/1994 X1 was discovered jointly by the world's greatest ever comet hunter, Rob McNaught, and by Ken Russell, both of the Anglo-Australian Observatory at that time. The comet has an orbital period of eighteen years and this is P/1994 X1's first observed return to the inner solar system; at perihelion it should have passed within 1.28 AU of the Sun and 0.86 AU from the Earth, while in northern Aquarius. A peak magnitude of 12 is possible and advanced amateurs may already have been able to image it in the autumn months of 2012. There were no visual observations of it when it was discovered in

1994 so it could have a few surprises in store. In January 2013, the comet crosses Cetus in an easterly trajectory and so should be at an altitude of thirty degrees, due south, as darkness falls from the UK. However, precisely how bright the comet will be is impossible to predict at the time of writing. Comet 2011 F1 (LINEAR) is predicted to reach tenth magnitude at perihelion (8 January 2013). However, it will not fully emerge from the twilight glare until February/March when it will be a distinctly Southern Hemisphere object, but hopefully still around tenth or eleventh magnitude, so it should be visible in medium to large amateur telescopes.

2011 L4 (PANSTARRS)

The comet C/2011 L4 (PANSTARRS) is looking like it might be the cometary highlight of 2013, although as comets can disappoint, disintegrate and be outperformed by newer discoveries, it is just too early to be sure. Nevertheless, the prospects for C/2011 L4 do look very encouraging. This comet is predicted to reach perihelion on 11 March at a mere 0.30 AU from the Sun. In addition, it has a healthy absolute magnitude (the brightness when 1.0 AU from Sun and Earth) of 5.5, hinting that it should be a naked-eye object for a number of months close to perihelion. Its implied healthy inherent 'size' should also mean it is less likely to disintegrate near to perihelion, as occurred with C/2010 X1 (Elenin) in 2011.

As with all comets that have a small perihelion distance, observing C/2011 L4 when it is close to perihelion will be extremely difficult as the small angular elongation from the Sun means that when the Sun is sufficiently below the horizon for the sky to be dark the comet will only be at an altitude of a few degrees. For the first few months of 2013, comet C/2011 L4 will not be elongated from the Sun by more than thirty-five degrees, as it brightens from ninth magnitude to, potentially, zero magnitude at perihelion. It will also be in the far south of the celestial sphere moving through Scorpius, Crater, Telescopium, Microscopium, Grus and Pisces Australis in January and February. It then travels into the northern celestial sphere, moving from Cetus to Pisces in early March. Precisely how early in March the comet will be available for observation in the evening sky after perihelion depends on just how bright it will be after that time. By 20 March, the comet will be in

Pisces, heading towards Andromeda just east of the Square of Pegasus. Shortly after 19h UT the Sun will be ten degrees below the western UK horizon when the comet, at a predicted magnitude of 2.0, will be ten degrees above that same horizon.

From that point on the comet soars north into a darker evening sky, although fading as it does so. By 5 April the comet will be passing a degree or so west of the Andromeda galaxy, roughly twelve degrees above the horizon by 20h UT, in a dark sky. By this time it should still be a fifth-magnitude object with a decent length of tail. The view is slightly better in the morning sky around this time, at 03h 30m UT, for observers who do not mind a chilly pre-dawn observing session, because the comet is several degrees higher in the sky for the same degree of twilight. At these northerly declinations comet 2011 L4 will be circumpolar from UK latitudes (in other words, the comet is permanently above the horizon) and its northerly altitude increases over the coming weeks as it tracks through Cassiopeia and Cepheus and just clips Draco. On 27 May, the comet reaches its northerly maximum when it passes just four degrees from the pole star Polaris, by which

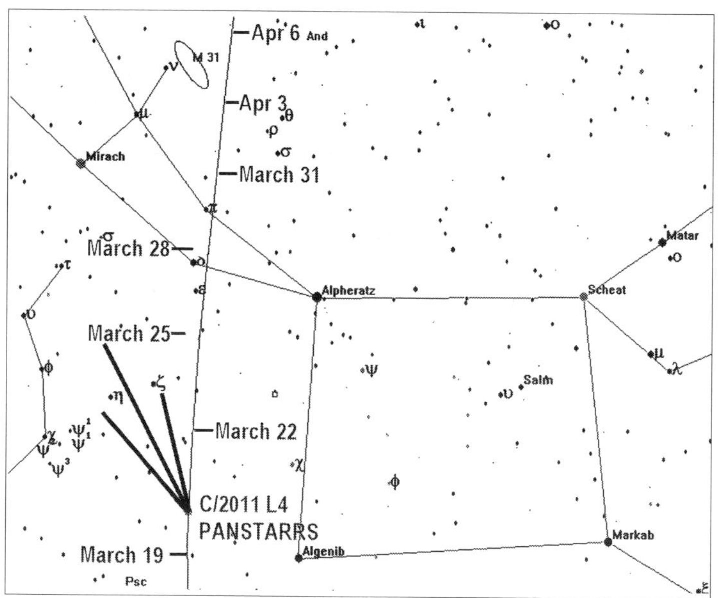

Figure 1. The track of comet C/2011 L4 (PANSTARRS) in March 2011

time it will have faded to magnitude 9. During the UK summer it will fade even further as it moves south through Ursa Minor and Draco. By the time it reaches Bootes in late July it will have faded to twelfth magnitude.

29P SCHWASSMANN-WACHMANN

Comet 29P/Schwassmann-Wachmann 1 is the most regularly outbursting comet in the sky. While it is not in the same explosive league as 17P/Holmes (famous for its half-million-fold outburst in October 2007), it has been observed flaring very regularly in recent years, possibly because it has been monitored far more closely than in the photographic era. When in its dormant state, 29P sits at around sixteenth to seventeenth magnitude: a challenging object even for CCD imagers. However, in recent years it has reached eleventh magnitude in outburst, putting it within easy visual range of large amateur telescopes or more modest telescopes equipped with CCD cameras. This year the comet reaches opposition on 25 April when it will be 5.2 AU from Earth and 6.2 AU from the Sun. Unfortunately, for UK observers, the comet will be low in the southern sky at -23 degrees, close to the border that divides southern Virgo from northern Hydra and some twelve degrees south of the bright star Spica. However, those observers in the Southern Hemisphere will enjoy the comet at a high altitude.

102P/SHOEMAKER

While not having prospects as exciting as C/2011 L4, the periodic comet 102P/Shoemaker should reach twelfth magnitude in September of this year and it will be reasonably placed in the night sky, travelling north-west as it clips the Alpheratz corner of the Great Square of Pegasus from mid-September to mid-October. 102P is scheduled to reach perihelion on 31 August at 1.97 AU from the Sun, seven years since its last and fourth observed return to the inner solar system, and it reaches opposition in mid-September. While a twelfth-magnitude comet is hardly exciting it will be well within visual range of the larger amateur telescopes and a routine target for medium aperture systems equipped with CCD imaging equipment.

2P/ENCKE

Comet 2P/Encke is one of the most famous comets in the night sky, mainly because of its remarkably small orbital period of just 3.3 years, its perihelion distance of 0.34 AU and the 61 previous returns to perihelion that have been observed. However, as Encke is not a large comet it reaches binocular brightness levels only by virtue of that small perihelion distance and so, unless the orbital circumstances are very favourable, it becomes a bright object only when it is immersed in twilight. Fortunately, its return in 2013 with a perihelion date of 21 November is a very good one from the perspective of Northern Hemisphere observers and, in this respect, the apparition is similar to that of 2003 when perihelion occurred some five weeks later, on 29 December. The astute reader will already have spotted that this ten-year gap between similar returns is not a coincidence, but the result of three orbits of 3.3 years equalling roughly ten years. However, unlike in 2003, Encke peaks in the morning sky this time around.

Encke should achieve twelfth magnitude by late September when it

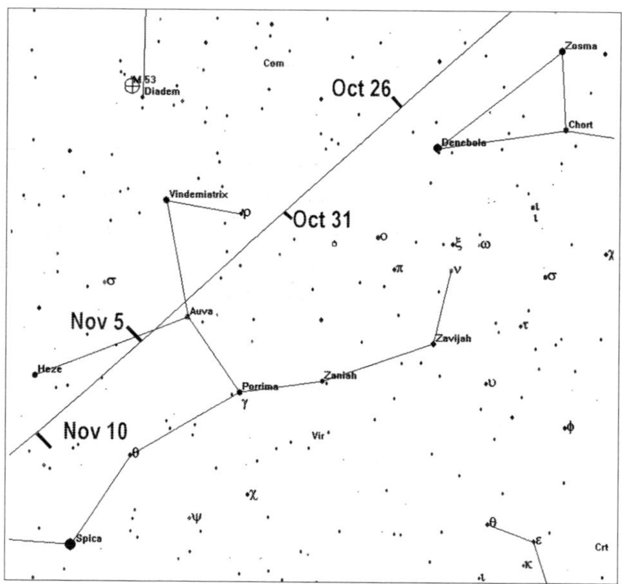

Figure 2. The track of 2P/Encke through south-eastern Leo and Virgo in October/November.

will be high in the northern celestial sphere in Auriga. During October it heads due east, passing through Lynx and Leo Minor and entering north-eastern Leo itself by the last week in October, by which time it should be an eighth-magnitude object in the pre-dawn sky. During November, closing in to perihelion, it crosses Virgo from north-west to south-east and, hopefully, should achieve sixth magnitude by the end of the first week of that month. However, by then it will only be fifteen degrees above the eastern UK pre-dawn horizon ninety minutes before sunrise and with twilight starting to interfere. By 15 November, the fifth-magnitude comet will be just eight degrees above the UK horizon with the Sun just twelve degrees down, around 06h UT. After perihelion the comet is a distinctly Southern Hemisphere object travelling through Libra, the head of Scorpius, southern Ophiuchus and Sagittarius.

Figure 3. An image of 2P/Encke captured at a similar return to this year's apparition in 2003, on 26 November. This is a composite of 14 x 20-second exposures with a Celestron 14 and SBIG ST9XE CCD. Note the comet's fan-shaped coma. (Image courtesy of Martin Mobberley.)

154P/BREWINGTON

As well as the return of comet 2P/Encke the autumn of 2013 should see a reasonable tenth-magnitude return for comet 154P/Brewington, on its third return to the inner solar system since Howard Brewington discovered it visually in August 1992, using a 0.40-m Newtonian. At the time of its discovery, Howard had already discovered three other comets including one other short-period visitor. From August to October, comet 154P will crawl slowly north-west through Aquarius brightening from thirteenth to (hopefully) tenth, or maybe even ninth, magnitude as it becomes better placed in the evening sky. On 7 November, the comet finally makes it out of Aquarius as it crosses into southern Pegasus, a few degrees north of alpha Aquarii (Sadalmelik). The comet reaches perihelion at 1.6 AU from the Sun on 12 December, but its closest approach to Earth will already have occurred on 24 September at 0.886 AU. The net effect of these two factors is that comet Brewington will probably peak in brightness in late November when it will be a healthy forty degrees above the UK south horizon as darkness falls.

The tables below list the right ascension (RA) and declination (Dec.) of, arguably, the six most promising comets at their peak as well as the distances, in AU, from the Earth and the Sun. The elongation, in degrees from the Sun, is also tabulated along with the estimated visual magnitude, which can only ever be a rough guess as comets are, without doubt, a law unto themselves!

P/1994 X1 (McNaught-Russell)

Date	RA (2000)			Dec.			Distance from Earth	Distance from Sun	Elongation from Sun	Mag.
	h	m	s	°	′′	′′′	AU	AU	°	
Jan 1	00	31	24.1	-07	29	20	1.015	1.334	83.8	14.8
Jan 11	01	10	16.7	-08	00	22	1.102	1.378	82.5	15.1
Jan 21	01	46	36.0	-07	54	57	1.202	1.433	81.2	15.5
Jan 31	02	20	20.5	-07	23	09	1.316	1.497	79.8	15.8
Feb 10	02	51	42.9	-06	34	02	1.440	1.568	78.1	16.2
Feb 20	03	21	02.8	-05	35	23	1.573	1.645	76.1	16.6

C/2011 L4 (PANSTARRS)

Date	RA (2000)			Dec.			Distance from Earth	Distance from Sun	Elongation from Sun	Mag.
	h	m	s	°	′	″	AU	AU	°	
Jan 1	17	16	00.9	-39	12	18	2.441	1.604	25.0	9.5
Jan 6	17	29	28.5	-40	14	21	2.322	1.517	27.3	9.1
Jan 11	17	44	35.5	-41	18	27	2.200	1.428	29.5	8.8
Jan 16	18	01	46.4	-42	23	37	2.074	1.336	31.4	8.3
Jan 21	18	21	33.4	-43	27	52	1.946	1.242	33.0	7.9
Jan 26	18	44	38.0	-44	27	34	1.817	1.146	34.3	7.4
Jan 31	19	11	51.6	-45	16	14	1.688	1.047	35.1	6.8
Feb 5	19	44	10.4	-45	42	40	1.561	0.946	35.2	6.2
Feb 10	20	22	19.7	-45	28	27	1.440	0.841	34.6	5.5
Feb 15	21	06	21.3	-44	05	31	1.327	0.733	33.0	4.8
Feb 20	21	54	46.8	-40	56	09	1.230	0.623	30.2	3.9
Feb 25	22	44	12.0	-35	19	42	1.154	0.513	26.3	2.9
Mar 2	23	29	48.2	-26	47	00	1.108	0.409	21.5	1.8
Mar 7	00	06	14.5	-15	21	22	1.098	0.327	17.1	0.9
Mar 12	00	28	31.0	-02	16	27	1.119	0.303	15.1	0.6
Mar 17	00	36	55.6	+09	58	56	1.154	0.353	16.9	1.3
Mar 22	00	38	02.0	+20	06	55	1.193	0.446	21.2	2.4
Mar 27	00	36	37.7	+28	26	23	1.233	0.554	26.1	3.4
Apr 1	00	34	32.1	+35	30	45	1.274	0.665	31.1	4.3
Apr 6	00	32	18.8	+41	44	01	1.315	0.774	35.9	5.0
Apr 11	00	30	02.9	+47	21	47	1.356	0.880	40.4	5.6
Apr 16	00	27	37.0	+52	34	06	1.397	0.984	44.8	6.2
Apr 21	00	24	45.3	+57	27	30	1.439	1.085	48.8	6.6
Apr 26	00	21	03.3	+62	06	12	1.481	1.182	52.7	7.1
May 1	00	15	51.3	+66	32	48	1.524	1.278	56.3	7.5
May 6	00	07	59.2	+70	48	37	1.569	1.371	59.6	7.8
May 11	23	55	09.3	+74	53	29	1.615	1.461	62.7	8.2
May 16	23	32	16.6	+78	44	41	1.664	1.550	65.6	8.5
May 21	22	46	19.8	+82	13	15	1.716	1.637	68.2	8.8
May 26	21	03	22.8	+84	47	55	1.770	1.722	70.5	9.1
May 31	18	10	51.4	+85	08	14	1.828	1.805	72.5	9.4
Jun 5	16	12	36.0	+83	05	40	1.889	1.887	74.3	9.6
Jun 10	15	20	14.3	+80	10	21	1.953	1.967	75.7	9.9

C/2011 L4 (PANSTARRS) *continued*

Date	RA (2000)			Dec.			Distance from Earth	Distance from Sun	Elong-ation from Sun	Mag.
	h	m	s	°	′	″	AU	AU	°	
Jun 15	14	55	31.9	+77	03	02	2.021	2.046	76.9	10.1
Jun 20	14	42	48.1	+73	55	37	2.093	2.124	77.8	10.4
Jun 25	14	36	05.1	+70	52	17	2.168	2.201	78.4	10.6
Jun 30	14	32	46.7	+67	54	43	2.246	2.277	78.7	10.8
July 5	14	31	36.0	+65	03	44	2.327	2.351	78.8	11.0
July 10	14	31	51.0	+62	19	43	2.411	2.425	78.6	11.3
July 15	14	33	06.4	+59	42	54	2.498	2.497	78.2	11.5
July 20	14	35	05.9	+57	13	21	2.587	2.569	77.6	11.7
July 25	14	37	38.7	+54	51	01	2.679	2.640	76.9	11.9
July 30	14	40	37.2	+52	35	43	2.772	2.710	75.9	12.0
Aug 4	14	43	56.5	+50	27	17	2.867	2.779	74.8	12.2
Aug 9	14	47	33.0	+48	25	34	2.963	2.848	73.6	12.4
Aug 14	14	51	23.5	+46	30	22	3.060	2.915	72.2	12.6
Aug 19	14	55	25.7	+44	41	32	3.158	2.983	70.8	12.7
Aug 24	14	59	37.4	+42	58	48	3.256	3.049	69.3	12.9
Aug 29	15	03	57.3	+41	21	58	3.354	3.115	67.7	13.1
Sep 3	15	08	24.3	+39	50	47	3.452	3.180	66.1	13.2

29P/Schwassmann-Wachmann

Date	RA (2000)			Dec.			Distance from Earth	Distance from Sun	Elong-ation from Sun	Mag.
	h	m	s	°	′	″	AU	AU	°	
Apr 1	14	02	40.2	-24	00	10	5.322	6.224	152.3	15.6
Apr 11	13	08	21.2	-23	46	55	5.264	6.223	161.7	15.5
Apr 21	13	53	44.3	-23	27	55	5.234	6.221	168.3	15.5
May 1	13	49	08.0	-23	04	27	5.234	6.220	166.9	15.5
May 11	13	44	49.6	-22	38	06	5.264	6.218	159.2	15.5
May 21	13	41	05.5	-22	10	44	5.322	6.217	149.7	15.6
May 31	13	38	08.4	-21	44	11	5.405	6.215	139.9	15.6
Jun 10	13	36	06.8	-21	20	06	5.510	6.213	130.1	15.6
Jun 20	13	35	06.6	-20	59	52	5.634	6.212	120.5	15.7
Jun 30	13	35	09.5	-20	44	28	5.771	6.210	111.1	15.7

102P/Shoemaker

Date	RA (2000)			Dec.			Distance from Earth	Distance from Sun	Elongation from Sun	Mag.
	h	m	s	°	′	″	AU	AU	°	
Aug 1	00	37	56.4	+08	01	54	1.309	1.986	116.8	13.0
Aug 11	00	38	54.7	+11	44	22	1.217	1.976	124.5	12.8
Aug 21	00	36	17.2	+15	30	36	1.140	1.971	132.5	12.7
Aug 31	00	29	47.7	+19	12	46	1.081	1.968	140.6	12.6
Sep 10	00	19	34.2	+22	38	36	1.043	1.970	147.9	12.5
Sep 20	00	06	32.4	+25	34	07	1.027	1.975	152.9	12.5
Sep 30	23	52	21.9	+27	48	22	1.035	1.983	153.7	12.5
Oct 10	23	39	07.6	+29	17	47	1.067	1.995	150.1	12.6
Oct 20	23	28	46.2	+30	08	33	1.119	2.011	143.8	12.8
Oct 30	23	22	26.8	+30	32	17	1.190	2.030	136.6	13.0

2P/Encke

Date	RA (2000)			Dec.			Distance from Earth	Distance from Sun	Elongation from Sun	Mag.
	h	m	s	°	′	″	AU	AU	°	
Sep 1	04	02	43.1	+33	26	20	1.156	1.577	93.2	14.8
Sep 6	04	18	54.5	+35	04	21	1.052	1.511	94.3	14.3
Sep 11	04	37	54.6	+36	49	29	0.951	1.443	94.9	13.8
Sep 16	05	00	49.2	+38	40	32	0.854	1.374	94.9	13.2
Sep 21	05	29	12.8	+40	33	08	0.763	1.302	93.9	12.6
Sep 26	06	05	12.2	+42	16	15	0.678	1.228	91.8	12.0
Oct 1	06	51	10.1	+43	25	37	0.604	1.151	88.0	11.3
Oct 6	07	48	23.6	+43	17	02	0.542	1.072	82.4	10.6
Oct 11	08	54	18.2	+40	51	32	0.499	0.990	74.6	9.9
Oct 16	10	01	11.7	+35	32	15	0.479	0.906	65.1	9.3
Oct 21	11	00	43.6	+27	50	22	0.486	0.819	54.8	8.6
Oct 26	11	49	23.2	+19	11	47	0.522	0.729	45.3	8.0
Oct 31	12	28	14.0	+10	53	42	0.585	0.637	37.4	7.4
Nov 5	13	00	08.4	+03	30	02	0.671	0.544	31.3	6.7
Nov 10	13	28	17.3	-02	59	27	0.781	0.456	26.6	5.8

2P/Encke *continued*

Date	RA (2000)			Dec.			Distance from Earth	Distance from Sun	Elongation from Sun	Mag.
	h	m	s	°	′	″	AU	AU	°	
Nov 15	13	55	59.2	-08	49	01	0.911	0.382	22.7	5.0
Nov 20	14	26	28.9	-14	08	53	1.056	0.339	18.7	4.6
Nov 25	15	01	04.8	-18	48	58	1.199	0.348	14.6	5.0
Nov 30	15	37	30.8	-22	28	15	1.327	0.403	10.8	6.2
Dec 5	16	12	49.9	-25	02	49	1.440	0.484	8.0	7.6

154P/Brewington

Date	RA (2000)			Dec.			Distance from Earth	Distance from Sun	Elongation from Sun	Mag.
	h	m	s	°	′	″	AU	AU	°	
Sep 1	22	35	23.0	-13	30	50	0.944	1.951	174.7	11.1
Sep 11	22	23	21.6	-11	43	33	0.904	1.894	165.0	10.6
Sep 21	22	12	15.8	-09	36	11	0.887	1.841	153.5	10.2
Oct 1	22	03	50.2	-07	13	04	0.891	1.792	142.4	9.8
Oct 11	21	59	20.3	-04	39	22	0.912	1.747	132.2	9.6
Oct 21	21	59	21.1	-01	59	10	0.947	1.708	123.1	9.4
Oct 31	22	03	49.7	+00	45	06	0.991	1.674	115.2	9.2
Nov 10	22	12	28.5	+03	32	49	1.042	1.647	108.3	9.1
Nov 20	22	24	49.7	+06	23	59	1.098	1.627	102.3	9.0
Nov 30	22	40	23.5	+09	18	12	1.159	1.614	97.2	9.1
Dec 10	22	58	46.3	+12	14	50	1.224	1.608	92.8	9.1
Dec 20	23	19	35.4	+15	12	22	1.293	1.610	88.9	9.3
Dec 30	23	42	31.5	+18	08	20	1.367	1.620	85.5	9.5

Minor Planets in 2013

MARTIN MOBBERLEY

Some 600,000 minor planets (also known as asteroids) are known. They range in size from small planetoids hundreds of kilometres in diameter to boulders tens of metres across. More than 300,000 of these now have such good orbits that they possess a numbered designation and 17,000 have been named after mythological gods, famous people, scientists, astronomers and institutions. Most of these objects live between Mars and Jupiter but some 8,000 have been discovered between the Sun and Mars and more than 1,300 of these are classed as potentially hazardous asteroids (PHAs) due to their ability to pass within eight million kilometres of the Earth while also having a diameter greater than 200 metres. The first four asteroids to be discovered were Ceres (1), now regarded as a dwarf planet, Pallas (2), Juno (3) and Vesta (4), which are all easy binocular objects when at their peak, due to them having diameters of hundreds of kilometres.

In 2013, most of the first ten numbered minor planets reach opposition at some point during the year. The dwarf planet (1) Ceres reached opposition late in 2012, on 16 December, when it was in north-eastern Taurus, just a few degrees north of the Crab Nebula, at magnitude 6.7. However, it will still be very well placed in that constellation as 2013 begins even if it will have peaked in brightness. The same can be said for (4) Vesta which was the brighter target in mid-December 2012 when it, too, was in Taurus. Again, it is still well placed as 2013 starts, in that same constellation and not far from the eye of the bull, the bright star Aldebaran.

Although slightly fainter, the minor planet (9) Metis is also very well placed in January 2013, near the Auriga/Gemini border. In May and June (6) Hebe will be a ninth-magnitude target close to the Ophiuchus/Serpens/Hercules border regions and from June to August (5) Astraea will be travelling through northern Sagittarius at around magnitude 11. The supply of binocular-magnitude summer asteroids doesn't end there either, because (3) Juno, (8) Flora and (7) Iris are all

Figure 1. Vesta imaged by the Dawn spacecraft in 2011. (Image courtesy of NASA/JPL-Caltech/UCLA/MPS/DLR/IDA/UMD.)

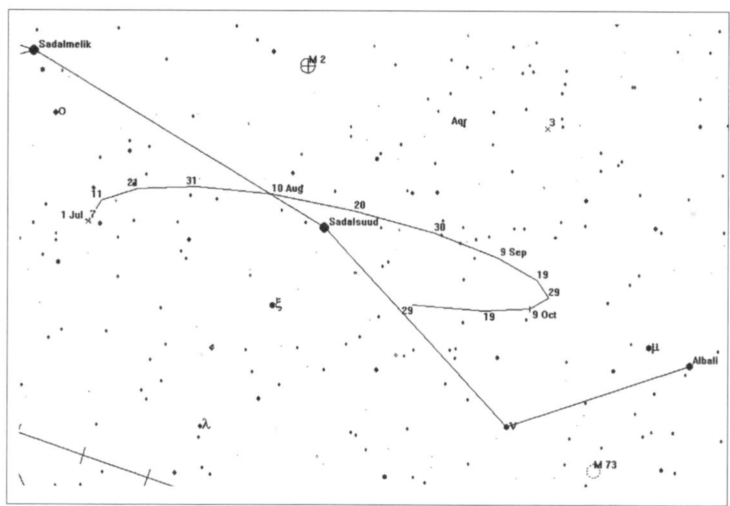

Figure 2. The track of eighth-magnitude Iris passing through Aquarius from July to October. Chart is 18 degrees wide.

well placed from June to August in eastern Aquila, north-eastern Sagittarius and Aquarius respectively. Finally, from October to December (10) Hygeia and (2) Pallas can easily be tracked down in Aries and Hydra.

Ephemerides for the best-placed bright minor planets at (or close to) opposition in 2013, listed in calendar order.

Dwarf planet (1) Ceres

Date	RA (2000)			Dec.			Distance from Earth	Distance from Sun	Elongation from Sun	Mag.
	h	m	s	°	′	″	AU	AU	°	
Jan 1	05	31	11.6	+26	00	49	1.704	2.659	162.8	7.1
Jan 11	05	22	35.2	+26	27	35	1.750	2.652	150.7	7.4
Jan 21	05	16	28.6	+26	50	12	1.821	2.645	139.2	7.6
Jan 31	05	13	23.8	+27	10	20	1.911	2.638	128.4	7.8
Feb 10	05	13	28.7	+27	29	16	2.016	2.632	118.3	8.0

(4) Vesta

Date	RA (2000)			Dec.			Distance from Earth	Distance from Sun	Elongation from Sun	Mag.
	h	m	s	°	′	″	AU	AU	°	
Jan 1	04	44	40.1	+18	15	49	1.663	2.569	151.3	6.9
Jan 11	04	37	59.1	+18	37	26	1.737	2.567	139.7	7.1
Jan 21	04	34	18.3	+19	03	52	1.831	2.565	128.8	7.3
Jan 31	04	33	45.4	+19	34	44	1.941	2.563	118.6	7.5
Feb 10	04	36	11.6	+20	09	11	2.061	2.560	109.1	7.6
Feb 20	04	41	22.1	+20	45	55	2.188	2.557	100.3	7.8
Mar 2	04	48	57.1	+21	23	27	2.317	2.554	92.1	7.9

(9) Metis

Date	RA (2000)			Dec.			Distance from Earth	Distance from Sun	Elongation from Sun	Mag.
	h	m	s	°	′′	′′′	AU	AU	°	
Jan 1	06	52	16.7	+28	27	39	1.137	2.117	174.4	8.5
Jan 11	06	41	05.4	+29	05	38	1.154	2.124	166.9	8.7
Jan 21	06	31	32.7	+29	29	32	1.197	2.131	155.2	9.0
Jan 31	06	25	05.1	+29	40	33	1.263	2.139	143.9	9.3
Feb 10	06	22	24.9	+29	41	41	1.347	2.147	133.3	9.5
Feb 20	06	23	41.1	+29	35	42	1.446	2.156	123.6	9.8

(6) Hebe

Date	RA (2000)			Dec.			Distance from Earth	Distance from Sun	Elongation from Sun	Mag.
	h	m	s	°	′′	′′′	AU	AU	°	
May 1	16	38	08.5	-00	04	07	1.865	2.756	145.6	9.9
May 11	16	30	53.4	+00	45	10	1.803	2.741	153.0	9.7
May 21	16	22	03.8	+01	20	09	1.764	2.726	157.2	9.6
May 31	16	12	32.0	+01	36	24	1.752	2.710	156.1	9.6
Jun 10	16	03	15.6	+01	31	24	1.764	2.693	150.2	9.7
Jun 20	15	55	11.7	+01	04	55	1.801	2.676	142.1	9.8

(5) Astrae

Date	RA (2000)			Dec.			Distance from Earth	Distance from Sun	Elongation from Sun	Mag.
	h	m	s	°	′′	′′′	AU	AU	°	
Jun 10	19	52	42.4	-16	15	24	2.069	2.935	141.8	11.5
Jun 20	19	46	28.5	-16	34	27	2.006	2.947	152.8	11.3
Jun 30	19	38	27.8	-17	00	41	1.968	2.958	164.0	11.1
July 10	19	29	21.0	-17	31	57	1.956	2.969	174.4	10.9
July 20	19	20	02.3	-18	05	30	1.972	2.979	170.8	11.0
July 30	19	11	27.1	-18	38	49	2.016	2.989	159.8	11.3
Aug 9	19	04	22.7	-19	10	02	2.086	2.998	148.6	11.5

(3) Juno

Date	RA (2000)			Dec.			Distance from Earth	Distance from Sun	Elong-ation from Sun	Mag.
	h	m	s	°	′	″	AU	AU	°	
Jun 30	21	05	54.5	-02	36	05	1.951	2.792	138.2	9.6
July 10	21	01	09.0	-02	50	49	1.851	2.767	148.1	9.4
July 20	20	54	28.7	-03	24	41	1.773	2.741	157.8	9.2
July 30	20	46	27.7	-04	17	04	1.720	2.715	165.5	9.0
Aug 9	20	37	52.9	-05	25	15	1.693	2.688	166.0	9.0
Aug 19	20	29	43.2	-06	44	10	1.693	2.661	158.6	9.0
Aug 29	20	22	54.6	-08	07	39	1.719	2.634	148.6	9.2
Sep 8	20	18	11.6	-09	29	43	1.767	2.607	138.3	9.3
Sep 18	20	16	04.1	-10	45	16	1.834	2.580	128.2	9.4
Sep 28	20	16	42.9	-11	50	52	1.916	2.552	118.6	9.6

(8) Flora

Date	RA (2000)			Dec.			Distance from Earth	Distance from Sun	Elong-ation from Sun	Mag.
	h	m	s	°	′	″	AU	AU	°	
Jun 30	20	17	14.2	-19	48	10	1.269	2.237	156.2	9.3
July 10	20	08	45.4	-20	48	32	1.215	2.219	167.8	9.0
July 20	19	58	25.9	-21	53	41	1.185	2.201	178.8	8.7
July 30	19	47	38.0	-22	56	36	1.181	2.183	167.7	8.9
Aug 9	19	37	52.6	-23	51	19	1.200	2.165	155.8	9.1
Aug 19	19	30	34.4	-24	34	11	1.241	2.147	144.4	9.3

(7) Iris

Date	RA (2000)			Dec.			Distance from Earth	Distance from Sun	Elong-ation from Sun	Mag.
	h	m	s	°	′	″	AU	AU	°	
July 20	21	54	47.8	-04	26	56	1.333	2.255	147.2	8.7
July 30	21	48	18.7	-04	20	29	1.256	2.228	157.5	8.4
Aug 9	21	39	39.7	-04	31	36	1.202	2.202	167.1	8.1
Aug 19	21	29	52.0	-04	58	07	1.172	2.176	170.3	7.9
Aug 29	21	20	15.8	-05	35	21	1.166	2.150	162.4	8.1
Sep 8	21	12	12.0	-06	17	14	1.183	2.125	151.7	8.3
Sep 18	21	06	49.6	-06	57	10	1.221	2.099	141.1	8.5
Sep 28	21	04	46.5	-07	29	51	1.275	2.075	131.0	8.7

(10) Hygeia

Date	RA (2000)			Dec.			Distance from Earth	Distance from Sun	Elong-ation from Sun	Mag.
	h	m	s	°	′	″	AU	AU	°	
Oct 8	02	31	12.8	+20	01	41	2.551	3.468	152.4	10.7
Oct 18	02	24	17.2	+19	31	48	2.505	3.472	163.7	10.5
Oct 28	02	16	35.7	+18	53	27	2.487	3.476	173.9	10.3
Nov 7	02	08	50.8	+18	09	40	2.499	3.480	170.4	10.4
Nov 17	02	01	47.6	+17	24	37	2.541	3.483	159.2	10.6
Nov 27	01	56	02.8	+16	42	34	2.611	3.486	147.8	10.8

(2) Pallas

Date	RA (2000)			Dec.			Distance from Earth	Distance from Sun	Elong-ation from Sun	Mag.
	h	m	s	°	′	″	AU	AU	°	
Oct 28	09	06	01.9	-14	51	30	2.208	2.143	73.2	8.8
Nov 7	09	20	38.0	-16	30	50	2.112	2.138	78.0	8.7
Nov 17	09	33	49.6	-18	05	27	2.013	2.134	83.1	8.6
Nov 27	09	45	23.8	-19	32	07	1.911	2.131	88.7	8.5
Dec 7	09	55	03.2	-20	46	46	1.808	2.130	94.8	8.3
Dec 17	10	02	29.6	-21	44	10	1.704	2.131	101.5	8.2
Dec 27	10	07	25.5	-22	18	12	1.602	2.133	108.8	8.0

NEAR-EARTH ASTEROID APPROACHES

A list of some of the most interesting numbered and named close-asteroid approaches in 2013 is presented in the table, along with a couple of very close encounters with the un-numbered objects 2005 WK4 and 2000 KA. It should be borne in mind that the visibility of close-approach asteroids is highly dependent on whether they are close to the solar glare and in which hemisphere the observer is based, but by the very nature of their proximity they move rapidly across the sky. The table gives the minimum separation between the Earth and the asteroid in AU (1 AU = 149.6 million km) and the date of that closest approach and also the constellation in which the object can be found when closest. The brightest magnitude achieved and the corresponding date and constellation for that condition are also given. Some of the objects listed are as faint as magnitude 17 or 18 at their best and so present a real challenge, even to advanced amateur astronomers using CCDs. Scores of other tiny asteroids with provisional designations will come within advanced amateur CCD range during 2013, many of them undiscovered as this *Yearbook* goes to press.

Some interesting numbered and named close-asteroid approaches in 2013 along with a couple of very close encounters with provisional designation objects (2005 WK4 and 2000 KA).

Asteroid desig./ name	**Min. sep. (AU)**	**Date closest 2013**	**Const.**	**Peak mag.**	**Peak date**	**Const.**
(99942) Apophis	0.09666	Jan 9.49	Pyx	15.8	Jan 21	Pup
(3752) Camillo	0.1478	Feb 12.25	Vir	13.3	Feb 13	Com
(7888) 1993 UC	0.1260	Mar 20.07	And	14.3	Mar 28	Cep
(96315) 1997 AP10	0.1178	Mar 28.20	Pup	14.7	Mar 21	Vel
(7753) 1988 XB	0.1183	July 10.55	Crt	16.6	Jun 20	Lib
2005 WK4	0.025	Aug 9.21	Ari	13.9	Aug 12	Cet
(6037) 1988 EG	0.1847	Aug 12.17	Tau	17.9	Aug 4	Psc
(52760) 1998 ML14	0.05624	Aug 24.44	Vir	15.0	Aug 13	Lup
(4581) Asclepius	0.1188	Aug 25.63	Ser	18.0	Aug 6	Oph
(39565) 1992 SL	0.1798	Sept 23.44	Sgr	17.2	Sep 27	Cap
(152664) 1998 FW4	0.04500	Sept 24.76	Aur	15.3	Sep 22	Ari
(6063) Jason	0.07903	Nov 11.55	Car	12.7	Nov 8	Lep
(85774) 1998 UT18	0.1724	Nov 21.67	Leo	17.9	Nov 15	Cnc
2000 KA	0.022	Nov 24.41	Leo	15.4	Nov 23	Cnc

Accurate minor-planet ephemerides can be computed for your location on Earth using the MPC ephemeris service at: www.minorplanetcenter.org/iau/MPEph/MPEph.html.

Meteors in 2013

JOHN MASON

Meteors (popularly known as 'shooting stars') may be seen on any clear moonless night, but on certain nights of the year their number increases noticeably. This occurs when the Earth chances to intersect a concentration of meteoric dust moving in an orbit around the Sun. If the dust is well spread out in space, the resulting shower of meteors may last for several days. The word 'shower' must not be misinterpreted – only on very rare occasions have the meteors been so numerous as to resemble snowflakes falling.

If the meteor tracks are marked on a star map and traced backwards, a number of them will be found to intersect in a point (or within a small area of the sky) which marks the radiant of the shower. This gives the direction from which the meteors have come.

Bright moonlight has an adverse effect on visual meteor observing, and within about five days to either side of Full Moon, lunar glare swamps all but the brighter meteors, reducing, quite considerably, the total number of meteors seen. In 2013, the April Lyrids, Orionids, Leonids and Geminids will all suffer interference by moonlight. Observations of the April Lyrids will be hampered by a waxing gibbous Moon, only three days from full, which sets in the early morning hours. For the Orionids, the Moon is a waning gibbous in Taurus/Gemini very close to the radiant, and the peak of the Leonids coincides with a full Moon in Taurus. Remember, though, that it is often possible for visual observers to minimize the effects of moonlight by positioning themselves so that the Moon is behind them and hidden behind a wall or other suitable obstruction.

The Geminids are now the richest of the annual meteor showers, but this year's peak coincides with a waxing gibbous Moon, only three days from full, in Aries/Taurus, so high observed rates are most likely in the pre-dawn hours of 14 December. Geminid maximum is expected at around 01h, but past observations show that bright Geminids become more numerous some hours after the rates have peaked, a consequence of particle-sorting in the meteoroid stream.

There are many excellent observational opportunities in 2013, with the annual showers that peak in late July or early August being particularly well placed with regard to the Moon this year. The Delta Aquarids, Piscis Australids, Alpha Capricornids, Iota Aquarids and the rise to maximum and peak of the Perseids may all be observed under dark skies. Indeed the 'high summer' period of 2013 looks like being a bumper time for meteor observers! Watches carried out any time at the end of July and in early August are likely to yield good observed rates. The peak of the Perseids occurs at about 16h on 12 August, so the best rates are likely in the pre-dawn hours of 12 August and during the night of 12/13 August. The Taurids (in early November), will also be observable with little or no interference from moonlight. The shower has shown unexpected activity in the past – most recently in 2005 – so it is hoped that there will be good coverage of the shower this year.

The following table gives some of the more easily observed showers with their radiants; interference by moonlight is shown by the letter M:

Limiting Dates	Shower	Maximum	Radiant RA		Dec.	
			h	m	°	
1–6 Jan	Quadrantids	3 Jan, 12h	15	28	+50	
19–25 April	Lyrids	22 April, 11h	18	08	+32	M
24 Apr–20 May	Eta Aquarids	5–6 May	22	20	-01	
17–26 June	Ophiuchids	20 June	17	20	-20	M
July–August	Capricornids	8, 15, 26 July	20	44	-15	
			21	00	-15	
15 July–20 Aug	Delta Aquarids	29 July, 6 Aug	22	36	-17	
			23	04	+02	
15 July–20 Aug	Piscis Australids	31 July	22	40	-30	
15 July–20 Aug	Alpha Capricornids	2–3 Aug	20	36	-10	
July–August	Iota Aquarids	6–7 Aug	22	10	-15	
			22	04	-06	
23 July–20 Aug	Perseids	12 Aug, 16h	3	04	+58	
16-30 Oct	Orionids	21–24 Oct	6	24	+15	M

20 Oct–30 Nov	Taurids	5, 12 Nov	3	44	+22	
			3	44	+14	
15–20 Nov	Leonids	17 Nov, 19h	10	08	+22	M
Nov–Jan	Puppid-Velids	early Dec	9	00	-48	
7–16 Dec	Geminids	14 Dec, 01h	7	32	+33	M
17–25 Dec	Ursids	22–23 Dec	14	28	+78	

Some Events in 2014

ECLIPSES

There will be four eclipses, two of the Sun and two of the Moon.

15 April: Total eclipse of the Moon – Australia, New Zealand, Pacific Ocean, North and South America.

29 April: Annular eclipse of the Sun – Southern Indian Ocean, Antarctica and Australia.

8 October: Total eclipse of the Moon – Central and Eastern Asia, Australia, New Zealand, Pacific Ocean, North and South America.

23 October: Partial eclipse of the Sun – Northern Pacific Ocean and North America.

THE PLANETS

Mercury may be seen more easily from northern latitudes in the evenings about the time of greatest eastern elongation (31 January) and in the mornings about the time of greatest western elongation (1 November). In the Southern Hemisphere the corresponding most favourable dates are 21 September (evenings), 14 March and 12 July (mornings). The planet may also be spotted from both hemispheres in the evenings around the time of greatest eastern elongation on 25 May.

Venus passes through inferior conjunction on 11 January and is visible in the mornings from January until September. It attains its greatest brilliancy in February and reaches greatest western elongation (47°) on 22 March. *Venus* passes through superior conjunction on 25 October and will become visible in the evenings towards the end of the year.

Mars is at opposition on 8 April in Virgo.

Jupiter is at opposition on 5 January in Gemini.

Saturn is at opposition on 10 May in Libra.

Uranus is at opposition on 7 October in Pisces.

Neptune is at opposition on 29 August in Aquarius.

Pluto is at opposition on 4 July in Sagittarius.

Part Two

Article Section

ALMA: The World's Most Complex Telescope

STEPHEN WEBB

INTRODUCTION

It's the closest thing to setting foot on Mars without leaving Earth. On the coast of northern Chile lies the Atacama Desert and one of the most extreme environments on our planet. Atacama's barren and rugged landscape covers roughly the same area as England. It is the highest desert in the world, with a peak 6,893 m above sea level, but perhaps more intriguing is the fact that Atacama can lay claim to be the driest desert in the world. Its extreme aridity arises from the interplay of several geological, atmospheric and oceanographic features. First, clouds from the east are blocked by the mighty Andes. Second, very little moisture from the Pacific Ocean can reach the desert because of the blocking effect of the Chilean Coastal Range, a 3,100 km-long mountain range that runs parallel to the Andes. Third, a cold ocean flow called the Humboldt Current and a subtropical anticyclone called the South Pacific High combine to suck moisture from the air in that region. The result is that some places in the Atacama Desert have never seen rainfall for hundreds, perhaps thousands of years. Unsurprisingly, few people choose to live there. But for some it's a great place to work.

Astrobiologists are drawn to the Atacama Desert: it's ideal for testing Mars exploration strategies. Despite the extreme aridity, life is not impossible here. Some parts of the desert occasionally see the Camanchaca, the local name for a marine fog, which contains sufficient moisture for specific flora to scrape a living. Drier parts of Atacama may appear lifeless, but in late 2011 researchers from institutes in Spain and Chile used a 'life detector chip' to discover an oasis of bacteria and archaea two to three metres under the surface of one of these lifeless regions. Perhaps this tells us that the next generation of Mars explorers should search for signs of life under the surface of the Red Planet?

Film directors appreciate the location, too. Recent films shot partly on location here include *The Motorcycle Diaries*, the James Bond thriller *Quantum of Solace* and the visually stunning *Nostalgia for the Light*. It is astronomers, however, who most value this harsh place and it is easy to understand why: the high altitude, the paucity of cloud cover, the exceptional dryness, the lack of light pollution and radio interference from towns and cities . . . this constellation of properties makes the Atacama Desert one of the best places on Earth to conduct astronomical observations. From a location such as this, astronomers can get their best view of many important objects in the southern sky, including the centre of our Galaxy and the Magellanic Clouds.

OBSERVATORIES AT ATACAMA

The Atacama Desert is the astronomy capital of the world. It is home to several world-leading observatories, such as La Silla, Paranal, Cerro Tololo, Cerro Pachón and Las Campanas, and many of the telescopes here are outstanding. For example Paranal Observatory, which is at an altitude of 2,635 m above mean sea level, plays host to the Very Large Telescope (VLT). The VLT consists of four 8.2 m telescopes that can observe independently in the visible and infrared but can also, along with four smaller auxiliary telescopes, be combined to form an optical interferometer. When working in interferometric mode the VLT is currently the largest optical telescope in existence. Las Campanas is home to the twin 6.5 m Magellan telescopes while La Silla is currently home to nine telescopes, one of which (in conjunction with the HARPS spectrograph) is the world's foremost hunter of exoplanets. However, of all the observatories in Atacama, pride of place must belong to the Llano de Chajnantor Observatory.

It's quite a journey to reach Chajnantor. From Santiago it's a two-hour flight to Calama, followed by an eighty-minute bus drive to San Pedro de Atacama, the nearest town, followed by a forty-minute drive in a four-wheel-drive pick-up truck to the ALMA base camp. The 4WD pick-up is mandatory. To make further progress, and reach the heights of the Chajnantor plain (Figure 1), requires travel in a properly certified vehicle. The journey is worth it. Llano de Chajnantor Observatory is home to several different and independent facilities, including the University of Tokyo Atacama Observatory (which, at 5,640 m above sea level, is the highest permanent astronomical observatory in the world) and the Atacama Cosmology Telescope (which is the third-highest observatory). By far the largest observatory at Llano de Chajnantor, though, is a revolutionary new telescope, the most complex telescope in the world: the Atacama Large Millimeter/sub-millimeter Array – ALMA, for short.

Figure 1. The Martian-like landscape of the Chajnantor plain lies 5,000 m above sea level in the Atacama Desert. This is where ALMA is being constructed. The pyramid shape to the left of centre is the Licancabur volcano. (Image courtesy of ALMA (ESO/NAOJ/NRAO).)

BUILDING A TELESCOPE AT THE TOP OF THE WORLD

Constructing ALMA has been a costly affair. ALMA is nowhere near as expensive as the James Webb Space Telescope, the story of which is perhaps appropriate for a future volume in this series, but when ALMA becomes fully operational in 2013, it will be by quite some margin the most expensive ground-based telescope ever built. The projected costs of $1.3 billion are being met by an international partnership of Europe (in the shape of the European Southern Observatory (ESO), which is an organization for astronomical research supported by fourteen European countries plus Brazil); North America (through the US

National Science Foundation in cooperation with the National Research Council of Canada and the National Science Council of Taiwan); East Asia (the National Institutes of Natural Sciences of Japan, in cooperation with the Academia Sinica in Taiwan); and Chile. Science operations are led by the ESO (for Europe); the National Radio Astronomy Observatory (for North America); and the National Astronomical Observatory of Japan (for East Asia). ALMA is thus a truly global endeavour! But why should a ground-based telescope, even one of extreme complexity, be so expensive? And what is the expected scientific pay-off for all this expenditure?

To answer the first of those questions we need to know a little more about ALMA, and what astronomers want it to do.

When astronomers try to observe the universe using electromagnetic waves with a wavelength somewhere between a few hundred micrometres and a few millimetres – in other words, in that region between the far infrared and the microwave parts of the spectrum – they immediately hit a problem: such waves are heavily absorbed by water vapour in Earth's atmosphere. That's why a telescope for observing at such wavelengths must be constructed on a high, dry site; the 5 km-high plateau at Chajnantor is thus an ideal home for ALMA, which will observe all the wavelengths in the region of 0.3–9.6 mm that can reach Earth's surface. Chajnantor's harsh environment and relative inaccessibility, however, present obvious technical challenges when developing the world's most ambitious and complex ground-based astronomical project. Simply constructing the observatory in this location guarantees it will be expensive.

Choosing to observe at the relatively long millimetre and submillimetre wavelengths gives rise to another problem for astronomers, this time to do with resolving power. The resolution that a single telescope can achieve depends on the ratio of the wavelength of observed light to the diameter of the telescope's aperture. Take one of the VLT telescopes as an example. It has a diameter of 8.2 m and might observe in the infrared at a wavelength of 2 μm. In that case, and by using adaptive optics to correct for atmospheric disturbances, one of the VLT telescopes has a maximum resolution of about ten millionths of a degree. Now consider an antenna for observing in the millimetre range. A typical antenna, of the type in use at ALMA, might have a dish with a diameter of 12 m: that gives it an advantage of about 50 per cent compared to the VLT mirrors. However, the wavelength at which the antenna observes is hundreds of times longer than 2 μm: this more than outweighs the antenna's size advantage. A single 12 m antenna observing at a wavelength of 1 mm has a resolution of about five thousandths of a degree – much worse than the VLT mirror. Indeed, to have a comparable resolution at millimetre wavelengths, a single antenna dish would need to have a diameter of several kilometres; that is impossible with current technology. The solution to this difficulty, of course, is to do what the VLT does: observe with more than one instrument and link them together as an interferometer. The resolution of an interferometer depends on the maximum separation between its antennae, not the size of the individual antennae. ALMA is thus not a single antenna but an *array* of antennae (the highest such array in the world); signals from the antennae are brought together in such a way that the array acts as if it were a single telescope. The ALMA antennae can be arranged in patterns spread over distances ranging from 150 m (in which configuration ALMA will be able to get a broad overview of the objects it is observing) to 16 km (in which configuration ALMA will be able to 'zoom in' on particular details). When the configuration is at maximum separation, ALMA's resolution will far exceed that of the Hubble Space Telescope operating at visible wavelengths. Interferometers can thus give impressive results – but inevitably they cost more to build than single telescopes.

When complete, ALMA will consist of sixty-six antennae. The main array will have fifty antennae each with 12 m-diameter dishes; the compact array, which allows the imaging of extended sources, will consist of four antennae with 12 m-diameter dishes and twelve with 7 m-diameter dishes. These parabolic dishes must collect submillimetre and millimetre radiation and focus it into a detector, in much the same way as the mirror of an optical telescope brings visible wavelengths to a focus. In order to deliver on the promise of superacute vision the ALMA dishes need to be the most precise ever made. Furthermore, even though they are exposed to bright sun, freezing temperatures and wind gusts in excess of 150 km per hour, they must be designed to keep their parabolic shape to within a third of the thickness of a human hair. In addition, the antennae must be capable of being pointed with an angular accuracy of 0.6 seconds of arc – the apparent width of a one-pence coin at a distance of three miles (Figure 2). Meeting these unprecedented technical requirements, of course, adds further to the cost.

Figure 2. When in place and observing at the Array Operations Site, the ALMA antennae must all point to precisely the same point in the sky. This photo, taken in May 2011, shows fourteen of the 12 m antennae. (Image courtesy of ALMA (ESO/NAOJ/NRAO) and J. Guarda (ALMA).)

Perhaps even more challenging is the requirement to move these 100-ton antennae. The antennae are built at the Operations Support Facility (OSF), which is 28 km away and more than 2 km below the Chajnantor plateau (Figure 3). To take their place in the array they must be put on transporter vehicles capable of climbing up a 10 per cent slope to the observing site. So these transporters must be beasts, right? Well, yes, but there's more. ALMA's versatility will come from its ability to be put into different array configurations: when the antennae are close together, they have improved sensitivity for studying extended sources and when they are far apart, they provide that high-spatial resolution mentioned earlier. So not only must the transporters be capable of lifting a 100-ton antenna and carting it up a mountain, they must then be able to place it on a concrete foundation pad to within an accuracy of a few millimetres. The antennae transporters are thus engineering feats in themselves.

There are two of these twenty-eight-tyred giants, named Otto and Lore, both of which are 20 metres long, 10 metres wide, 6 metres high and 130 tons in weight. Otto and Lore are both powered by two diesel

Figure 3. The ALMA antennae are assembled at the Operations Support Facility and then transported to the Array Operations Site. This photo of the Operations Support Facility shows two 12 m-diameter antennae and one 7 m-diameter antenna. (Image courtesy of ALMA (ESO/NAOJ/NRAO).)

engines, each of which generate more power than a pair of Formula 1 car engines and yet are capable of working in the rarefied atmosphere of the Chajnantor plateau. Their top speed of 12 km per hour when carrying an antenna may seem less impressive, but it's fast enough for this work and it required the development of a custom braking system!

Even when the ALMA array is in a stable configuration, Otto and Lore will have work to do: antennae will occasionally need to be picked up from the array and taken back down to the OSF for repair. These vehicles may appear to be mere industrial carriers, but without them the construction of ALMA would have been impossible. Figures 4–6 show Otto and the journey it makes in transporting an antenna from the Operations Support Facility up to the Array Operations Site. Figure 7 illustrates how Otto and Lore will be able to move the antennae into different array configurations.

The final technical challenges occur once the antennae are in place. Millimetre and sub-millimetre radiation entering the atmosphere above Atacama will encounter water molecules. Even in these dry

Figure 4. The size of the ALMA transporters is apparent when you see a Smart car performing an undertaking manoeuvre. Otto and Lore were built by the German company Scheuerle Fahrzeugfabrik GmbH and named in honour of Otto Rettenmaier, the owner, and his wife. (Image courtesy of ESO/H. H. Heyer.)

Figure 5. The transporter Otto taking the sixteenth antenna from the Operations Support Facility up to the Array Operations Site. This particular antenna was a milestone for the project because it enabled early science operations to begin. (Image courtesy of ALMA (ESO/NAOJ/NRAO) and S. Stanghellini (ESO).)

Figure 6. After its journey up from the Operations Support Facility, the sixteenth ALMA antenna is about to be lowered into place on one of the concrete foundation pads. (Image courtesy of ALMA (ESO/NAOJ/NRAO) and S. Rossi (ESO).)

Figure 7. An aerial view of the Chajnantor plateau, looking to the south-west. The photo was taken in March 2011, at which time seven of the antennae were in place. Also visible is road-laying work. Otto and Lore will transport the antennae along these roads when astronomers want the array to be put into a different configuration. (Image courtesy of ALMA (ESO/NAOJ/NRAO) and W. Garnier (ALMA).)

conditions, and at those wavelengths that can reach ground level, water molecules can partially absorb, scatter and delay the radiation. It's critically important for successful interferometry that atmospheric effects are measured and corrected for, so the ALMA team had to develop sensitive vapour radiometers. These radiometers have been placed in seven weather stations that continuously measure the amount of line-of-sight water vapour in the atmosphere. Once waves from a cosmic source have made it to an antenna, the dish brings them to a focus, at which sits a state-of-the-art receiver cryostatically cooled to just a few degrees above absolute zero. The receiver turns the waves into an electrical signal, which is then digitized and passed along up to 15 km of optical fibre to a central computer or correlator. This process happens for all ALMA's antennae, and signals from the different antennae are brought together and combined. However, in order for interferometry to work, all the antennae and associated electronics must work in perfect harmony – to a precision of a trillionth of a second. The path followed by a signal, from antenna to computer, must be known to an accuracy of a hair's width. Finally, the central correlator must provide a supercomputer level of performance in order to achieve what ALMA is designed to do: form images of astronomical objects.

There is one further challenge involved in the construction and subsequent maintenance of ALMA. It's a challenge that's much more important than the technical difficulties mentioned above since it can be a matter of life and death: how can human beings perform the complicated actions required of them at ALMA given the harsh environment in which the work takes place?

Earlier I mentioned how difficult it is to reach ALMA. One element in the journey I didn't mention is the need for a medical check-up when you reach the OSF at 2,900 metres. Blood pressure and blood oxygen-level readings are taken. Only if you pass the test will you be allowed to continue to Chajnantor (and then only if you sign a release form acknowledging the dangers of spending time at 5,000 metres). You'll be told to dress appropriately: gloves, a hat that covers the ears, wind pants and a parka or a down jacket. And you'll need sunglasses, plenty of suncream, and lots of drinking water (Figure 8).

Even so, if you spend more than a couple of days at ALMA, you're likely to feel your cheeks tighten and your lips chap. After a couple of weeks, which is the limit before you'll be ordered to return to sea level,

Figure 8. Doing work at an altitude 5,000 metres is not easy! (Image courtesy of ALMA (ESO/NAOJ/NRAO).)

your head is likely to ache and your nose and the skin on your hands might bleed. Dizziness, and what ALMA astronomers call 'jelly legs', is a common complaint. You have a good chance of suffering from some sort of altitude sickness. The main problem is that the air is thin at this altitude: oxygen levels are about half of those at sea level. Unless you carry an oxygen cylinder every step will be an exertion, every breath will be tiresome. Most dangerous of all, your thought processes will slow: imagine suffering from extreme fatigue and yet having to drive Otto or Lore, or to give instructions on how to position one of the giant ALMA antennae to millimetre-levels of precision.

The construction of ALMA thus presents huge logistical, technical and personal challenges to the international teams of astronomers and engineers who are involved. Small wonder that ALMA costs so much. But will the science it allows be worth it? The answer is a resounding: yes!

SCIENCE WITH ALMA

The mission team expects the ALMA array to be complete in 2013 (Figure 9). Even with only sixteen antennae in place, a project milestone reached in the summer of 2011, ALMA is capable of doing exceptional science: as noted below, astronomers are already using ALMA for early science observations. Nevertheless, the completed array will be immensely more powerful than it is at present. ALMA will be by far the world's largest and most sensitive telescope for observing at millimetre and sub-millimetre wavelengths. And there is a lot to see.

Figure 9. An artist's impression of how ALMA will look when completed. The image shows one of the transporters moving an antenna into place for one particular array configuration. (Image courtesy of ALMA (ESO/NAOJ/NRAO) and L. Calçada (ESO).)

The clouds of hydrogen from which the very first stars formed will have produced certain emission lines; those particular wavelengths, after the subsequent expansion of the universe has stretched them by a factor of about ten, will be in the region that ALMA can observe. So, one of the most interesting tasks for ALMA will be to study the earliest stars and galaxies that came into existence after the Big Bang. ALMA may also be able to see the first star deaths: early massive stars will have aged quickly, producing heavier elements that were dispersed throughout protogalaxies via supernova explosions and gamma ray bursts. Again, ALMA can detect certain types of emission from the host protogalaxies because the expansion of the universe redshifts the radiation into the region in which ALMA is sensitive. Thus ALMA should help us to understand how stars formed and 'switched on' after the cosmic dark ages, and how protogalaxies evolved.

ALMA will also deepen our knowledge of how stars form in the present-day universe. The general process is understood: a star forms from the gravitational collapse of a protostellar nebula. However, astronomers have no clear evidence for such collapse. To observe infalling material on scales the size of the solar system requires an instrument with high sensitivity and resolution; the instrument must be able to measure velocities with precision; and the instrument must observe at particular wavelengths (namely those at which the collapsing protostellar nebula emits and at which the surrounding clouds of gas and

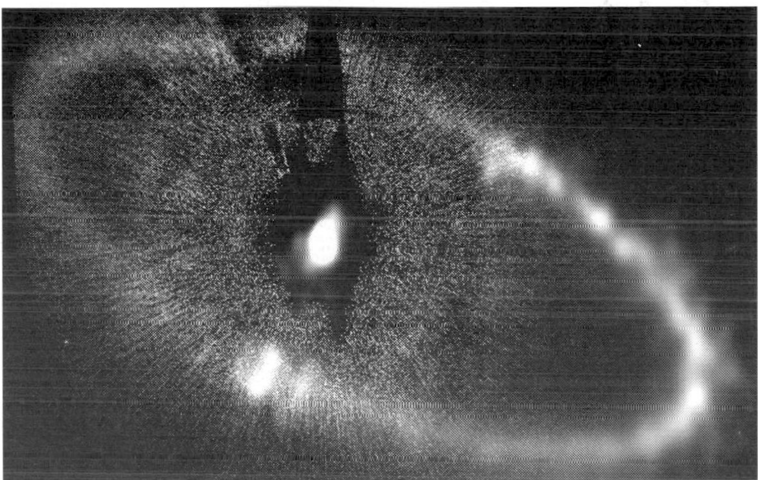

Figure 10. The bright U-shaped structure to the bottom right of this image is a part of a dust ring around Fomalhaut as seen by just sixteen ALMA antennae. The rest of the image shows an earlier picture obtained by Hubble. Even in its early, limited phase ALMA shows exquisite sensitivity and resolution. This particular image lends weight to the theory that Fomalhaut's dust disk has been shaped into a ring by the gravitational 'shepherding' effect of two planets either side of the ring. (Image courtesy of ALMA (ESO/NAOJ/NRAO). Visible light image: the NASA/ESA Hubble Space Telescope.)

dust are transparent). For quite some time ahead, ALMA will be the only instrument with these characteristics. It will be ALMA that unlocks the secrets of stellar formation.

ALMA will also help us learn more about planetary formation, since it will illuminate the structure of protoplanetary disks and tell us how such disks evolve and give rise to planets such as our own. Figure 10, which was released in April 2012, shows the first science result from ALMA. It demonstrates how effective the array will be in this endeavour. ALMA will also be a significant instrument in the search for exoplanets.

The clouds of interstellar gas and dust from which stars form are also the sites of complex astrochemistry. When molecules in these clouds collide they emit radiation associated with their quantized rotational levels, and these will typically be in the millimetre or submillimetre range: perfect for ALMA. There is a host of interesting astrochemistry for ALMA to study, since astronomers have already identified more than 160 different molecular types in such clouds. Molecular hydrogen is, of course, the most commonly occurring molecule, followed by carbon monoxide (which can be used as a tracer of interstellar clouds in distant galaxies). Clouds contain not just diatomic molecules, however; some molecules contain as many as thirteen atoms and there is evidence for anthracene (twenty-four atoms), buckminsterfullerene (sixty atoms) and 70-fullerine (seventy atoms). A polycyclic aromatic hydrocarbon (PAH) such as anthracene is particularly interesting because it has been suggested that PAHs played an important role in the early stages of the development of life. There's an enticing prospect that ALMA will tell us something of how life itself began, and how widespread it is likely to be in the universe.

It is not just the birth of stars that ALMA will investigate. ALMA will also observe old, highly evolved stars and study how they lose mass by blowing stellar winds into space. Stellar winds from old, cool stars are a prime source of dust grains in interstellar space, and ALMA will pick up the thermal emission from such grains. The array will measure the angular sizes of circumstellar envelopes, structures formed by stellar winds from old stars. These measurements will be combined with information about nearby stars, whose distances are known by independent means. Since ALMA can pick out circumstellar envelopes with ease, astronomers will be able to use this technique to measure accurate distances to certain stars as far away as 10 kiloparsecs.

Closer to home, ALMA will be used to study the Sun. The processes that give rise to radiation with millimetre and sub-millimetre wavelengths are different to those that give rise to visible light or radio waves, and take place in different regions of the Sun. Thus ALMA will provide astronomers with a new way of studying our star. We can expect to learn much more about solar flares, which can accelerate electrons to very high energies; about the formation, structure and evolution of solar prominences and filaments, which are currently not well understood; and about the conditions in the lower parts of the solar atmosphere where the temperature is relatively cool, just below the jump to the three million-degree temperatures of the corona.

As if all this were not enough, ALMA will be one of the prime ground-based telescopes for studying the atmospheres of other objects in the solar system; not just the extensive atmospheres of the planets, but even bodies such as Io, Titan and nearby comets. From our closest neighbours in space to the most distant objects imaginable, the peerless sensitivity and unmatched angular resolution of ALMA will study cosmic objects in a way no telescope has done before.

THE FUTURE AT ATACAMA

The Atacama Desert is set to consolidate its position as the world's premier location for telescopes. In a few years' time, at the Cerro Pachón observatory, construction of the Large Synoptic Survey Telescope (LSST) will commence. The LSST will have an 8.4-m mirror. This is not exceptional by today's standards, but it will possess an extremely wide field of view. When combined with an extremely sensitive 3,200 Megapixel camera capable of short exposure times, the result is an optical telescope that can study the universe in ways previously not thought possible. The LSST will provide a time-lapse movie of the universe, and highlight objects that move or change rapidly – supernovae, near-Earth asteroids, Kuiper belt objects. Even more exciting than the telescope's design is the fact that the deluge of data it produces will be made available to the public.

Not far away from Cerro Pachón, at the Las Campanas observatory, the twin 6.5 m Magellan telescopes will soon have a big brother for company. By 2020, the Giant Magellan Telescope (GMT) should be complete. The GMT will consist of seven 8.4-m-diameter mirror segments arranged so that it has the equivalent resolving power of a 24.5-m primary mirror and a light-gathering capacity that is about 15 times greater than one of the single Magellan telescopes. It will be a wonderful facility. Indeed, the GMT will be the largest optical telescope ever built. But it won't hold that record for long.

The LSST and GMT will undoubtedly perform wonderful science, but they don't rival ALMA in either complexity or cost. The ESO plan an instrument that *will* exceed ALMA in these respects. The Cerro Armazones near Paranal is to be the site for the European Extremely Large Telescope (E-ELT). This mind-boggling instrument will dwarf even the GMT. The E-ELT will consist of 798 hexagonal mirror segments arranged to form a primary mirror with a 39.3-m diameter. A telescope with such a vast collecting area will have the capability of addressing some of the most enduring mysteries of the universe. The successful construction of the E-ELT will require a variety of technical, managerial and engineering problems to be overcome. However, until the E-ELT becomes operational, ALMA will remain the most sophisticated and expensive ground-based telescope ever constructed. By any measure, ALMA is a testament to the skill, ingenuity, imagination – and even courage – of the scientists, engineers and technicians who are building it.

FURTHER READING

For more information about progress on the array, and for general news about ALMA, your first port of call should be the observatory's website (www.almaobservatory.org). The site features a webcam pointed at the Array Operations Site so that you can see what this piece of the Atacama looks like. The three films that were mentioned in the Introduction above feature scenes set in this desert environment; they are worth watching if you can't make the trip to Chile!

The piece of astrobiological research referred to in the Introduction appears in: Parro, V. *et al.*, 'A microbial oasis in the hypersaline Atacama subsurface discovered by a life detector chip: implications for the search for life on Mars'. *Astrobiology* 11 (10), pp. 969–996, 2011.

In addition to ALMA, a host of new telescopes and observatories are being developed. Some of them operate in even harsher environments than ALMA (neutrino telescopes are buried in Antarctic ice or anchored to deep sea beds; dark matter detectors are typically placed deep underground). Some of them are trying to initiate new types of astronomy (gravitational wave telescopes hope to study the universe via a so-far undetected signal carrier; cosmic-ray observatories are intended to take forward charged-particle astronomy). And new space-based telescopes will observe at all parts of the electromagnetic spectrum. My recent book *New Eyes on the Universe: Twelve Cosmic Mysteries and the Tools We Need To Solve Them* (Springer, 2012) provides an overview of all these developments.

The *Dawn* Asteroid Mission

DAVID M. HARLAND

INTRODUCTION

For almost two centuries, asteroids were merely star-like points in the sky, but then spacecraft began to inspect some of these objects in detail. Initially, these encounters were 'target of opportunity' flybys by spacecraft heading for other destinations, but recently spacecraft have been launched on missions specifically to study asteroids, and in some cases to enter into orbit around them to conduct detailed studies. This article provides some context for the *Dawn* mission that is currently underway.

ASTEROIDS

Johann Daniel Titius of Wittenberg noted in 1766 that there was a progression in the sizes of the orbits of the six planets that were then known. Their relative distances from the Sun matched the values derived by adding 4 to the series 0, 3, 6, 12, 24, 48 and 96, except that the entry for 24 was mysteriously vacant. In 1772, Johann Elert Bode of Berlin, reviving a speculation by Johann Kepler, proposed that there was an as-yet-undiscovered planet corresponding to this gap in what became known as the Titius-Bode 'law'. On 21 September 1800, Franz Xaver von Zach convened a meeting of several astronomical friends at Johann Hieronymus Schröter's observatory near Hanover to discuss mounting a search for this 'missing' planet. It was agreed that the object would probably reside near the plane of the ecliptic and letters would be sent to colleagues urging that over the coming year they methodically search a series of zones of sky in an effort to locate it. The group nicknamed themselves the 'celestial police'.

On 1 January 1801, however, Giuseppe Piazzi in Sicily, who was not in the group and was nine years into the task of compiling a star catalogue, saw a star-like object moving slowly across the sky from night to night. After tracing its westward motion for several weeks, he watched it come to a halt and then reverse its path, just as do the outer planets as a result of the changing line of sight as Earth overtakes an object further out. Unfortunately, he then lost the object in the twilight as it slipped behind the Sun. However, the mathematician Carl Friedrich Gauss managed to process the meagre data to calculate the object's orbit (thereby proving that it was located in the gap between Mars and Jupiter) and after it emerged from superior conjunction it was recovered by von Zach on 31 December. Because at 27.7 AU[1] the object's mean distance from the Sun was in excellent agreement with the Titius-Bode prediction, the mystery was considered to have been resolved and the object was named Ceres in honour of the patron goddess of Sicily.

While searching for comets on 28 March 1802, Heinrich Olbers in Bremen was surprised to find a *second* object in the same part of the sky, and in a similar orbit. William Herschel, being unable to resolve either object as a disk using his most powerful telescope, suggested that these star-like objects be termed 'asteroids' rather than planets. Piazzi countered that 'planetoids' was a better term, because they were clearly unrelated to the stars. In an attempt to explain why there were several objects, Olbers suggested that the planet which had originally orbited at that distance from the Sun had been shattered by a collision with a comet. After Karl Ludwig Harding, one of Schröter's assistants, discovered the third one on 2 September 1804, Olbers found the fourth (his second) on 29 March 1807. After a few years without another sighting, the 'celestial police' disbanded itself. But in 1830, Karl Ludwig Hencke started his own systematic search, and was rewarded with his first asteroid in 1845 and his second two years later. This opened the flood gate and by 1850, there were ten. Initially, they were eagerly sought after, but with the advent of photography the pace of discoveries accelerated and they were derided as being no more than the 'vermin' of the sky.

It was evident from the start that the asteroids were 'minor' planets at best, and in 1853, Urbain Le Verrier showed by consideration of perturbations that the total mass of the 'asteroid belt' was no more than

[1] One Astronomical Unit (AU) is the average radius of the Earth's orbit around the Sun, which is approximately 150 million km.

that of Mars. (In fact, we now know that their total mass is less than 0.1 per cent of the mass of Earth.)

Table 1: The first four asteroids discoveries

No.	Asteroid	Year
1	Ceres	1801
2	Pallas	1802
3	Juno	1804
4	Vesta	1807

ASTEROIDS AND METEORITES

Individual asteroids are classified by their spectral characteristics, with the majority falling into three groups: C-class, S-class, and M-class that are generally identified with carbon-rich, stony, and metallic compositions respectively. (More complex taxonomic classifications have been developed, but they needn't concern us here.)

Meteorites that fall to Earth have traditionally been divided into three categories: stony meteorites are rocks mainly composed of silicate minerals; iron meteorites are largely composed of metallic iron-nickel; and stony-iron ones contain both metallic and rocky material.

Only 6 per cent of meteorites are iron or stony-iron meteorites. Most iron meteorites are thought to have come from the cores of asteroids that were once molten, with the denser metal separating from silicate material and sinking to the centre; after the body solidified, it was broken up by collisions with other asteroids.

Stony meteorites are classified as either chondrites or achondrites. The majority are chondrites, named for the small round particles they contain. The chondrules are composed mostly of silicate minerals that appear to have been melted whilst freely floating in space. Chondrites are typically 4.55 billion years old, and are thought to be the 'building blocks' of the planets. Achondrites do not have chondrules, and are similar to magnesium- and iron-rich terrestrial igneous rocks. They are believed to represent the crustal material of thermally differentiated asteroids.

Table 2: Asteroids inspected by spacecraft (prior to *Dawn* reaching Vesta)

Asteroid	Spacecraft	Year	Size (km)	Class
951 Gaspra	*Galileo*	1991	18.2 × 10.5 × 8.9	S
243 Ida*	*Galileo*	1993	53.6 × 24.0 × 15.2	S
253 Mathilde	*NEAR-Shoemaker*	1997	66 × 48 × 46	C
433 Eros**	*NEAR-Shoemaker*	1998	34.4 × 11.2 × 11.2	S
9969 Braille	*Deep Space 1*	1999	2.1 × 1 × 1	S(Q)
5535 Annefrank	*Stardust*	2002	6.6 × 5.0 × 3.4	S
25143 Itokawa	*Hayabusa*	2005	0.54 × 0.30 × 0.21	S
2867 Steins	*Rosetta*	2008	6.67 × 5.81 × 4.47	S(E)
21 Lutetia	*Rosetta*	2010	121 × 101 × 75	M

* *Galileo* found Ida to possess a small satellite 1.4 km in diameter named Dactyl.

** *NEAR-Shoemaker* made a flyby of Eros in 1998 and orbited it 2000–2001.

Using a more detailed taxonomic scheme the stony spectral class of Braille is classified as Q-class and that of Steins as E-class.

Most of the asteroids visited thus far by spacecraft (Figure 1) would seem to be rubble piles with bulk densities so low as to indicate that

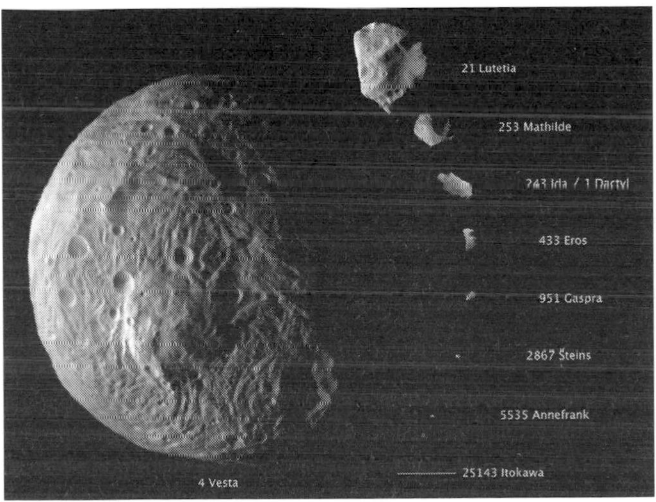

Figure 1. The relative sizes of the asteroids visited by spacecraft prior to the *Dawn* mission, illustrating that Vesta (depicted using imagery from that mission) is much larger. (Image courtesy of NASA/JPL-Caltech/JAXA/ESA.)

they contain a high proportion of voids. Several may well comprise several bodies in contact. Their surfaces are covered by blankets of regolith material which smooth out their features.

THE *DAWN* MISSION

In 2000, NASA invited further proposals for its Discovery programme, and one of the missions selected in 2001 was *Dawn*, which was to exploit proven ion-propulsion technology and a fortunate asteroidal alignment to visit Vesta in 2011 and Ceres in 2015.[2] But this was to be no mere flyby mission, since the spacecraft was to go into orbit around each asteroid in turn in order to undertake morphological and gravity mapping, and to determine the precise mass, shape and composition of

Figure 2. Although Vesta is by far the largest asteroid yet visited by spacecraft, it is the smallest body in this figure. Ceres, the largest object in the main asteroid belt, is also shown. (Image courtesy of NASA/JPL-Caltech/UCLA.)

[2] In 2006, the International Astronomical Union decided to reclassify Ceres and Vesta (together with such outer Solar System bodies as Pluto and Eris) as 'dwarf planets'.

its surface. As the first- and third-largest members of the main belt respectively, Ceres and Vesta are intermediate in size between the smallest asteroids with the 'pristine' composition of the solar nebula and the thermally evolved composition of the rocky planets (Figure 2), so they were expected to provide insight into the process of planetary formation. The name of the mission was chosen to reflect this investigation into the 'dawn' of the Solar System.

CERES

Ceres orbits the Sun at an average distance of 2.77 AU. Being 899 x 959 km, it is the largest member of the asteroid belt. It has been classified as a relatively rare type that bears similarities to the C-class. Little was discovered about it until the 1970s due to the intrinsic difficulty of observing such a small body and the fact that its spectrum was fairly flat, but then photometry by both terrestrial and space-based instruments measured its surface reflectance, and it was studied using radar. The breakthrough was ultraviolet spectroscopy obtained in the early 1980s that showed an absorption band believed to be caused by hydrated minerals. In addition to carbonates, Ceres is evidently covered with a dry clay-like material in a layer at least several centimetres thick. In fact, Ceres is similar in density to the water-rich Jovian moons Callisto and Ganymede. It has been calculated that water ice should be present at a depth no greater than several tens of metres from the surface, and spectral emissions by the hydroxyl radical have been seen from Earth. The discovery in the mid-1990s of a small number of 'main belt comets', seemingly ordinary asteroids which develop comas and tails at any point of their (almost circular) orbits, lent support to the theory that some asteroids can retain significant reservoirs of volatiles that can be exposed by impacts. Recently, observations by the ten-metre telescope of the Keck Observatory in Hawaii and by the Hubble Space Telescope (Figure 3) have resolved details on the surface of Ceres as fine as fifty kilometres (giving an image with a resolution equivalent to a naked-eye view of the Moon).

These observations refined estimates of its shape and measured its rotation period at just over nine hours. Two roundish spots situated at opposite longitudes in the Northern Hemisphere could mark major impacts. One spot, 180 km wide, showed a bright centre that could be a

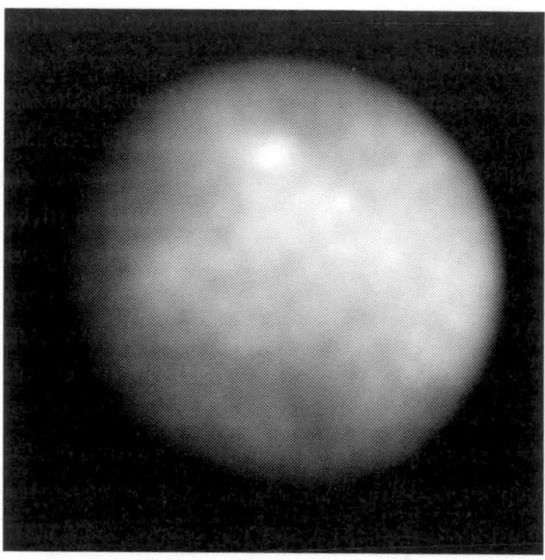

Figure 3. Ceres imaged by the Hubble Space Telescope on 23 January 2004. Astronomers enhanced the sharpness of the Advanced Camera for Surveys (ACS) images to bring out features on Ceres's surface, including brighter and darker regions that could be asteroid-impact features. Ceres's round shape suggests that its interior is layered like those of terrestrial planets such as Earth. Ceres may have a rocky inner core, an icy mantle, and a thin, dusty outer crust inferred from its density and rotation rate of 9.1 hours. (Image courtesy of NASA, ESA, J. Parker (Southwest Research Institute), P. Thomas (Cornell University), L. McFadden (University of Maryland, College Park), and M. Mutchler and Z. Levay (STScI).)

central peak. The other one has been informally named Piazzi in honour of the asteroid's discoverer. High-resolution observations also confirmed the lack of surface relief, which was consistent with an ice-rich crust, possibly containing more water than all the terrestrial oceans.

VESTA

Although the fourth member of the main belt to be identified, Vesta is the brightest and is actually visible to the naked eye against a very dark sky. It orbits at an average heliocentric distance of 2.34 AU. Its spectrum shows a heterogeneous surface and the presence of basaltic rock, indicating that it must have differentiated to some degree during its history. This discovery, made in the early 1970s, was significant because it provided the first link between an asteroid and the family of achondrites

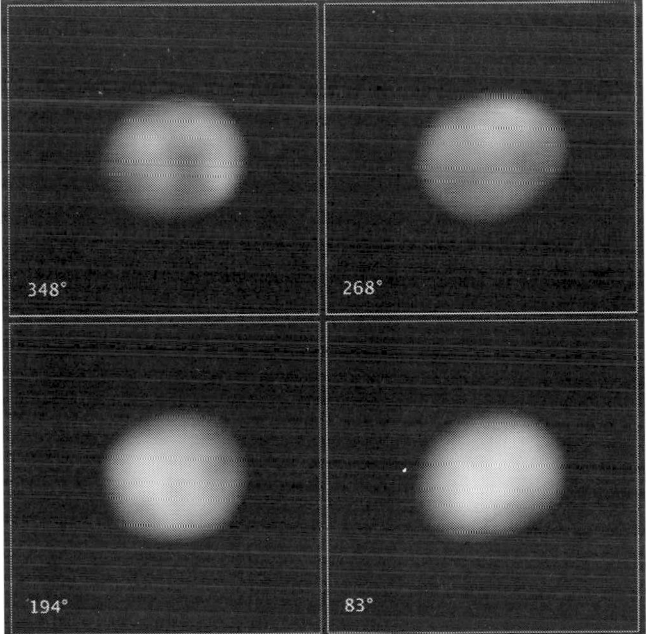

Figure 4. Vesta imaged by the Hubble Space Telescope on 25 February 2010 from a range of 211 million km with a resolution of 40 km. The irregular shape shows what is evidently a huge crater excavated at the south pole. The asteroid is some 530 km across its widest diameter, and the crater is about 460 km in diameter. (Image courtesy of NASA/ESA/STScI/UMd.)

known as the Howardite-Eucrite-Diogenite (HED) meteorites that have igneous compositions indicating varying degrees of thermal processing. Moreover, several other asteroids have been found to have a similar spectrum to Vesta, and together they constitute the taxonomic V-class.[3] It was speculated that a considerable amount of material was ejected from Vesta by a major impact a billion years ago, and that in the intervening years some of the larger fragments have been perturbed into near-Earth asteroids and smaller debris has fallen to Earth as meteorites.

In the 1980s, attempts were made to correlate the variability of Vesta's spectrum with its rotation in order to make crude maps of its

[3] The small V-type near-Earth asteroids include 3551 Verenia, 3908 Nyx and 4055 Magellan.

surface composition in terms of classes of meteorites and types of minerals. Hubble Space Telescope observations in the 1990s indicated Vesta to be ellipsoidal, 578 x 560 x 458 km in size, rotating in about 5 hours and 20 minutes. The non-spherical shape was mainly the result of a crater some 460 km wide and 13 km deep that removed at least half of its southern hemisphere in an impact with an 80-km object that must have come very close to shattering the asteroid (Figure 4). This impact was consistent with the presumed origin of the V-class asteroids and the HED meteorites.

The mass of Vesta was fairly well determined by the perturbations it exerted on the orbits of other asteroids. As early as 1968 it was noted that every eighteen years Vesta approached within 6 million km of the smaller asteroid 197 Arete. Vesta was also involved in the first observation of an asteroidal occultation when it passed in front of a background star in 1958. By precisely timing the duration of the occultation as viewed by different observers, it was possible to directly measure the shape and size of the occulting object. In 1984, this technique was applied to Ceres. Vesta's density proved to be twice that of Ceres, which was consistent with it having a rocky rather than an icy constitution.

THE *DAWN* SPACECRAFT

A mission to orbit one main belt asteroid would be very difficult using conventional chemical propulsion owing to the large amount of propellant required to rendezvous with and enter orbit around such a body. (This was achieved by *NEAR-Shoemaker* with Eros, but it is a near-Earth asteroid.) To attempt two such asteroids would have been out of the question. By sharing part of its scientific and engineering teams at JPL and Orbital Sciences Corporation, the ion-propelled *Dawn* mission sought to capitalize on the lessons learned from *Deep Space 1*, which pioneered this type of propulsion. The chief engineer and mission director for the *Dawn* mission is Dr Marc D. Rayman of JPL (Figure 5).

Apart from its propulsion system, the spacecraft, which was based on the bus of a communications satellite, was fairly conventional. It was an aluminium and carbon composite honeycomb box 1.64 x 1.27 x 1.77 metres in size, around a cylindrical thrust tube that enclosed tanks. The xenon fuel tank was one metre across. Attitude determination was

Figure 5. Marc Rayman of JPL, the 'chief engineer' for the *Dawn* mission, standing alongside the spacecraft during ground testing. (Image courtesy of Marc D. Rayman.)

to be achieved by a fully redundant system consisting of two star trackers, triple gyroscopic and accelerometer platforms, and no fewer than sixteen solar sensors. Attitude control would employ a combination of four redundant reaction wheels, ion engine thrust and a dozen hydrazine thrusters. There was a fixed high-gain antenna on one side of the bus and a set of three low-gain antennae positioned to maintain communication at times when precise pointing was not practicable. The spacecraft would require a lot of power for its ion propulsion. A further complication was the need to continue firing even at large heliocentric distances. The bus carried a pair of solar wings mounted on gimbals. They were capable of delivering 10 kW of power at 1 AU, but just 12 per cent of this at the orbit of Ceres. With its accordion-like wings fully deployed, *Dawn* had a total span of almost twenty metres, making it one of the largest interplanetary vehicles ever launched by NASA.

The ion propulsion system would perform all of the manoeuvres during the cruise, rendezvous, orbit insertion, shaping and escape phases of the mission. To provide the required total change in velocity, the vehicle would have to thrust for the equivalent of two-thirds of the eight-year duration of its mission. It was decided to provide the ion propulsion system with three identical thrusters, but only one would be operated at any given time. They were all placed on the base of the bus, one on the axis and the others offset and angled to deliver their thrust through the centre of mass.

The three scientific instruments were mounted on one side of the bus, in body-fixed positions. The Max Planck Institute for Solar System Research in Germany developed the framing camera. There were two identical units, each with a 1,024 x 1,024 pixel CCD providing a 5.5-degree field of view. The filter wheels had a clear slot and seven filters in the visible and infrared ranges to investigate the mineralogy of Vesta. Only a few of the filters would be useful for the flatter spectrum of Ceres. The design of this instrument was based on those carried by *Venus Express*, *Rosetta* and its *Philae* lander. The gamma-ray and neutron spectrometer was the only US-built instrument on board. It was based on hardware flown on the *Lunar Prospector* lunar orbiter and on *Mars Odyssey*, and was to measure the elemental composition of the surfaces of the two target asteroids and detect water ice and other volatiles. Italy contributed a visible and infrared mapping spectrometer derived from instruments on *Cassini*, *Rosetta* and *Venus Express*. This provided good spectral resolution from the near-ultraviolet through the visible range into the near-infrared, and was to identify minerals on the surfaces of the asteroids with good spatial resolution. The combined observations of the spectrometers should facilitate identification of a wide variety of hydrated minerals and enable the role of water in the geological evolution of Ceres to be characterized. (It might then be possible to deduce that certain meteorites came from Ceres.) And, of course, the Doppler monitoring of the spacecraft's radio signal would chart the gravity fields of the asteroids to determine their masses, moments of inertia, localized mass concentrations, internal structure, etc.

The spacecraft's launch mass was 1,218 kg, including 425 kg of xenon for the ion propulsion system and 46 kg of hydrazine for the attitude thrusters; sufficient for a total velocity change over the full duration of the nominal mission of about 11 km/s, which, remarkably, was comparable to that provided by the Delta II 7925H launch vehicle. The low-thrust propulsion trajectory allowed considerable flexibility in the launch window. The target of a launch in June 2006 would ensure arrival at Vesta in July 2010 for an eleven-month reconnaissance, after which the spacecraft would depart and reach Ceres in August 2014. A delay in launch beyond 15 October 2007 would rule out attempting a two-asteroid profile until the 2020s. It narrowly missed being cancelled by development issues and budget crises, and the launch was delayed to 2007. After several delays due to technical issues and weather, the mission finally got under way on 27 September.

TO VESTA

After coasting for forty minutes in a low parking orbit, the second and third stages of the launch vehicle fired in succession and *Dawn* was released into a 1.00 x 1.62 AU solar orbit. The spacecraft fired its thrusters to cancel its inherited spin, waited for its xenon fuel to settle, and then opened its solar panels and turned them to face the Sun. Its systems were tested over the next two months. A test of the central ion thruster on 6 October monitored its performance at several different throttle levels. The angled thrusters were then tested. For the engine tests, the spacecraft thrusted towards Earth and the resultant Doppler shift was measured to calculate the acceleration imparted. While engineers checked the engines, scientists tested their instruments. The camera was first activated on 18 October and calibrated by taking pictures of star fields. On 17 December, after all of the tests had been successfully concluded, *Dawn* oriented itself to point its central engine in the optimal direction for its ion-propelled mission. As in the case of *Deep Space 1*, the cruise would be a succession of long periods of thrusting interrupted on a weekly basis for telemetry dumps (during which thrusting would be suspended so as to point the high-gain antenna at Earth) interleaved with longer periods of ballistic flight for housekeeping and other tasks. On 8 August 2008, the spacecraft reached aphelion beyond the orbit of Mars. On 31 October, the engine was switched off to initiate about seven months of ballistic cruising that would include a Mars flyby. In ten months of thrusting it had consumed 72 kg of xenon and achieved a 1.81-km/s velocity change which placed it into a 1.22 x 1.68 AU orbit. The vehicle spent the end of 2008 in solar conjunction with respect to Earth.

The Mars flyby on 18 February 2009 as *Dawn* headed in from aphelion was at an altitude of 542 km. It approached from the night-side, with the Sun occulted by the planet. It sped by at a relative speed of 5.31 km/s, with the closest point of approach above the volcanic region of Tharsis. On crossing over to the day-side, *Dawn* was to image regions over which the *Mars Express* orbiter would pass an hour later.

This would enable the performance of the cameras on both vehicles to be cross-checked. The gamma-ray and neutron spectrometer was to take calibration spectra of Mars that could be compared with those of other missions orbiting the planet. *Dawn* was to continue to image the planet on the outbound leg for a week to calibrate optical navigation for the approach to Vesta. Unfortunately, a hitherto unsuspected flaw in the software of the attitude control system caused the spacecraft to enter safe mode soon after closest approach, halting the instrument calibration sequences. However, the main objective of the flyby was the gravity-assist, which successfully deflected *Dawn* into a 1.37 x 1.84 AU orbit and steepened the plane of its orbit relative to the ecliptic by more than 4 degrees to match the 7.1 degrees of Vesta. The 2.6 km/s of velocity increment from the flyby was equivalent to an additional 100 kg of xenon for the ion propulsion system.

Dawn reached perihelion on 16 April 2009, shortly after new flight software was uploaded from Earth. On 8 June (after a seven-month coast during which the ion engines had been operated for only about ten hours) thrusting was resumed and in November the vehicle entered the asteroid belt. In early 2011, during the final leg of the flight to Vesta, orbital mapping operations were rehearsed. In March, thrusting was halted and the instruments were checked out. The cameras were tested after a long period of inactivity by taking pictures of star fields. The thrusters were also calibrated and their minute push measured. After a week of tests, the vehicle resumed thrusting for the encounter.

AT VESTA

The interplanetary cruise ended on 3 May, and the approach phase began. The first navigation images of Vesta were taken on that day from a range of 1.21 million km; the target was still a bright dot only a few pixels wide. One week later, the mapping spectrometer took its first spectra. The gamma-ray spectrometer was also activated in early May. By mid-June the imagery from *Dawn* began to rival the best obtained by the Hubble Space Telescope and the frequency of imaging was increased to twice weekly. At that time the craft was only as far from Vesta as the Moon is from Earth. Approach movies started to show hints of surface features and texture variation on the asteroid, observed at an angle and a phase relative to the Sun inaccessible from Earth. The pictures were already sharp enough to show the central peak of the south polar crater. Vesta appeared to be less spherical than other bodies in the solar system of similar size, but Vesta was known to be made of rigid rock rather than ice. These distant observations were used to refine knowledge of the orientation of its spin axis to ensure that the vehicle would be able to enter the desired polar orbit. The visible and infrared spectrometer also started to obtain data.

On 1 July, the camera took more pictures of Vesta. The probe was then well south of the equator and had a direct line of sight of the central mound of the polar crater, which looked like a 'gigantic belly button'. The mound rose about 22 km above the surrounding terrain and was 200 km across at its base, making it the second tallest mountain in the solar system after Olympus Mons on Mars. Otherwise the southern hemisphere appeared remarkably smooth, with only a few craters in the north, and smooth terrain around the equator. Possibly, the giant crater formed only in the last 2.5 billion years, and showered the rest of the surface with debris that hid older features. On 9 and 10 July, *Dawn* took three sequences of seventy-two images of the space around Vesta to search for any satellite (Figure 6). Although satellites

Figure 6. A view of Vesta taken by *Dawn* on 9 July 2011 from a range of 41,000 km, six days before the spacecraft was captured by the asteroid's gravity, with a resolution ten times better than the Hubble Space Telescope imagery. (Image courtesy of NASA/JPL-Caltech/UCLA/MPS/DLR/IDA.)

were deemed likely for such a large, spheroidal body, none larger than several metres in size were detected. (A number of other main belt asteroids were detected moving in the background.)

In contrast to a conventionally propelled mission, there was no critical, fast-paced orbit-insertion manoeuvre. Instead, the ion thrusters gradually matched the vehicle's heliocentric speed with that of its target, so that it gently came under the asteroid's gravitational influence. *Dawn* smoothly went from orbiting the Sun to orbiting Vesta as it continued to thrust. This occurred at 4.48 UTC on 16 July. The precise time of the insertion could not be determined beforehand, since it depended on the relatively poorly known mass of the asteroid (Figure 7).

When *Dawn* entered orbit around Vesta at an altitude of 16,000 km, it was at a relative speed of a mere 27 m/s (about 100 km/h). Vesta was 188 million km from Earth, and would be in opposition to Earth later in the month. The ion thrusters had been on for 70 per cent of the mission thus far, and had provided an overall velocity change exceeding 6.6 km/s, consuming 252 kg of xenon. Confirmation that *Dawn* had indeed entered orbit came more than twenty-four hours later when the probe suspended thrusting and turned to communicate with Earth.

Figure 7. A depiction of the *Dawn* spacecraft in proximity with Vesta, using imagery of the asteroid obtained by the spacecraft. (Image courtesy of NASA/JPL-Caltech.)

Along with the engineering data returned on this occasion were sequences of images taken earlier in the day.

The first image returned from orbit looked straight down at the south polar crater and its central mound (Figure 8). The floor of the crater was crossed by parallel ridges and grooves, making it resemble the Uranian satellite Miranda. Moreover, as with that moon, there were kilometres-tall cliffs. Even more impressive, grooves up to 15 km wide with steep walls and flat bottoms ran parallel to the equator, spanning two-thirds of the circumference of the asteroid. Individual troughs were up to 380 km long. These could have been formed by the south polar impact, which crushed the surface, making overlapping compressional wave-like tectonic features and ridges. However, the grooves looked more cratered than the southern crater, as if they were older. Or perhaps the floor of the giant crater had been resurfaced, making it appear younger. Conspicuously absent were smooth deposits of ejecta or pools of rock melted by the impact. There were also enigmatic fresh-looking small craters with dark streaks on their sides (Figure 9). Other deposits of dark material could hint at ancient volcanic activity.

Multispectral images taken at the same time implied compositional differences at the surface. The 475-km polar crater was officially named Rheasilvia in honour of Rhea Silvia, the mythical Vestal Virgin mother

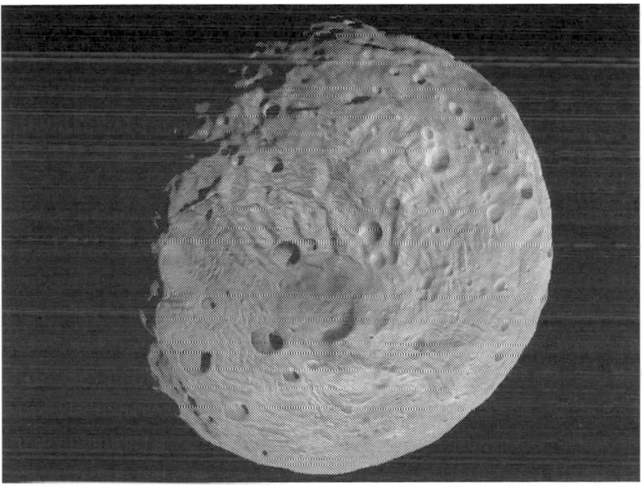

Figure 8. The south polar region of Vesta looking down on the Rheasilvia impact basin. (Image courtesy of NASA/JPL-Caltech/UCLA/MPS/DLR/IDA.)

Figure 9. A young crater within the Rheasilvia impact basin. It is 15 km in diameter, has layering visible in its wall and external material has spilled into the crater. (Image courtesy of NASA/JPL-Caltech/UCLA/MPS/DLR/IDA.)

of Romulus and Remus, the founders of Rome. Other features would be named after Vestal Virgins of Roman history and famous Roman women. Three days were dedicated to acquiring images, spectra and rotation movies. *Dawn* then flew over the night-side and one week after orbit insertion it was over the day-side on the northern hemisphere, where it was winter. The surface poleward of 52°N was in darkness (Figure 10). The visible portion of the northern hemisphere appeared to be more cratered than the southern terrains seen so far, and included three shallow, flat-floored craters in a pattern that resembled a snowman, but crater counts implied that the two hemispheres were about the same age.

On 25 July thrusting was suspended at a height of 5,200 km in order to conduct scientific observations over the next four days. Full rotation movies were obtained, with *Dawn* flying over a range of latitudes. On 2 August, *Dawn* reached the survey orbit altitude of 2,700 km, in which it officially started scientific observations nine days later. This activity was to be undertaken during seven sixty-nine-hour orbits, over twenty days. Observations were mainly carried out using the camera and infrared spectrometer, while the gamma-ray and neutron detectors collected background data, the asteroid being still too distant to pro-

Figure 10. The northern hemisphere of Vesta with the pole in darkness. (Image courtesy of NASA/JPL-Caltech/UCLA/MPS/DLR/IDA.)

vide a significant signal. Data were collected and stored on board over the day-side and then relayed to Earth over the night-side. This and subsequent orbits were designed to prevent the craft from entering the shadow of Vesta, so that the solar panels would continue to generate power. The two-way radio link between Earth and the spacecraft was used to precisely measure the Doppler shift and thereby map the gravitational field and internal structure of the asteroid. The mass in particular was established to be within a few per cent of that determined from Earth. Gravity tracking confirmed Vesta to be a differentiated body, with a denser metallic core spanning 40 per cent of its diameter. The core was probably associated with a fossil magnetic field, but *Dawn* was not equipped to detect it – the planned magnetometer had been deleted because of budget constraints. Although the infrared spectrometer had a glitch that prevented data collection on two out of seven orbits, it made over 3 million spectra covering 63 per cent of the visible surface. On the other hand, the entire illuminated surface was imaged by the camera at least five times, producing a total of more than 2,800 images. Remote-sensing observations showed a geological variety unseen on any other minor planet visited by spacecraft, but failed to uncover any unambiguous volcanic features.

On 1 September, *Dawn* resumed thrusting to deliver a velocity change of 65 m/s and descend to the 680-km orbit with a period of about 12.3 hours in which most of the imaging would be undertaken, including colour and stereo imaging by orienting the vehicle to point its camera sideways instead of straight down. It thrusted through eighteen revolutions, pausing frequently for orbit determination (because Vesta's irregular gravity field was not yet precisely known), until it reached this high-altitude mapping orbit on 18 September. After two short corrections and imaging sequences, *Dawn* restarted observations on 29 September. It completed six ten-revolution mapping cycles, acquiring an almost complete stereoscopic topographic map of Vesta, as well as more than seven thousand pictures and fifteen thousand spectral frames covering all but the northern polar regions. Topographic reconstructions revealed the south polar crater to have an '8' shape, suggesting that it is actually two craters superimposed, with Rheasilvia superimposed on an older basin about 100 km in diameter (Figure 11).

Observations also showed the presence of a second, broader set of troughs, north of the equator with an older and more degraded appearance. The transition to the low-altitude mapping orbit began on 2 November, and would take more than five weeks to complete, *Dawn* pausing every three days in order to perform orbit measurements. It briefly entered safe mode on 4 December while turning to the attitude to

Figure 11. A processed view of the south polar region of Vesta, showing the mountain at the centre of the Rheasilvia impact basin. The landscape is portrayed as if it were flat and the vertical scale is 1.5 times that of the horizontal scale to emphasize the topography. (Image courtesy of NASA/JPL-Caltech/UCLA/MPS/DLR/IDA/PSI.)

fire the engine, and finally restarted observations on 12 December. The low-orbit phase would be the longest one, with the vehicle on average 210 km above the irregular surface of Vesta (it was to have been 10 km lower but controllers decided to raise it). The low orbit had a period of four hours twenty-one minutes on average and was tailored to obtain high signal-to-noise gamma-ray and neutron spectra and gravitational field data using one of the low-gain antennae. Of course, remote-sensing instruments also returned images and data at high resolution.

It was concluded that, owing to its size, Vesta is a 'transitional' object between an evolved planet and an unevolved asteroid. In general, its geology is more like larger rocky bodies such as the Moon than the asteroids. The multispectral investigation of the surface supported the hypothesis that the HED meteorites derived from Vesta, in all likelihood blasted out by the Rheasilvia impact. The Howardite meteorites appear to represent surface breccia, the Eucrites a surface basalt flow and the Diogenites an igneous rock from a deep magma chamber.

After surveying in this low orbit, the spacecraft resumed thrusting to progressively widen its orbit for a second high-altitude mapping orbit phase. Observations obtained from this orbit provided a companion set of data and images to those obtained during the first high-altitude mapping orbit phase, completed in October 2011. A key difference was that the angle at which sunlight hit Vesta had changed, illuminating more of its northern hemisphere. The principal science observations undertaken in this new orbit were obtained with the framing camera and the visible and infrared mapping spectrometer.

Following this final science data-gathering phase, *Dawn* then spent almost five weeks spiralling out from the giant asteroid to the point at which Vesta lost its gravitational hold on the spacecraft, on 5 September 2012. *Dawn* then turned to view Vesta as it departed and acquired more data. Then the spacecraft set its sights on the dwarf planet Ceres, and began a two-and-a-half-year journey to investigate the largest body in the main asteroid belt.

TO CERES

Dawn will pursue about three-quarters of a solar orbit, most of which will be spent thrusting in order to reach Ceres in February 2015. During this cruising phase, it will pass relatively near a number of asteroids and, if funds are available to support an encounter, it might detour to inspect one or more of these objects. One flyby candidate is Arete, the object that first enabled Vesta's mass to be determined.

On approaching Ceres, *Dawn* will undertake a similar ballet to that at Vesta: first entering an orbit at an altitude of 5,900 km with a period of about 10 days, reducing to 1,300 km and 17 hours, and finally to 700 km and 9 hours. The primary mission is scheduled (by funding) to conclude in July 2015. If Ceres is indeed found to have an environment potentially favourable to life, the main end-of-mission option is to leave the spacecraft in a 'quarantine' orbit around it that should not crash on Ceres for at least half a century. Unlike *NEAR-Shoemaker* at Eros, it will not be possible to soft-land *Dawn* on Ceres owing to the asteroid's stronger gravity. Depending on the amount of propellant remaining and the trajectory options, it might be possible to manoeuvre the spacecraft out of Ceres's orbit and perform a flyby of another target of opportunity.

Digital Meteor Imaging

NICK JAMES

Meteors, or shooting stars, are visible as occasional, brief streaks of light on most nights when the sky is clear and there is no Moon. They are caused by tiny particles, called meteoroids, which enter the atmosphere at high speed (11–72 km/s) and burn up due to friction. The streak of light that results is called the meteor trail. Brighter meteors often leave behind an ionized tube of atmospheric molecules which emit light for a period after the meteor has passed. This residual effect is called the meteor train or, sometimes, persistent train. The incidence of meteors varies throughout the year, peaking at the times of the major meteor showers, which are listed elsewhere in this yearbook. At other times the rates are lower and meteors that don't belong to showers are called sporadics.

VISUAL METEOR OBSERVING

Astronomers are interested in the characteristics of the meteoroids that cause shooting stars. It is known that, in most cases, these particles come from comets and so the size distribution and frequency of the meteoroids can tell us something about the original comet and the subsequent motion of the particles in space. Amateur meteor observers can help to determine shower and sporadic meteoroid size distributions and rates by observing the sky and estimating meteor brightness (magnitude) and times of appearance.

The trails of all the meteors in a shower radiate from a particular point on the sky called the radiant. This is a perspective effect resulting from the parallel trajectories of all of the meteoroids in the shower. Occasionally, a much larger meteoroid enters the atmosphere and this causes a very bright meteor called a fireball or bolide. These events can, very rarely, lead to a meteorite fall. Multiple station observations of the same meteor can be used to determine the true trajectory using triangulation if the start and end of the trail can be accurately measured. This allows us to determine the original orbit of the meteoroid and, in the case of a bright fireball, it would allow determination of the fall area for any possible meteorite(s).

Meteor observing is still an area of astronomy where the visual observer can make a significant contribution to science with minimal expense. Such observing requires only warm clothes, an accurate watch and great dedication. Group meteor watches can also be great fun in the warmer summer months when they provide an excellent social activity for local clubs and societies. This is why the Perseid meteor shower each August is so well observed. Things tend to be rather different in the winter and showers such as the Quadrantids, which peak in early January, are under-observed since only the hardiest and most dedicated observers can stick it out through the long, cold nights. The rest of this article discusses meteor-observing techniques which don't need the observer to be out throughout the night.

METEOR IMAGING WITH STILL CAMERAS

For many years meteor observers have used still cameras to capture meteor trails. These days film-based cameras have been replaced by their modern digital equivalents but the techniques are broadly similar, although the individual exposure times are generally much shorter. With a tripod-mounted camera, lens (usually a wide-angle) at full aperture, a high ISO setting, and the camera aimed at an elevation of about 50° above the horizon, the observer hopes that a bright meteor will flash through the field of view while the shutter is open. Digital SLRs (DSLRs) are very efficient at collecting background light from the sky, particularly at a setting of ISO 1600, so exposures should generally be kept relatively short – no more than five minutes' duration in a really dark, rural location, and probably only ten to thirty seconds from a more typical observing site. With some DSLRs, the camera can be operated using a programmable timer attached to the shutter control to take repeated exposures one after the other for as long as required, provided the battery is fully charged beforehand. An example of a meteor captured in this way is shown in Figure 1. This approach is

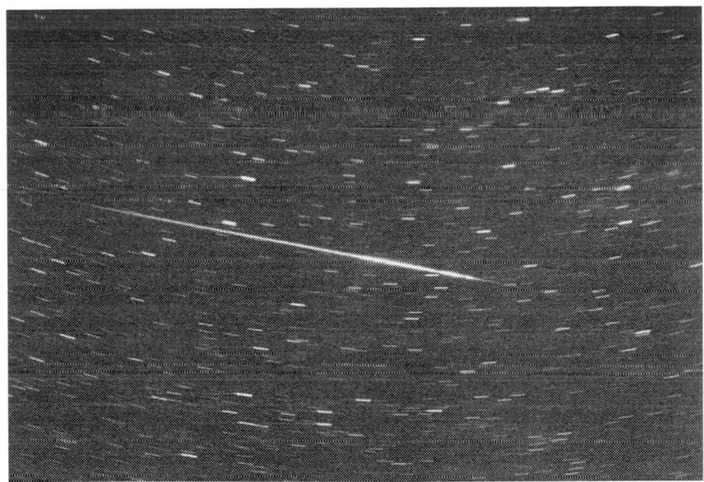

Figure 1. A Leonid meteor captured on film using an SLR on 18 November 1999, 01:56:13 UTC. (Image courtesy of Nick James.)

effective but, even with high ISO settings, only bright events are detected and there is no information about speed, duration or the exact time of appearance.

IMAGING WITH TV CAMERAS

An alternative to still cameras is to use a TV system which produces full-frame rate (25 frames per second) video of the night sky. This allows the speed, duration and time of appearance of meteors to be determined and enables continuous monitoring of the sky for long periods. In the past, TV cameras were not sufficiently sensitive for meteor work and they had to be used with specialized image intensifiers. This was a successful combination pioneered in the UK by the late Andrew Elliott. Andrew obtained spectacular results over a number of years (Figure 2) but the use of image intensifiers has significant disadvantages for long-term, automated monitoring of the sky. The intensifiers are expensive, need special power supplies and can easily be damaged by bright light. It is particularly difficult to set up a permanent intensifier camera-based system since the intensifier has to be protected from illumination during the day.

Figure 2. A bright Perseid caught on a TV/Image Intensifier system. (Image courtesy of the late Andrew Elliott.)

As amateur astronomers we frequently benefit from technological developments in other fields and in the last few years TV cameras have become available which are suitable for use in meteor observing without the use of intensifiers. These are based on highly sensitive detector chips. A limiting factor in the sensitivity of a TV camera is the detector read-out noise. This noise swamps the very small signal from faint objects and limits what can be detected at TV frame rates. A measure of the performance of a sensor is its Quantum Efficiency (QE). This is the probability that a photon of a particular wavelength will be captured by the camera and turned into usable signal. Sony pioneered the development of a type of CCD detector called the Exview HAD (Hole Accumulation Diode) sensor. This collects more photons by concentrating them on to the small sensitive area of the pixel using integral micro-lenses. The depth of the pixel detection layer is also increased so that the camera's sensitivity is extended into the near infrared. All of this leads to a significant increase in photon capture and a larger signal. Overall, the sensitivity of this type of detector is at least four times that of previous cameras.

MODERN TV CAMERA SYSTEMS

TV cameras designed for the security market based on this technology are incredibly sensitive. With suitable lenses, they can detect stars in real-time video down to second or third magnitude over a wide area of sky. Such cameras are ideal for monitoring the sky for meteors and other transient events. A camera which is particularly suitable for this type of work is the Watec 902H2 Ultimate (Figure 3). This camera has a sensitivity of 0.0001 lux at f/1.4. Such a camera can provide good-quality security video in faint moonlight or can be used to detect meteors and other objects in the night sky. These cameras are quite expensive new (approximately £250) but can occasionally be picked up on eBay for a fraction of this price.

Figure 3. The Watec 902H2 TV camera. (Image courtesy of Nick James.)

In addition to the camera the lens is a key component in the imaging system. The lens needs to be as fast as possible to improve sensitivity. There is a trade-off between very short focal lengths which give a large field of view but which have a small physical aperture, and longer focal-length lenses which will detect fainter meteors but which have a narrow field of view. For my system I wanted a large field of view so I use a 3.8mm, f/0.8 lens which provides a sky coverage of 96° x 72°. Again, these C-mount TV lenses can often be obtained from eBay at very good prices. Since I wanted my installation to be permanent, I use a conventional CCTV housing to protect the camera and lens (Figure 4). This is waterproof and has a heater to prevent dewing in the winter.

Figure 4. A waterproof, heated CCTV housing containing my meteor video system. (Image courtesy of Nick James.)

It also has a fan to keep the camera cool in the summer. This is pointed at an elevation of 45° so it covers elevations from 10° to 80°.

The camera's power supply is critical to the overall system performance. Any ripple on the DC voltage will be visible as dark and bright bands rolling vertically through the frame. Most cheap 12V DC supplies are not sufficiently clean and external filtering is required. Alternatively, the camera system can be run off a battery supply which is recharged during the day. This is the approach I use with a PC managing the charge/discharge of the battery via a USB controlled relay.

PROCESSING THE ACQUIRED VIDEO STREAM

It's all very well collecting hours of sky video every night but it wouldn't be practical to sit through it all looking for meteor events. Thankfully, this function can be performed by suitable software utilities and there are a number of packages available. Two of the most popular are Metrec and UFOCapture.

Metrec is probably the program most used by serious meteor observation systems. It is very powerful but it had two significant problems for my project. First, it requires a specific video capture card, made by Matrox, which is now obsolete; and second, it runs in real time on the captured video stream and so the processing has to be done on a computer physically near the camera or video has to be replayed from an external recorder. UFOCapture is an alternative program which is also very capable. It will work with a wider range of video capture devices but still has a number of limitations that meant that it didn't do exactly what I wanted. My objective was to develop a complete stand-alone meteor detection system that would run on low-end, cheap PC hardware, so I decided to develop my own software from scratch.

In my system, composite video from the camera is digitized by a Hauppauge PVR250 hardware MPEG2 encoder card and stored on network disk at a rate of around 3GB per hour. This means that around 40GB of video is recorded on a long winter night. Modern disks of 1-2 TB allow several weeks of recordings to be stored. The Hauppauge card is available very cheaply (under £20) from several sources and it converts the video stream in hardware so that there is no need for the local PC to have a powerful processor. The video is recorded in standard CCIR TV format, 720 x 576 pixels, interlaced, at 50 fields (25 frames) per second. Frames in the video stream are time-stamped by the card.

Analysis of the video stream is done in non real-time. My software is written in the C programming language and runs on a low-end Linux PC. It reads each decoded frame from the MPEG video and forms a moving average of the twelve frames either side of the current frame. This moving average is then subtracted from the current frame to generate a difference frame. If the video image is static the difference frame would have pixels which were all zero. Any new object or movement would show as groups of pixels brighter or fainter than zero. In reality the video is noisy and so the difference image contains pixels above and below zero even in the case of a static input. The software performs a statistical analysis of each difference frame to determine the noise level and then sets a threshold to minimize the number of false detections. Any pixels above this threshold are considered to be real detections and are passed on to the next stage of processing.

Adjacent pixels above the detection threshold in each difference frame are grouped together to be processed as a single object. Each difference frame may contain a number of detected objects which could be aircraft, satellites, birds or meteors. The software compares detected objects in adjacent difference frames to see how they are moving across the sky. Aircraft and satellites tend to move quite slowly and birds tend not to move in straight lines and so the software can generally reject these candidates. This is particularly important for me since one of my cameras points in the direction of an approach path for London's Heathrow airport and, in the evening, I can have many aircraft strobes in the field of view at the same time. An example of this is shown in Figure 5 which is a 20-second stack of frames showing five aircraft in the field of view.

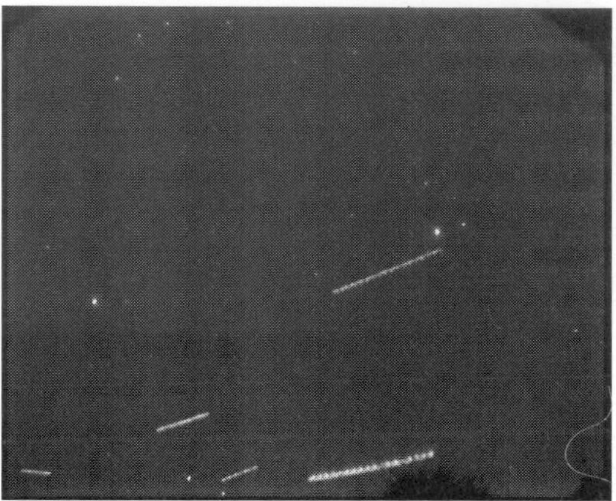

Figure 5. A 20-second stack of video frames showing aircraft trails in the field of view in the evening sky. (Image courtesy of Nick James.)

Detected objects which meet the criteria for a meteor trail are recorded in a log file along with estimated magnitude, angular speed across the sky and the time of appearance obtained from the video timecode. This log is used by another software program to extract segments of video around the detection time for manual checking and image stacking. Even though individual frames only show stars to second magnitude or so, the stacks can go much deeper. For stacking I generally use a statistical technique which corresponds to a mean for static objects and a maximum for moving objects. This makes moving objects, such as meteors, stand out well from the background objects. The stacked images can go surprisingly deep. Ten seconds of stacked video (250 frames) is enough to show the Milky Way and stars to fifth magnitude (Figure 6).

AUTOMATED METEOR OBSERVING

The system described above is very effective. It never gets cold, tired or bored and it can monitor the sky continuously during dark hours.

Figure 6. A 10-second stack of video frames showing the Milky Way and stars to magnitude 5 on 22 April 2012, 01:55 UTC. (Image courtesy of Nick James.)

During the 2011 Geminids it detected 105 shower meteors on maximum night despite moonlight and drifting cloud. It has also detected a number of very bright fireball events. Some example detections are shown in Figures 7–9. These results demonstrate the effectiveness of the system and I would like to encourage others to set up similar automatic cameras around the country based on a common

Figure 7. A bright Lyrid meteor in Bootes on 22 April 2012, 01:09 UTC. (Image courtesy of Nick James.)

Figure 8. A bright sporadic meteor in Leo on 18 November 2011, 05:25 UTC. The bright object to the right of the trail is the Moon. (Image courtesy of Nick James.)

Figure 9. A bright meteor travelling from Cygnus to Hercules on 12 April 2012, 03:50 UTC. The last quarter gibbous Moon is behind the tree at bottom right. (Image courtesy of Nick James.)

design. Such networks are already active in the US and parts of Europe, and the NASA CAMS network (jttp://cams.seti.org) is a very good example of what can be achieved.

An alternative to video observing is to use a camera capable of longer exposures which are repeated throughout the night. This approach has been used very successfully by Peter Meadows. He employs a DMK AU03 camera with an Opticstar 2.8 to 12.0 mm, f/1.4 lens giving a maximum field of view of 97° x 73° (Figure 10).

Figure 10. Peter Meadows's still camera (a DMK AU03) used for meteor imaging. (Image courtesy of Nick James.)

The camera points roughly SE and exposures of 10-second duration are taken automatically throughout the night until about an hour before dawn. Images are reviewed manually the following day and generally show stars to magnitude 6. A homemade dew heater is connected to a 12v battery and the camera is connected to a laptop located within a nearby shed. An example of a meteor captured using this system is shown in Figure 11.

The images obtained by Peter's still camera are complementary to those obtained by the video and we have had many occasions where the two cameras capture the same event. A particularly good example of this was a very bright Leonid fireball that was seen on the morning of 19 November 2011. The video, running at 25 frames per second,

Figure 11. A bright Perseid captured by Peter's system on 11 August 2010, 00:42 UTC. (Image courtesy of Peter Meadows.)

allowed us to determine that the peak magnitude of this fireball was around -11.3 (Figure 12).

It also allowed us to determine the precise time of appearance, the angular velocity and the initial train decay rate. The still imager provided detailed information on the shape and decay of the train over the next three minutes as shown in Figure 13.

As with many areas in astronomy, meteor observing is now benefiting from the availability of high-quality imaging equipment and these recent developments mean that we can contribute significant numbers of meteor observations without having to sit outside in the cold throughout the long winter nights.

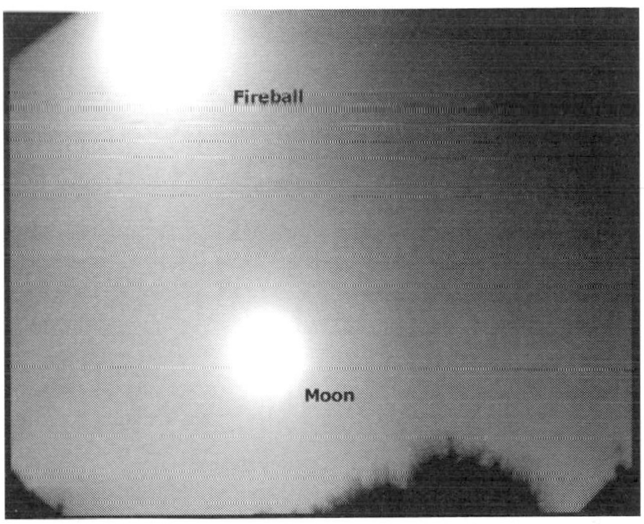

Figure 12. A single video frame of a Leonid fireball in misty conditions on 19 November 2011, 04:07:45 UTC. At this point the fireball is brighter than the Moon. (Image courtesy of Nick James.)

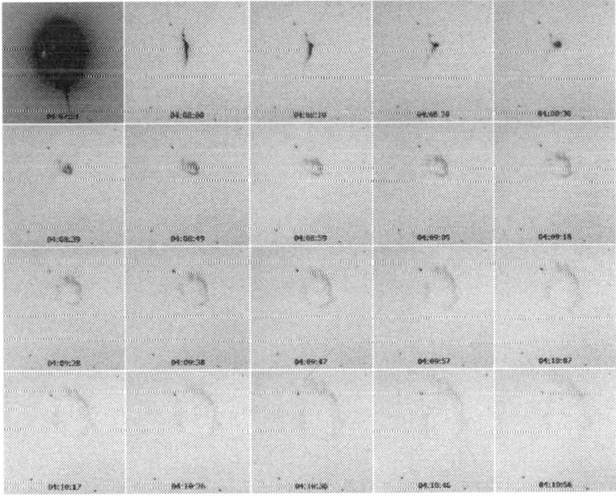

Figure 13. Development in the train of the fireball of Figure 12 for 3 minutes after the event. (Image courtesy of Peter Meadows.)

Will Hay: Entertainer and Amateur Astronomer

MARTIN MOBBERLEY

Eighty years ago the popular entertainer Will Hay (1888–1949) discovered a massive white spot in the atmosphere of the planet Saturn while using his 6-inch Cooke refractor from the back-garden of his home at 45 The Chase, Norbury. That house, in South London, still exists today, roughly midway between Streatham Common and Norbury Park, and it carries a blue English Heritage plaque signifying that the 'Comic Actor and Astronomer' lived there. Hay was the first to admit that he had been very lucky and, at the British Astronomical Association (BAA) meeting in October 1933, he stated that he regretted that 'a more regular and constant observer had not seen the white spot first'. There were certainly many amateur astronomers around who were far more frequent planetary observers than Will in the 1930s, and the entertainer knew many of them. These included the larger-than-life BAA Saturn Section Director Captain Maurice Anderson Ainslie (nicknamed 'The Skipper') and Will's best astronomical buddy Dr W. H. Steavenson (known as Steave to his friends), who lived less than two miles north-east from Hay at West Norwood. Other amateur astronomers Hay mingled with included Dr Rowdon Marrian Fry, Dr Gerald Merton, Frank Holborn, Guy Porter, Revd T. E. R. (Theodore) Phillips and Dr Reggie Waterfield.

SERENDIPITY

It is tempting to think that Will Hay's discovery of Saturn's white spot was purely a huge fluke – with his busy life in the world of stage and radio entertainment he can surely have been a casual observer at best. However, while his greatest astronomical achievement was indeed a lucky find, there is no doubt that Hay was utterly fascinated by the night sky and, if not for the exponential increase in his film work throughout the 1930s, who knows what else might have been achieved? The BAA journals of the 1930s show that Hay not only used, but also constructed and refined, a mechanical chronograph used for astronomical timings and he had even mastered the tricky technique of off-set guiding when photographing comets with a five-inch aperture f/5 Zeiss triplet lens. This astrograph was piggybacked on to his refractor.

The history of astronomy is littered with chance discoveries and I cannot resist mentioning a few. Perhaps the most amusing example, and one which has a tenuous link to the 1933 white spot, was the discovery of comet 1946 K1 (Pajdusakova-Rotbart-Weber). Anton Weber independently discovered the spot seen by Hay but, thirteen years later, in an event which would have seemed perfectly at home in a Will Hay film, Weber co-discovered comet 1946 K1 while sitting on the outside toilet in his Berlin home! His home was located in the Steglitz district of Berlin and due to wartime damage there were gaping holes in the lavatory building. Through one of these, on 31 May, he spotted a diffuse cometary fuzz in Lyra, which he was able to confirm with binoculars minutes later.

Another discovery with a coincidental link to Will Hay was that of nova Lacerta 1936. On 18 June of that year, in the complex border region between Cepheus and Lacerta, a second-magnitude nova erupted, just one day before a total solar eclipse occurred across the Mediterranean. The nova was first discovered by Gomi, from Tokyo, but was independently spotted by Dr Nielsen observing from the P&O cruise ship SS *Strathaird* in the Aegean Sea. That ship had been chosen by Reggie Waterfield of the BAA for the total solar eclipse expedition and one of the other passengers was none other than Will Hay. I've digressed a bit here but simply to illustrate that where astronomical discoveries are concerned, especially visual ones, there is often an element of luck and coincidence involved. Who can forget the two British Christmas comets of 1960 and 1980, for example? The former discovered on Boxing Day by Mike Candy with a brand-new comet-sweeper and the latter, twenty years later, by Roy Panther on a Christmas Day sweep through Lyra.

As we shall see shortly, when Will discovered the white spot on Saturn in 1933, the planet was far from being well-placed in his London sky. It was 22:35 GMT on 3 August and a gibbous Moon, two days from full, was low down and transiting the south meridian in

Sagittarius. Saturn, at a tree-scraping declination of -17° 30 m, would have been at a dismal altitude of 17° when Will noticed the spot. The planet was in Capricornus, transiting at midnight, when its altitude would only have been a few degrees higher. Such an unfavourable altitude may well have dissuaded other British observers from making an observation and, therefore, the discovery of a lifetime. Planets that low down are invariably bathed in spurious colour and distort wildly as the light passes through such a huge mass of our atmosphere.

Precisely how lucky Hay had been in 1933 was something that I had wondered about for many years, but in 2005, I teamed up with the late and much-missed Ken Goward of the Society for the History of Astronomy to investigate his observing career in more detail. During three years of digging and delving we discovered (with much help from the Royal Astronomical Society librarian Peter Hingley) that Will Hay's observing logbook had recently been handed in to the RAS. This book proved to be a real insight into the amount of time Will had been able to spend observing during the 1930s. The results from much of our research were published in the April 2009 *BAA Journal* (vol. 119, no. 2).

In an era when visual work dominated amateur astronomy, the observing logbook was the universal method in which observations were recorded. Many logbooks were no less than treasured works of art. Essentially every logbook was a diary, in which everything was recorded, sometimes in meticulous handwriting, along with copious notes as to what was seen and the prevailing atmospheric conditions. Sometimes, as in Patrick Moore's numerous logbooks, amusing insights into the workings of Spode's law (in modern parlance, Sod's law!) were noted, such as when cloud intervened, trees got in the way at the worst possible time, or when a problem developed with the telescope. Sadly, in the twenty-first century this very human approach seems largely to have been lost with amusing comments simply replaced with cold, hard captions relating to digital exposures and equipment.

If it were not for Will's energy and perfectionist streak being directed towards his comedy routines surely he would have been one of the greatest British amateur astronomers of his generation, but his entire period of film-making (twenty in ten years!) followed immediately after his white spot discovery and so his subsequent astronomical endeavours were almost wiped out.

GETTING SERIOUS

At some point leading up to 1932, Will Hay seems to have decided to get serious about amateur astronomy. Born on 6 December 1888, Will would have turned forty-four at the end of that year and so he had left it quite late to start making regular observations. However, throughout the previous decade his schedule had been ludicrously hectic. He had entertained live audiences at different venues up and down the country continuously, often moving to another theatre every two weeks. In addition, Will had taken his bumbling schoolmaster 'Fourth Form at St Michael's'-style of act to America, South Africa and Australia. In 1926, he had moved from Ramsgate to Norbury and his new London address had a back garden long enough for an observatory or two. He was also a wealthy man by 1932 and, although he was famously 'tight' with money, he could afford to purchase some serious astronomical equipment. Furthermore, the hobby which had engrossed him since moving to London, namely flying, seemed less of an attraction after a number of fatal accidents (and the death of a friend) had forced him to re-evaluate the safety of aviation. Now based in London, Will was within easy reach of all the main London theatres and the BBC radio studios as well as the Gainsborough and Ealing film studios. So, with the night sky having fascinated him throughout his whole life, and with a stable address, the time must have seemed right to purchase some powerful equipment and keep a proper observing logbook.

ACQUIRING SOME EQUIPMENT

Will Hay's observing logbook opens on 28 April 1932 and it is clear that he already owned some useful equipment, including a 6½-inch reflector. It transpires that he applied for membership of the British Astronomical Association around the same time the logbook was created because, at the BAA meeting of 29 June 1932, held at Sion College on the Victoria Embankment, Hay's name appeared along with five other candidates, on the list of those accepted as members by the BAA Council earlier in the day. He was listed as William Thomson Hay of 45, The Chase Norbury SW16 and the existing members who supported his proposal for membership were none other than Dr W. H. Steavenson and Captain Maurice Anderson Ainslie.

Hay, like so many other BAA members of that era, had sought the advice of Dr Steavenson, when acquiring his equipment. Steavenson (1894–1975) was more than five years younger than Will, and had lost the sight in his right eye due to a childhood accident, but as a teenager he had come within a hair's breadth of co-discovering comet Quenisset in 1911. He was a mere thirty-two years old when he was elected as the President of the British Astronomical Association in 1926, in an age when the Association was dominated by Victorian-era greybeards who were either military men or members of the clergy! Even today his record of being the youngest ever BAA President still stands. By profession Steave was a surgeon and so, technically, was an *amateur* astronomer, but despite this, in 1957, he even became the President of the Royal Astronomical Society. Will and Steave became firm friends, helped enormously by the fact that Steave lived within two miles of Will's Norbury address, at 70 Idmiston Road, West Norwood, SE27, where his gigantic 20½-inch Newtonian was erected.

The first piece of kit that Steave helped Will to acquire was a large 12½-inch f/7.2 Newtonian made by the telescope maker George Calver in 1895 (Figure 1). The instrument, already thirty-seven years into its

Figure 1. Will Hay and his 12½-inch f/7.2 Calver Newtonian in 1933. (Image courtesy of Mirrorpix.)

working life in 1932, had previously been owned by a Mr Paxton and on his death his female relatives had sold the instrument. The telescope is still in use today. In May 1934, Hay gave it to another BAA colleague and medical man, Dr Rowdon Marrian Fry who, via an elderly Dr Steavenson, passed it to Dr Robert Paterson in 1967. More recently, in June 2010, the Newtonian was passed on to another amateur astronomer, Dale Holt. So, 118 years after its original date of manufacture, Hay's big Calver reflector is still in use.

However, only two months after Hay's observing logbook was created, a splendid 6-inch f/15 Cooke refractor appeared for sale in the showroom of Broadhurst Clarkson. On 17 June, just twelve days before he was elected as a BAA member, Hay purchased the refractor, no doubt guided in his choice by Steave himself, whom everyone in that era consulted about telescopes. Even the eleven-year old Patrick Moore would consult Steavenson two years later when he too joined the BAA and purchased his three-inch brass-tubed refractor.

THE COOKE INSTALLED

Will Hay's observing log records that three days after he purchased the Cooke refractor Steave delivered it in person and spent 20 and 21 June 1932 helping Will erect the instrument in the garden at number forty-five. Quite what the state of the refractor's observatory was at that stage is unclear. Hay also had a partially constructed circular-base observatory which housed the 12½-inch Calver. The twelve foot-diameter ring, awaiting a dome, was situated in the middle of the back garden which stretched some seventy feet to the north-east of the house itself. The run-off roof building for the Cooke refractor (completed by the summer of 1933) was located at the north-eastern end of the garden, immediately behind the Calver's partly built observatory. When the run-off roof was fully extended to the north-east it would almost have touched the fence separating Hay's property from the first one along the north-western side of Hillcote Avenue. From studying the details of Hay's run-off roof observatory it appears that the south-western gable folded down in operation to allow better access to the southern sky and low-altitude planets, once the sloping roof had slid away to the north-east. When stowed, the refractor's tube would have lain under the highest part of the roof. The run-off roof observatory, according to

Hay's notes, was designed by him but constructed by a company called Overends. A solid iron plinth some six feet in height and roughly nine or ten inches in diameter was used to carry the Cooke and its equatorial head (Figures 2 and 3).

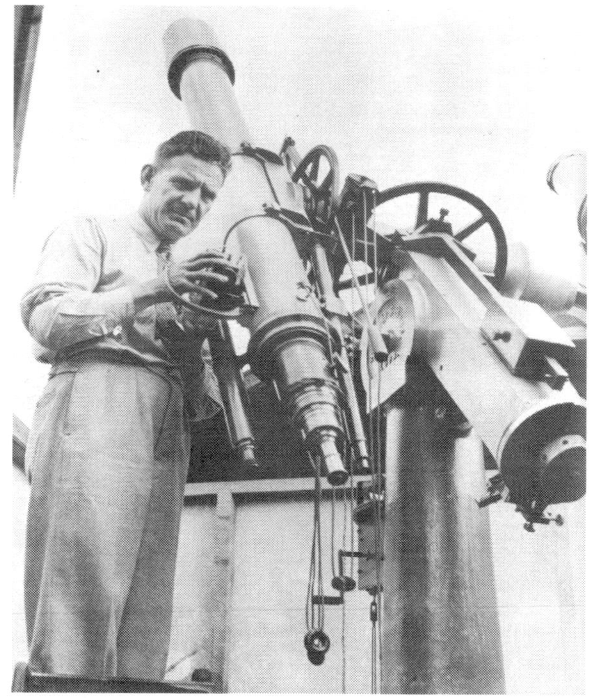

Figure 2. Will Hay, around 1933/34, standing on a chair while adjusting a camera he mounted piggyback on his Cooke refractor. (The original photographer is unknown.)

It is clear when one looks at the arrangement in Will's back garden that the house itself would have seriously restricted the viewing of low-altitude objects to the south and south-west from the Calver's location and may have prevented Saturn being viewed at all in the 1930s. Even from the position of the Cooke refractor the planet would have been a bit of a roof-scraper. Ironically, if Hay had completed the dome for the Calver Newtonian before 1933 it would have partially blocked the south-western view from the Cooke refractor, possibly preventing Saturn being observed that year and thereby changing history!

Figure 3. Will Hay with the Cooke refractor, days after he discovered the white spot on Saturn with that instrument in August 1933. (Image courtesy of Mirrorpix.)

A YEAR OF OBSERVING

On page seven of Hay's observing log he noted down first light with the Cooke on 22 June 1932 in a session lasting from 22h 45m to 00h 30m UT. What did he observe on that first night with his big new toy? Well, he chose objects that would quite possibly be first-light targets for any modern observer getting a new telescope in June, namely: the Ring Nebula M57 in Lyra, the globular cluster M13 in Hercules, the Owl nebula M97 in Ursa Major, and the double stars Zeta Bootis and Otto Struve 371 in Lyra. Finally, he moved on to Saturn and recorded that the Cassini division was strongly marked despite mediocre seeing.

Will's next recorded observing session was two and a half weeks later, on the night of 9/10 July 1932. For this session his logbook records that he once again viewed Saturn with the 6-inch Cooke at a magnification of 300x. He even records the RA, Dec. and altitude of the planet to arc second accuracy in his notes where he records that it transited his London horizon at 01h 08m 00s at an altitude of 18 degrees 36 minutes and 34 seconds. Well, they say he was a perfectionist with his comedy routines and this obviously applied to his observing as well! On that night, more than a year prior to his white spot discovery, he wrote:

> *Image boiling at times but very clear at intervals. The image on the whole was very distinct and Cassini's division was strongly marked. There seems to be a faint division or fine shading in Ring A tho' it may have been an illusion. Ring B seemed to be shaded on its inner margin. The crêpe ring was well marked where it crosses the planet but was also plainly visible against the dark sky. At first I thought this appearance was illusory but it persisted throughout the clear intervals.*

As for any modern-day amateur using a new telescopic toy, Will's first year with the Cooke refractor was quite an active one and, despite his hectic performing schedule and family life (he was the father of Gladys, William and Joan, born in 1909, 1913 and 1917) and the UK weather, he seemed able to observe every few weeks. His name will be forever linked with the planet Saturn but Will observed variable stars, comets and the other planets whenever he could in that first year. He even had a Hilger spectroscope for observing solar prominences. In fact, the log also records that Will occasionally swept the sky for new comets!

It is clear from his notes that Will preferred the 6-inch Cooke refractor to the hefty 12½-inch Calver reflector despite the larger instrument's light-grasp advantage. Precisely why this was Will does not say, but any amateur astronomer reading his logbook can make some deductions. A heavy Newtonian with a seven foot-long tube can be a brute to move around the sky and, even with a rotating top end, it can be gruelling and a bit scary to use when the eyepiece is so high off the ground, if observing objects overhead. By comparison the eyepiece of the Cooke refractor would always have been at a more convenient height. In addition, Hay occasionally makes comments which hint that the mirror was in need of re-silvering, such as when he states he could 'only use x120 due to the bad condition of the mirror'. As the Newtonian still exists today it is clear that there is nothing fundamentally wrong with the Calver's optics and so the state of the mirror coating must be a major reason he so preferred the Cooke refractor. In the 1930s mirror surfaces were silvered, whereas nowadays they are coated with aluminium and a silicon oxide overcoat. A silvered mirror could need re-coating every six months if exposed regularly to the damp air and Will was a very busy man.

THE WHITE SPOT

After using the Cooke refractor regularly for one year, and on the penultimate observing session before Will would go down in astronomical history, namely on 7/8 July 1933, he observed Saturn and described the image as 'boiling', but with the satellites Titan and Rhea being 'very plain'. He also observed the 'Wild Duck' cluster M11, noting that it was 'easily visible despite light sky and Full Moon'. He closed the observing session down at 00:30:00 GMT.

On the night of 1/2 August 1933, two evenings prior to Will discovering a giant white spot on Saturn, there was no hint of what would occur during the next observing session. His notes recorded the following:

> *Observed Saturn and Ring Nebula Lyra with 12.6 Calver Reflector.*
>
> *Observed Saturn with 6" Cooke Refractor. Image good. Titan, Rhea, Tethys and Dione all visible.*
>
> *Observed variable stars T. Herculis and R. Pegasi. T. Herculis brightening rapidly. R. Pegasi very bright. Swept for Comets – no luck. Closed down 1h 10m 00. 1933, Aug 2nd.*

Professionally Will was performing his bumbling schoolmaster 'Fourth Form at St Michael's' act at the Trocadero, Elephant and Castle during the first week of August 1933, so he only had a short six-mile journey home to Norbury. On the evening of Thursday, 3 August, he pointed the big refractor towards Saturn. His logbook records the following historic entry:

> *August 3rd. G.M.T. 22^{H}. 35^{M}. 00. Observing Saturn with 6" Cooke Refractor. I was surprised to see a large bright area in the Equatorial region of the planet and just left of the central meridian. I rang up Dr Steavenson who observed the planet thro' his 20½" reflector and*

confirmed the existence of the bright area. I observed the spot from 1933 Aug 3 22^{H}. 35^{M} until it reached the limb of the planet at 1933 Aug 4th, 00^{H}. 10^{M}. 00. The large bright area seemed to be followed by other bright areas tho' not so conspicuous or well defined as the principal one. I also suspected similar bright patches near the N. Polar region of the planet. There was also a well-defined shading in Ring A which seemed to divide the Ring, and at times appeared to be almost a fine line. I am inclined to think it is an optical illusion caused by prolonged gazing and focusing on Cassini's division. I used power x175 on the 6" Cooke throughout the observation.

So, at 22h 35m on 3 August 1933, Will's place in astronomical history was secured (Figures 4a and 4b). He was already a household name, even though his film era was yet to materialize, but now he was seen rather differently from the confused schoolmaster he portrayed on stage.

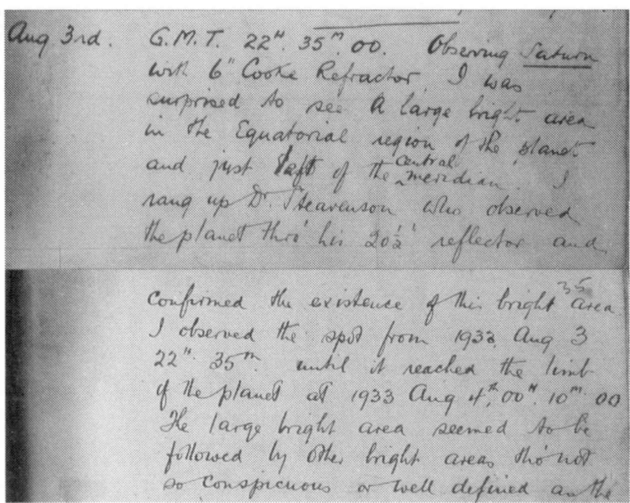

Aug 3rd. G.M.T. 22^{H}. 35^{m}. 00. Observing Saturn with 6" Cooke Refractor. I was surprised to see a large bright area in the Equatorial region of the planet and just left of the central meridian. I rang up Dr. Steavenson who observed the planet thro' his 20½" reflector and

Confirmed the existence of this bright area. I observed the spot from 1933 Aug 3 22^{H}. 35^{m} until it reached the limb of the planet at 1933 Aug 4th. 00^{H}. 10^{m}. 00 The large bright area seemed to be followed by other bright areas tho' not so conspicuous or well defined as the

Figure 4a. Will Hay's handwritten notes on the discovery of the white spot on that fateful night of 3 August 1933.

It was obvious that 'Hay the man' was a scholar, not a buffoon. Newspaper photographers and film crews were soon making a beeline to number 45, The Chase, Norbury, and Hay was rather startled at the

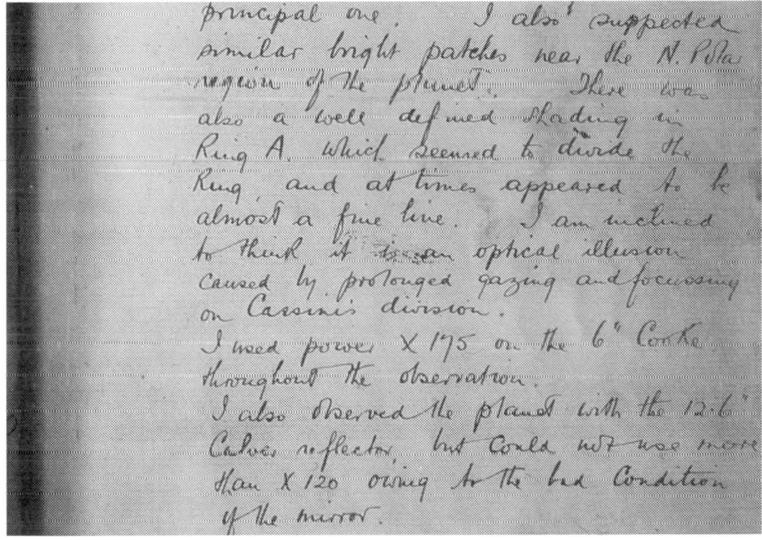

principal one. I also suspected
similar bright patches near the N. Polar
region of the planet. There was
also a well defined shading in
Ring A. Which seemed to divide the
Ring and at times appeared to be
almost a fine line. I am inclined
to think it ~~is~~ an optical illusion
caused by prolonged gazing and focussing
on Cassini's division.
I used power X 175 on the 6" Cooke
throughout the observation.
I also observed the planet with the 12·6"
Calver reflector but could not use more
than X 120 owing to the bad condition
of the mirror.

Figure 4b. Will Hay's handwritten notes from 3 August 1933 continued.

interest, appearing quite bemused, even embarrassed by the publicity. The front page of the *Daily Mirror* carried the headline:

COMEDIAN'S BIG DISCOVERY ON PLANET

and in a smaller sub-heading:

Spot as Clue to Saturn's Mysteries? –
WILL HAY BEATS AMERICA

Cinemas up and down the country showed a British Paramount News 'Exclusive' trailer headed:

'STAR' TURNS STAR-GAZER!
Norbury – Will Hay discovers spot on Saturn!
'Schoolmaster comedian' tells how he located white 'blemish' on planet.

A deep-voiced upper-class British Paramount News commentator spoke the following introduction: 'In real life, Hay was the opposite of his screen character' and the captions disappeared to reveal Will and Steave, immaculately presented in suits and ties with Steave smoking his pipe, inside the walled ring which surrounded the 12½-inch Calver. The six-inch Cooke was in the background and its run-off roof was moved back to reveal the instrument. The film crew had obviously filmed the first shot from an upstairs window of number 45 with the close-up shots filmed at ground level. They had obviously thought that the big Calver was the most impressive telescope, even though the discovery was made with the refractor (Figures 5a and 5b).

Figure 5a. The observatory arrangement at 45, The Chase, Norbury in August 1933. (From an old newsreel movie clip.)

Hay commented, 'There seems to be quite a lot of fuss being made about this spot which I discovered on Saturn last Thursday night,' and in the close-up he explains, 'but this spot, this new spot, allows us to see the globe actually rotating, which it does in about ten and a quarter hours.' The deep-voiced commentator then adds: 'He was an accomplished amateur scientist and engineer and wrote passages into the scripts of his films which his screen character could not possibly understand.'

Figure 5b. Hay and Steavenson alongside the Calver Newtonian just after the white spot discovery. (From an old newsreel movie clip.)

Needless to say, following the discovery, Will embarked on an intense period of observing the spot whenever possible, whatever his entertainment commitments. With Saturn well south of the celestial equator it would not remain visible for many more weeks. In fact, analysis of BAA records shows that Hay submitted more observations of his spot than any other BAA observer (Figures 6a and 6b).

Figure 6a. Will Hay produced a Saturn sketch from his white spot-discovery night observation. (Image by kind permission of the Royal Astronomical Society.)

Figure 6b. Will Hay produced a second Saturn sketch six days after the white spot discovery, showing it had become elliptical. (Image by kind permission of the Royal Astronomical Society.)

The BAA members who submitted observations of the 1933 white spot were: Hay, Steavenson, Phillips, Parr, Waterfield, Hargreaves, Strachan, Butterton (New Zealand) and Ainslie (the Saturn section director, observing from the Cape, South Africa, where he was visiting his daughter).

With the planet only being at a decent altitude above the UK horizon for a short period of time and with its globe rotating every ten hours, even a clear night was no guarantee that the spot could be seen, as it could be on the opposite side of the planet when Saturn crossed the meridian.

Six days after his discovery, Hay's views on astronomy were summarized in an interview with the *Daily Mail* as follows:

> *He believes astronomers to be the finest brotherhood on earth, the only men who see life in its true proportion. 'If we were all astronomers there'd be no more war,' he says. He has the broad, calm philosophy of a man who thinks habitually in terms of universes. All astronomers are philosophers, he will tell you.*

In total Will made fourteen observations of Saturn and two splendid colour sketches of the spot between 3 August and 14 September. He even observed it whilst at an engagement in Brighton on 18 August and from Bournemouth on 13 and 14 September, his last observations of the spot. A few selected extracts from his logbook, detailed below, during August and September 1933 hint not only at his observations, but also at his thought processes regarding the feature. P. E. stands for 'preceding edge', F. E. stands for 'following edge' and C. M. stands for 'central meridian'.

Aug 6th Dr W H Steavenson and I observed Saturn from 21 ^{H} *50* ^{M}*. Dr W. H. S. at the 12" Calver Reflector and I at the 6" Cooke Refractor. Cloudy periods prevented prolonged observation but the white spot was plainly visible and it crossed the central meridian at 22* ^{H} *12* ^{M} *(Dr W. H . S.) 22*^{H} *13* ^{M} *(W. H.). I estimated that the spot was at the 'discovery' position at 22*^{H} *29*^{M}*; which gives a rotation period (based on 7 rotations) of 10*^{H} *16*^{M} *17*^{S}*. The region of the Equatorial Zone following the spot seems to be darker than usual, and discontinuous. Seeing conditions fair. Moon just past full. Sky hazy with cloudy intervals. Image only good at intervals. Powers X 125 and X 175 on Cooke. X 100 & X 175 on 12.6" Calver.*

Aug 9 G.M.T. 00 ^{H} *00* ^{M} *00* ^{S}*. Continued observation on Saturn. The bright spot appeared coming around the limb and the preceding edge seemed less well defined than when first discovered on Aug 3rd. It was rather difficult to determine where the preceding edge of the spot commenced, and I imagined that the main spot was preceded by a smaller bright area from which it was separated by a very faint dark mark. As the whole spot came into view it was clearly noticeable that the following edge was extremely well marked and also that the Equatorial Zone following the spot was very much darker than usual, so much so that the dark belt seemed to extend right up to the Crape ring in one unbroken dark shading. The spot itself seemed to be slightly longer than before and it encroached into the dark belt on its northern edge.*

Aug 9 Observed spot on Saturn.

When I commenced observation at 21 ^{H} *55* ^{M} *preceding edge had already crossed the C. M. The image was not so bright as previous night. The preceding edge of spot was diffuse. The following edge seemed less well defined than at previous observation and the spot seems to be extending & losing its sharp outline. The area immediately following the spot is still dark but becomes brighter farther along in longitude.*

Aug 12 Observed Saturn. Sky very transparent but seeing conditions (for Planetary work) almost impossible. At times the Rings merged into the planet and the whole image was shapeless. I commenced observing at 21 ^{H} 45 ^{M} G. M. T. and I at once noticed that the spot was 'Early'. It was impossible to say accurately when the centre of spot was on the meridian but in the rare glimpses I could get of the spot as a whole I should estimate the transit of C. M. at about 21 ^{H} 48 ^{M}.

The spot has lengthened considerably and the P. E. is almost obliterated. The area following the F. E. is getting brighter. The actual spot, however, still remains conspicuous. I concentrated on the F. E. and timed its transit across the C. M. AT 22^{H} 10. It was certainly past the C. M. at 22^{H} 15^{M}.

Observed M.13 Ring Neb M12, M31. The clusters were brilliant. Have never seen M.13 better with this aperture. 6 Cooke X 175. Closed down 00^{H} 30^{M} G.M.T. Aug 13 1933.

Aug 13. I venture to suggest that the spot may have been caused in the following manner. Assuming that the Equatorial Zone consists of a bright layer which covers a dark layer situated at a lower altitude, a sudden concentration of bright matter towards a centre would probably disclose the dark layer at each side of the centre of disturbance. The concentration of bright matter would account for the abnormal brightness of the spot. As the concentrated bright matter subsided it would cause a lengthening of the spot in one direction or another, and if the subsidence adopted a lateral direction of movement as a whole in addition to the vertical one, it would cause an alteration in the apparent rotation period of the spot measured when the concentration was more or less stationary. The concentration would no doubt occur more rapidly than the subsequent subsidence, and would probably remain stationary for a period. The subsidence would no doubt be slower and spread out in any direction, again covering up the lower dark matter. I suggest that at the moment of its greatest concentration it 'over flowed' over the dark northern belt, giving the appearance of having 'bitten' into this area, as observed on Aug 9th, by Dr W. H. Steavenson, Rev T. E. R. Phillips and myself.

Aug 18 G.M.T. 21 ^{H} 30 ^{M}. Observed Saturn with 3 ½" Cooke at Brighton X 150. Altazimuth Stand. Cloudy intervals. Image fairly clear at times. Spot easily visible and seemed to be extended more than ever in the preceding direction. Following end not so well defined as previ-

ously seen. Estimated that the F. E. was on the C. M. at 21 ^{H} 50 M. II Class observation.

Aug 21st G.M.T. 23 ^{H} 10 ^{M} 00. Observed Saturn. Image not too steady. I imagined I saw a round bright patch on the E. Z. in longitude immediately following the F. E. of the large spot. Clouds interfered with observations and I could not be sure of the existence of this round patch.

Aug 23 G.M.T. 23 ^{H} 15 ^{M} 00. Observed Saturn until 1 ^{H} 00 ^{M} 00 Aug 24th G.M.T. The spot has become very elongated and the preceding end indistinguishable from the rest of the E Zone. The following end is still visible but seeing conditions were not good enough to permit of taking a reliable estimate of the time of its C. M. passage.

Hay's final observations of the white spot were made, while he was on tour, from Bournemouth and were recorded on page fifty-seven of his logbook as follows:

Sept 13. Observed Saturn at Bournemouth with Mr Strachan's Coudé 9" O. G. Also observed M13.

Sept 14. Observed Saturn at Bournemouth with Mr Strachan's 9" Coudé. The F. E. of spot had just crossed the C. M. at 19 ^{H} 10 ^{M} G.M.T.

W. Strachan was a noted BAA observer in the 1930s but died, aged sixty-one, on 16 March 1935, only nineteen months after Hay made his last white spot observation at Strachan's Bournemouth observatory. Strachan had been a cripple since infancy and so his nine-inch Coudé refractor with its fixed eyepiece position was the only instrument he could manage. After Strachan's death the parts of the refractor were bequeathed to Lionel Guest, but he too died shortly afterwards and bequeathed the lens to Captain Ainslie.

1934 AND BEYOND

So, how much observing did Will manage after Saturn and its spot had moved away into the western autumn twilight? Well, sadly, virtually nothing during the next couple of years. Either by sheer coincidence, or because of his added astronomical fame, almost immediately after the white spot excitement had faded (December 1933, to be precise) Will commenced his intense phase of feature film work, starting with serious character acting in *The Magistrate* and continuing in *Those Were the Days, Radio Parade of 1935* and *Dandy Dick*. Of course, his real talent was in comedy, both performing it and writing the scripts too, so *Boys Will Be Boys* and *Where There's a Will* quickly followed. However, it was only when he moved into the comedy that both mirrored his schoolmaster persona and teamed him up with the young actor Graham Moffatt and the dentally challenged 'old boy' Moore Marriott that his films really drew in the crowds. *Windbag the Sailor, Good Morning Boys, Oh, Mr Porter!, Convict 99, Old Bones of the River, Ask a Policeman, Where's That Fire?* and *The Ghost of St Michael's* were hugely popular. Lesser hits like *Hey! Hey! USA!* and *The Black Sheep of Whitehall* drew respectable audiences too. Rattling off twenty films in the ten years immediately following the white spot discovery meant that Will only had one serious rival on the silver screen of the 1930s, namely the young George Formby (1904–1961). In addition, Will's fame, wealth, sense of humour and intelligence attracted the ladies too, so that his marriage to his wife Gladys quickly deteriorated. Where could amateur astronomy fit into that lifestyle? Well, surprisingly, he

Figure 7a. Will Hay attended the 1935 BAA meal held on 24 April at Frascati's Restaurant in Piccadilly. Unfortunately, the only known picture of him at that meeting shows the side of his head, partly covered by a woman (possibly his wife?) staring directly at the camera in the extreme lower left of the image. W. H. Steavenson can be seen in the background just above Hay's head.

Figure 7b. Dr W. H. Steavenson and Will Hay at the 877th RAS Club Dinner on 13 March 1936, aged forty-one and forty-seven respectively. (RAS archive photograph taken by Dr Gerald Merton.)

Figure 7c. The only known decent-quality photograph of Will Hay at a BAA meeting/meal. This was the 1937 BAA meal held on 28 April at Frascati's Restaurant in Piccadilly. Will is looking straight at the camera and, once again, Steave can be seen in the background above Will's head.

did fit the time in now and again and occasionally attended meetings of the BAA and the RAS (Figures 7a, 7b and 7c), invariably accompanied by Steave who, despite being single and having a glass eye, often had a glamorous female companion with him too!

Over the twelve years covered by Will Hay's logbook there are some 124 nights of observations from 1932 to 1944, mostly with the six-inch Cooke refractor. The fourteen nights observing his white spot in August and September 1933, as well as the fourteen months preceding this time, mark Will's most intense period of serious observing. Admittedly, Will had smaller telescopes well before 1932 but he was not contributing observations to the BAA or the RAS prior to that year.

On page 56 of his logbook, dated May 1934, Will notes that a 6½-inch reflector he had owned for many years had been handed over to a Mr Lee and the big 12½-inch Calver was given to Dr R. M. Fry, a prominent BAA member. On the same page he also noted that his six-inch Cooke refractor had been removed from its old run-off roof observatory and installed in the big Calver's former twelve-foot diameter structure which he now refers to as a dome. He mentions that the Cooke's observatory will now be used as a workshop and a laboratory.

The logbook entries are very sparse at this point and it is easy to get confused as to whether the years 1933, 1934 or 1935 are being mentioned. However, after re-examining these pages it is clear that the sequence of events in Will's life, after the white spot era, go roughly as follows:

14 September 1933: final observation of the white spot from Bournemouth.

30 April 1934: Jupiter observed in poor seeing at a magnification of only x115.

May 1934: Cooke refractor moved to the former ring/dome observatory and both Newtonians disposed of.

8 June 1934: Jupiter observed in fair seeing at x175. (This was one day after the first screening of the film *Those Were the Days* in which Hay played the character Brutus Poskett).

Sadly, the brief use of the Norbury back garden's rearranged observatory would not last long and there is then a fifteen-month gap in

Hay's logbook with no entries at all from 8 June 1934 to September 1935. During this time Will would have been acting in three films, namely *Radio Parade of 1935*, *Dandy Dick* and *Boys Will Be Boys*. He was also continuing with his live performances and with radio broadcasts. Remarkably, in that same hectic year of 1935, his book *Through My Telescope* was published by John Murray, copies of which still change hands for a hundred pounds today, somewhat more than the 3s/6d price (17.5 pence) when new! Will had originally written the book around the time he first acquired the six-inch Cooke refractor, but hesitated to publish it at first in case it might be thought that he was seeking to advertise himself as a professional astronomer.

With his fame and wealth increasing exponentially, and his marriage deteriorating at the same rate, Will decided to buy a larger house just for himself, namely, The White Lodge at 6 Great North Way, Hendon, North London. Nowadays this property is surrounded by trees to shield it from the Great North Way dual carriageway where it passes Sunny Gardens Road. A footbridge passes over the Great North Way just to the east of the house. Will and his wife Gladys were already separated by a deed of separation and this progressed to a judicial separation in late 1935, the year his youngest child reached the age of eighteen. Will's reputation as 'a ladies' man' was now being reported in the newspapers of the time which probably stopped him being knighted.

Hay's logbook notes: 'Observatory dismantled September 1935'. Just below this entry Hay records that on 20 October 1935, the observatory was 're-erected at Gt. North Way Hendon' and '6" Cooke re-installed and adjusted'. Saturn was observed from the new observatory over the following nights with 'Seeing fair to very good'. The only other observation in October 1935 was an entry made on the 29th where he records further observations of Saturn and 'Seeing fair. Sky clear. High Winds. Good night for faint objects, Orion Nebula extremely good'.

What the logbook does not mention is that Hay also installed the chorus girl Peggy Bradford, in his new Hendon home in October 1935, along with his 6-inch Cooke! She would soon become his secretary.

The dome at the White Lodge was easily visible to anyone travelling along the Great North Way (Figure 8) and it appears to have been a much more substantial building than the circular Norbury silo shown in 1933 press photographs, despite the fact that his observing notes hint it was the same basic structure he had designed and erected at his previous address.

Figure 8. Will Hay outside his substantial dome at 6 Great North Way, Hendon, which housed his Cooke refractor. (The original photographer is unknown.)

One of the visitors to Hay's Hendon home was the stand-up comedian Tommy Trinder (1909–1989) and his tape-recorded comments have been replayed many times in BBC radio programmes and articles about Hay, specifically:

'One of the things he used to love to do was to take you to his home, and out in the backyard in Hendon he had this telescope; he was so enthusiastic about it all. He used to get you to look through this spy hole. All I ever saw looked like a steam pudding and Bill used to go into raptures over it, so much so he sounded more like Fanny Craddock than an astronomer!'

In the gaps in his relentless film work Will was able to make valuable observations of a number of astronomical objects from the new Hendon base and he reported them to the RAS and the BAA. He made particularly valuable photographs, observations and measurements of Comet Peltier 1936 K1 (originally designated 1936 II or 1936a) and Comet Van Gent 1941 K2 (originally designated 1941 VIII or 1941d).

The astrometric measurements of the latter comet were described in the *BAA Journal* along with the description of Will's cross-bar micrometer. A photograph of Comet Peltier (1936 K1), taken by Hay with his five-inch Zeiss triplet lens, even appeared in the national press (Figure 9).

Figure 9. An offset guided photograph of Comet Peltier 1936 K1 (1936a) from p. 78 of Hay's logbook. The exposure was made on 25 July 1936. The comet reached third magnitude. Hay used his five-inch aperture f/5 Zeiss lens, piggybacked on to the Cooke refractor, for his comet photography. (Photograph by kind permission of the Royal Astronomical Society.)

ECLIPSE CHASING

In March 1936, the BAA Mars Section Director, Dr Reggie Waterfield, described to members at the monthly BAA meeting his plans to travel to that year's 19 June total solar eclipse, the track of which, conveniently, started in the calm seas of the Mediterranean. Waterfield served on the RAS/Royal Society 'Joint Permanent Eclipse Committee' and the 1936 totality would be his third successful campaign. His plan involved sailing on a cruise ship, the P&O lines SS *Strathaird*, to the Island of Chios in the Aegean Sea. Hay signed up for the cruise, which fitted chronologically between the release of *Boys Will Be Boys* and filming for *Windbag the Sailor*. Whether the cruise ship experience had any bearing on that latter film I know not!

The cruise was a huge success, attended by many of the wealthier BAA members. As mentioned earlier, one of Hay's fellow travellers on that trip, Nielsen, made an independent discovery of the second-magnitude Nova CP Lac while on the ship. A photograph of totality taken by J. L. Haughton, on the same expedition, appeared in the December 1936 *BAA Journal* (vol. 47, no. 2) and it is reproduced here (Figure 10). A very similar picture appears in Hay's logbook, but the Haughton print better captured the reflection of the corona in the sea.

Figure 10. A photograph of the 1936 June total solar eclipse taken by Dr J. L. Haughton, who was one of Hay's colleagues on that eclipse cruise. (From the December 1936 *BAA Journal* (vol. 47, no. 2).)

Will Hay gave a brief presentation at the 25 November 1936 BAA meeting during which he showed a 16mm-format cine film of the 19 June eclipse, shot through a 150mm focal length f/4.5 lens. The film included scenes on board the cruise ship and must have been watched by a fascinated audience. At that same meeting Will delivered a talk on the application of a synchronous motor to his observatory chronograph. His observing logbook incorporates a dramatic photograph of lightning he took on the night following the Mediterranean total solar eclipse.

EVEN MORE FILMS

Following Will's successful eclipse trip and the release of *Windbag the Sailor*, his cinema success continued unabated. Audiences now craved more films with Will, Moore Marriott and Graham Moffatt as the three bumbling incompetents who somehow 'came good' in the end. In *Good Morning Boys*, released in April 1937, Hay again resumed his schoolmaster role, performing as Dr Benjamin Twist, but even this popular film would be eclipsed by the success of *Oh, Mr Porter!* released in October 1937. In this film, a classic of the years prior to the Second World War, Hay becomes the luckless stationmaster of a dodgy Irish railway station named Buggleskelly, in which his co-stars, Marriott and Moffatt, are on the fiddle. Somehow, despite their sheer incompetence, they foil a group of Irish terrorists! Despite the railway station being portrayed as being in Ireland it was actually filmed at Cliddesdon near Basingstoke. This would be Hay's most popular film and although *Convict 99*, released a year later, was also very popular, it could not match the previous film's success. The same goes for *Hey! Hey! USA!*, released in 1938, where Will acted without his two popular stooges. By this time Hay had a permanent girlfriend, the young Norwegian dancer Randi Kopstadt.

It was clear that although Will was highly popular with audiences, they only flocked to the cinema when the Hay/Marriott/Moffatt team were in the film together. So, the trio reformed for *Old Bones of the River*, *Ask a Policeman* and *Where's that Fire?*, which were all released in 1939. I have watched all of Hay's films that are available on DVDs and even watched *Where's That Fire?*, screened in front of a large audience at the British Film Institute when Graham Rinaldi launched his excellent biography of Will and kindly invited me along. That experience was quite an eye-opener. It will sound ludicrously obvious when I say this, but comedy films of the 1930s were designed to make cinema audiences laugh! 'That's a no-brainer' I hear you cry! Yes, of course, but it was only hammered home to me when I saw *Where's That Fire?* while sitting in an audience with several hundred other people. I had watched all the other popular Will Hay films on a DVD at home and, yes, they were amusing, and strangely infectious, with a few laugh-out-loud moments. However, played in front of a huge group of fans, they are totally different; cinema audiences laugh hysterically and in unison at well-timed gags and Will was the master of timing.

Another thing worth mentioning is possibly the only moment in any of Will's films that he actually mentions astronomy. It is in *Old Bones of the River* and the conversation goes like this:

PROF. TIBBETTS/WILL HAY: Anything wrong?

CAPT. HAMILTON/JACK LIVESEY: Well, nothing serious, but I wish in future that when you're teaching the native children you'd leave out references to the Sun, the Moon and the stars.

PROF. TIBBETTS/WILL HAY: What, no Astronomy?!

CAPT. HAMILTON/JACK LIVESEY: Yes. Well, I'd have to take you too literally. You know, you told them that the earth revolved from West to East.

PROF. TIBBETTS/WILL HAY: Well, so it does.

CAPT. HAMILTON/JACK LIVESEY: Yes, I know, but look at the result. This morning my Hussars stopped in their barracks waiting for the parade ground to come round to them!

PROF. TIBBETTS/WILL HAY: Hahahahaha! That's funny. That's ridiculous, isn't it? Why, their barracks are in the West.

Just before Capt. Hamilton collapses from his fever on to his desk he then repeats to Will Hay's character: 'And remember, in the future, no Astronomy . . .' CRASH!

THE WAR YEARS

Of course, the world of all those living in the British Isles changed dramatically in September 1939 with the declaration of war with Germany;

and the subsequent harder times, along with the bombing of London, affected film production. Nevertheless, Will's film career continued, even if Moore Marriott and Graham Moffatt no longer played a part in it. Between 1941 and 1943 five more films with Will Hay in the starring role were issued as well as a short wartime public information film about extinguishing incendiary devices entitled *Go to Blazes*, in which Will's daughter was played by a young Thora Hird. Those five films were: *The Ghost of St Michael's*, *The Black Sheep of Whitehall*, *The Big Blockade*, *The Goose Steps Out* and *My Learned Friend*. However, it was during the filming of *My Learned Friend* in late 1942, when Will was approaching fifty-four years of age, that he started feeling ill. He suspected the early warning signs of rectal cancer, which had taken his own father's life, and he was right. For most of 1943, Will was most unwell, even after an operation to halt the cancer early on. After a brief comeback on BBC radio in 1944 his health would take its penultimate nosedive.

Not surprisingly, Will's astronomical endeavours suffered greatly during this time. You needed to be fit to observe in the cold British nights and to push a six-inch refractor around the sky. In addition, during the Blitz, Will had removed his refractor from the dome, for fear of it being damaged in an air raid. Will had on one occasion told BAA colleagues that the skin of his dome was nothing more than 'unbleached linen treated with aircraft dope brushed well into the fabric and painted with protective paint . . . the whole dome cost £5!'

Will's observing notes reveal that his use of the Cooke thinned out as the Second World War approached and the last observation in his observing log, with the six-inch Cooke refractor, was made on 1 May 1942, where he merely recorded having observed Jupiter without further comment in this, his last year of good health. Regular observations of any kind appeared to end in November 1942 and the final two entries are in April and May 1944.

THE POST-WAR YEARS

In late 1945, with the Second World War finally over, and during a brief period when he was feeling physically stronger, Will was elected to serve on the BAA Council. However, a year later, he suffered a major stroke and the right side of his body and his speech were seriously affected; it became increasingly difficult for him to use any of his astronomical equipment. In 1947, after a six-month period of recuperation in South Africa, Will returned to London. His health was still poor and, after a further stroke, he instructed two friends, in 1948, to clear out his Hendon home so that he could move to somewhere smaller. His surviving observing logbook was eventually passed to the RAS in the first years of the twenty-first century, by the former London neighbours of Will's daughter Gladys (a Mr and Mrs Piper).

Hay died of a third stroke in his Chelsea flat in the early hours of Easter Monday 1949 (18 April) aged just sixty. Only three days earlier he had given a speech at a meeting of the show business charity organization *The Water Rats* and he had arranged a date with some friends at *The Savoy* on the day he died. He never made the appointment. Nine days later his death was announced to the BAA by the President, Dr Guy Porter, at the 27 April 1949 ordinary meeting, following the brief annual 'conversation meeting' and exhibition in the upper library at Burlington House, Piccadilly. The President's comments were recorded:

> 'Members will have heard with deep regret of the passing of Will Hay. William Thomson Hay became a member of our Association in June 1932 and his ability as an astronomer will be known to you all. Much prominence has been given to his discovery of the white spot on Saturn in 1933 and his descriptions of micrometers and a chronograph and other devices for observational work will be found in our journal. A painstaking observer, he was always ready to help those less experienced than himself; and, indeed, those of us who were privileged to have known him will remember him best in that way – as a great friend of all who were interested in astronomy.'

THE FATE OF THE SIX-INCH COOKE REFRACTOR

The destination of much of Will Hay's astronomical equipment is easy to trace. As already mentioned, the 12½-inch Calver Newtonian passed to Dr R. M. Fry, Dr Robert Paterson and Dale Holt. Some smaller items were bequeathed to the BAA and are still held today by the Association's curator of instruments. However, until recently, mystery surrounded the fate of the six-inch Cooke refractor that gave

Will his greatest scientific moment with the discovery of the white spot on Saturn in August 1933. Surely, someone must know where it ended up? Well, until 2011, no one I had contacted had a clue, but as I had advertised this mystery widely, I hoped someone, eventually, would come forward . . . and then I heard from a Mr David Wallis.

According to David, in 1949, when he was eighteen and an apprentice in the instrument trade, he assisted in dismantling Will's six-inch Cooke refractor when it was removed from his house in Hendon. The telescope was purchased by Charles Baker Ltd of 244 High Holborn and was put up for sale in their showroom, eventually being sold some months later, for £600, to a person who lived at Beaulieu in Hampshire. David spent some time overhauling the instrument before its delivery to the Beaulieu customer and was then sent with a more senior engineer to install the telescope in a purpose-built dome, located in the garden of a very large property, which had a walled garden. Unfortunately, after more than sixty years, David was unable to remember the address and, as he had travelled in the back of a windowless van, the route was invisible to him. Since establishing this new lead, I have explored various lines of enquiry to try to determine who the owner of the 1949 Beaulieu observatory actually was. BAA records have been searched and the University of York Borthwick Institute instrument archives department has been contacted. In addition, an advertisement was placed in a local Beaulieu newspaper and some very helpful Beaulieu historians (Tony Norris and John Beaumont) have given their time generously trying to locate a local astronomer from that era. At the time of writing, in early 2012, I still await a breakthrough, but I live in hope!

I leave the reader with a parting thought. If Will Hay had been in better health and had lived just another eight years, he could have been a guest on Patrick Moore's programme *The Sky at Night* in 1957. Maybe in a parallel universe somewhere it actually happened . . .

The Supernova That Won't Go Away

DAVID A. ALLEN

First published in the *1991 Yearbook of Astronomy*. Revised in June 2012.

On 23 February 1987, my life was changed as surely as if a volcano had erupted beneath my house. The event – the eruption itself – was, need I remind you, Supernova 1987A. In common with many Southern-Hemisphere astronomers, I set aside much of the research I had been engaged in so as to study this unique and exciting event. The narcosis of the first naked-eye supernova in almost four centuries floated me into areas of science that were totally foreign to me. I was challenged to take data of a quality that befitted their likely useful life of several centuries; I was challenged also to understand what those data meant, what secrets they bore of the inner workings of one of nature's most dramatic phenomena.

Still basking in the euphoria of a supernova in my own lifetime, I wrote of my experiences in the *1989 Yearbook of Astronomy*. There I tried to describe the frenetic early weeks, the succeeding hectic months, and then the gradual slowing of the pace. I ended with the comment that I was, at that stage, beginning to resume the research I had had to set aside. How wrong could I be? How naïve? Now, two years later, I come before you again, humbly, to correct that misinformation.

It had seemed a reasonable claim. Supernova 1987A was the explosion of a giant star, an event that propelled several times the mass of the Sun outwards at velocities approaching one-tenth that of light. The ejected gas gave out the light we recorded, and expansion adjusted its physical and chemical state. That adjustment was rapid at first, for the gas was dense and its constituent atoms jostled one another. But after a while the density fell due to the expansion, atoms scarcely met one another, and the pattern became set. Initially, measurements made on successive days could differ considerably; but as I draft these words, three years after the eruption, there is scarcely any change from one month to the next, other than the inexorable halving of the supernova's brightness in response to the radioactive decay of the cobalt it produced. Why, then, does SN1987A still command so much of my time? Why does adrenalin still flow every time I observe it?

Basically, because it has not simply faded into oblivion as I had expected. In retrospect, I can see that the first hint of its persistence came with the announcement, in March 1988, that light echoes had been found around it. To understand fully the implications of this it is useful to introduce just a tiny bit of quite simple physics.

ECHOES IN LIGHT

Just like sound, light can bounce off suitable objects. If you clap your hands in front of a distant cliff, you hear first the sound of the clap, and then its echo. The latter arrives later because that sound has taken a longer route. In the same way, if you were to let off a flash of light, you would witness the reflection from the cliff some time after you saw the initial flash. In this case the effect of the light echo is to make the cliff briefly brighten, and the experiment is best visualized taking place at night.

Because light travels so fast, the interval between the flash and its echo would be too small to detect without sensitive equipment, being only a few millionths of a second. In the vastness of space, however, where distances are measured in thousands or millions of light years, a delay of a year or more is perfectly possible – and it is not only possible, it is actually happening right now.

Imagine, if you will, a cliff near the supernova, as my cartoon (Figure 1) shows. Of course, there aren't really vast rock walls around stars, and my 'cliff' will turn out to be rather less solid; the light from the explosion of SN1987A travelled directly to us, covering a distance we believe to be about 170,000 light years. For simplicity, I'll assume the number is exactly 170,000. Suppose the cliff lies two light years away from the supernova, to the side. Then light from the flash would have travelled for two years to reach the cliff and for a further 170,000 years from the cliff to us. It would therefore arrive two years after the flash of the supernova itself. The supernova would appear and fade, to be followed two years later by a fainter flash from the direction of the cliff. A light echo, in fact.

Figure 1 shows a cloud too. Let's say that the cloud lies six light years above the supernova and eight light years in front. If I have correctly recalled Pythagoras's famous theorem, this puts it ten light years from the supernova, and 169,992 light years from us. The light would thus travel for 10 + 169,992 = 170,002 light years before it reached us, again appearing two years after the supernova. We would see the cliff and the cloud light up at the same time, even though they are in very different places. I could choose many spots to place a cliff, a cloud, or the like so

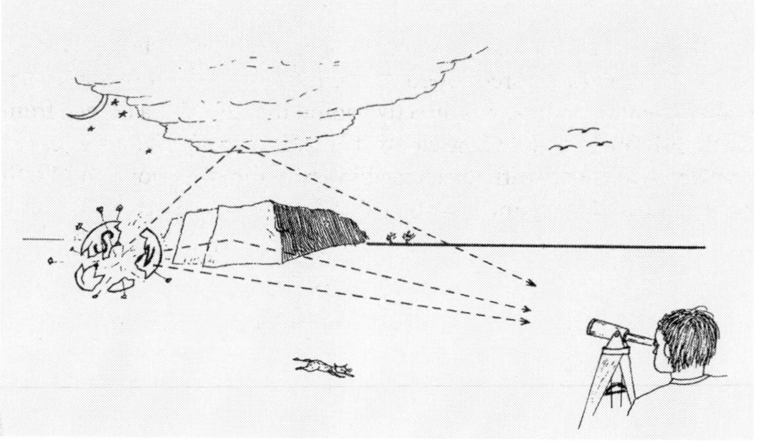

Figure 1. This cartoon shows how light echoes travel to a distant observer. (Diagram courtesy of David A. Allen.)

that the light echo arrived two years late, in early 1989. An obvious spot is one light year directly behind the supernova: the light then travels away from us for twelve months, bounces back to pass the supernova after a further twelve months, and only then begins its 170,000-year pilgrimage.

I have had to make the second object a cloud for one simple reason. If I had put a cliff there, it would have lain almost between us and the supernova. The far side, facing the supernova, would have been lit up, but the face we see would have remained dark. However, light passes through a tenuous cloud by a process given the technical name of forward scattering. When a cloud is backlit, the light rattles off the little particles and continues towards us. The light echoes I am about to describe arise in clouds of gas and, more importantly, tiny grains of dust that scatter light very efficiently.

I may have misled you by portraying the supernova as a flash of light. It actually brightened steadily from late February until mid-May 1987, and then began a slow decline. Each echo must do the same thing – in this case, two years later – thus a view twenty-seven months after the outburst, in May 1989, would reveal the supernova rapidly fading, whereas the echoes would all be at their brightest, and fairly constant, reflecting the light output of SN1987A two years previously. Another year later, supernova and echoes each would have faded, but neither would have gone out.

I listed three points from which we can receive an echo delayed by two years, and these are plotted in Figure 2, a diagram of somewhat higher scientific rigour. A is directly behind the supernova as seen from Earth; B is the cliff and C the cloud. These lie on a slender curve called an ellipse (for those with a technical interest, the supernova and Earth occupy the two foci). Indeed, any object capable of reflecting light and

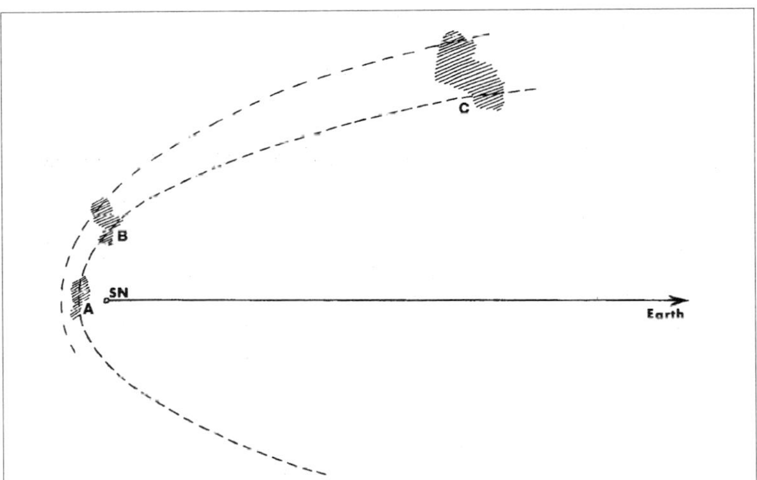

Figure 2. The inner broken line shows the location of any cloud capable of sending an echo back to Earth two years after the eruption of the supernova. As time passes, the broken line expands: the outer line shows the situation one year later again. (Diagram courtesy of David A. Allen.)

lying on the ellipse would also send back a light echo two years after the event. I have compressed on to a flat piece of paper what is actually a three-dimensional situation, and you should envisage not an ellipse but the cigar-shaped surface made by spinning this diagram about its long axis, the centre line that joins the supernova and Earth. The ellipse is not a real, visible thing. It is merely a set of places in space with the common property of introducing a two-year delay into the path of Earth.

Consider now the view from Earth. Obviously, position A lies directly behind the supernova so it is actually hidden, at least until the supernova fades. Position B appears off to one side, by about 2.4 seconds of arc – the angle subtended by two light years at the distance of SN1987A. Position C is six light years to the side, so it appears three times as far removed. Indeed, the greater the distance *in front* of the supernova that the reflecting object lies, the further to the side of the supernova will its echo appear on the photograph of the region.

This is a tremendously valuable fact. Normally, when we study photographs of the sky they appear flat: we are beguiled into thinking that everything lies side by side at the same distance from Earth. In reality, of course, the stars, gas and other features are spread out enormously along our line of sight. In most cases we cannot tell which are in front or behind, or by how much. In the case of the light echo we can. We have only to measure how much to the side of the supernova an echo appears, and we can determine how far in front of the supernova it arises. Only the date is needed – or rather the time since the light left the supernova. For those who like formulae, the distance D in front is given by: $D = 0.34d^2/t$ light years. Where d is the angle in seconds of arc between SN1987A and the echo, and t is the time in years since mid-May 1987, the epoch of greatest brilliance.

As time passes, then, the echoes change. The dotted line in the diagram shows the situation after three years have elapsed. The ellipse is now wider, so that the echo of cloud C appears further from the supernova. At this time there remains some faint echo light from that part of cloud C that was bright the year before. Observations made two and three years after the eruption, therefore, would show us the exact location of cloud C at the two points where the ellipses cross it. A series of such observations would map out the entire shape of cloud C in *three dimensions*, something that cannot be done in any other way. Gradually, every portion fades away, but so long as there is still a dust cloud for the current ellipse to cross, a bright echo will persist. This way the memory of the supernova can linger on long after the event itself becomes unobservable.

RINGS

It was never my intention to become involved in the light echoes. I was content to study just one facet of the phenomenon, one which would cease about the end of 1990 when SN1987A became too faint for the Anglo–Australian Telescope (AAT). Thus when our photographic expert, David Malin, produced the first picture of the echoes to be taken with the AAT, I looked it over with only casual interest.

As I knew from the previous announcements, the photograph, reproduced here as Figure 3, showed two rings of light around the supernova. But David is a clever chap. Whereas others had managed to photograph only partial rings, David had revealed complete circles of light. He had done this by subtracting away the stars and nebulosity that conceal and confuse the region. He hadn't actually obliterated the stars from the sky, you understand. He merely used an older photograph taken before SN1987A appeared, and by superimposing a positive of that one and a negative of the recent plate, he had made the unwanted bits cancel out fairly well. You can see the stars as white blobs surrounded by dark rings, while the nebulosity, which is quite bright hereabouts, has virtually vanished altogether. This pre-supernova photograph has been one of our most valuable assets.

I knew, too, when I first saw the photograph, what the two rings meant. Had the dust grains filled an extensive region in front of the supernova, they would have produced a large haze of light. The rings told us that most of the gas and dust had been tidily swept into piles, thin sheets that lay between us and 1987A. If you could set off from the supernova towards Earth, you would travel through clear space for about three hundred light years before coming to a smoggy region not more than a few tens of light years thick, and responsible for the inner ring. Continuing onwards, you would traverse clear space again until you were almost a thousand light years out, when you'd encounter another patch of smog: the outer echo.

As I glanced at the photograph, I was immediately struck by something odd. Take a moment to look yourself: do you notice anything?

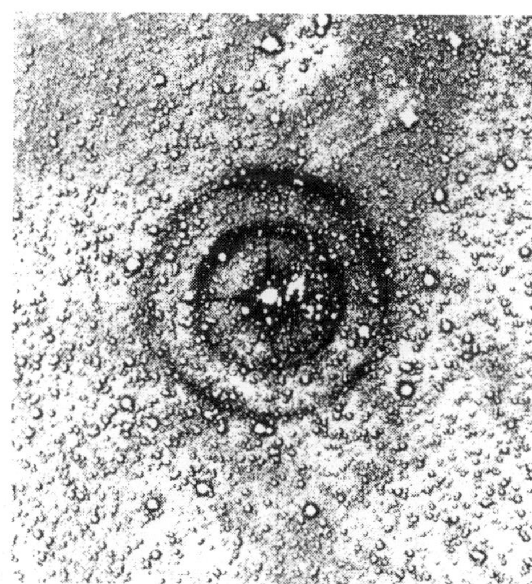

Figure 3. This is the difference between two photographs; features on the more recent one appear in the negative. In theory, the stars should have been cancelled out completely, but due to different atmospheric turbulence they were larger on the newer photograph, so have yielded dark rings with white centres. SN1987A is the black object in the middle, with stars superimposed in white. The central of these is the star that became the supernova. Like all bright stars, SN1987A has the four 'diffraction spike' arms, this time seen in black. The two dark rings are the light echoes. The photograph was taken on 15 July 1988, 507 days after the supernova outburst was first seen. See also Figures 4 and 5 below. (Image courtesy of David Malin and the Anglo–Australian Observatory.)

Here's a clue: the banks of dust should make circular echoes that have the supernova at their centre. Look again and you will see that the outer echo is off-centre. The gap between it and the inner ring is greater to the upper left than to the lower right.

I went home and began some complicated algebra. It took me several hours that night before I had worked it all out. I ended up with the equation given in the previous section, though this is actually a gross simplification of the true situation, a simplification made possible only because of the extreme distance of both SN1987A and the dust it illuminates. My algebra showed that the two clouds lay a very long way in front of the supernova – about 400 and 1200 light years.

The algebra had been worked out before. And I could have just dug through the literature instead. But deriving it for oneself gives a much better understanding of what is going on. In this case, I quickly realized that the outer echo is off-centre because the sheet of dust is tilted. The upper left portion is further in front of the supernova than the lower right by about 150 light years. From even a single photograph we were learning about the three-dimensional distribution of otherwise-invisible material in our neighbouring galaxy, the Large Magellanic Cloud.

Warrick Couch was also excited by the photograph. He, David and I began a proper analysis, as well as an observing programme, to follow the changes with time and so to map out the dust completely. In a scientific paper based solely on that first photograph we concluded that we are probably seeing the front and back faces of a gigantic bubble blown by a collection of hot stars that lie to the upper left and in front of SN1987A. Together the stars form the cluster known as NGC 2044, discovered by Sir John Herschel during his survey of the southern sky from South Africa. Hot stars can blow such bubbles either by a steady shedding of excess gas or by the eruption of one of them as a supernova. In either case, the outflowing material pushes the surrounding matter ahead of it like a snow plough, building up an ever denser ripple as it goes.

We now have several photographs, of which the best three are reproduced to the same scale in Figures 3, 4, and 5, and show how the rings have expanded. Some years must pass before we can show for certain that they are indeed two cuts through a single bubble. In the interim the photographs reveal plenty more of interest. Note how both rings have become split into two around one side. This seems to suggest that one bubble has been expanding into another one from a slightly different centre. The last photograph also shows a fainter, partial ring outside the other two to the right. The sheet responsible for this echo appears to be even more tilted, and we don't know its origin.

There are blobs appearing between the two rings as well, so the bubble isn't totally hollow, One of these, to the lower left, became for a while just about the brightest echo feature, though it is only small. The brighter the echo, the more dust grains there are in the cloud where it originates. So this cloud is particularly dense. We have no idea why such a cloud resisted the snow-plough action.

If you scrutinize Figures 4 and 5, you will also see a rather blobby echo immediately around the image of the supernova. There was no way to see this on the first photo because the supernova itself was still too bright, producing the burnt-out black circle in the negative. The fading of SN1987A has at last enabled us to see material in close

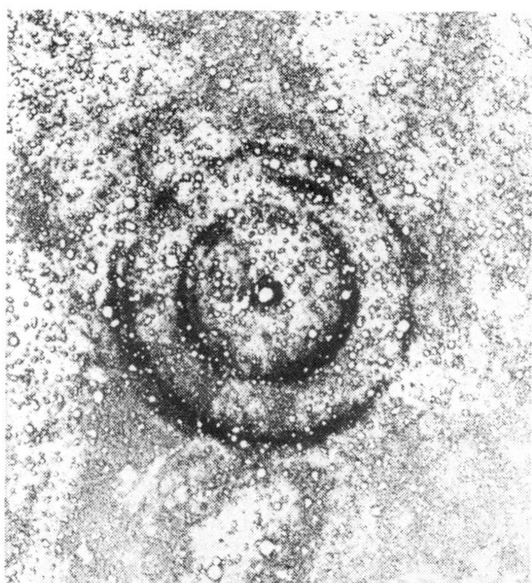

Figure 4. A photograph of the rings, made in the same way as Figure 3, dated 6 February 1989, 713 days after the supernova first appeared. (Image courtesy of David Malin and the Anglo–Australian Observatory.)

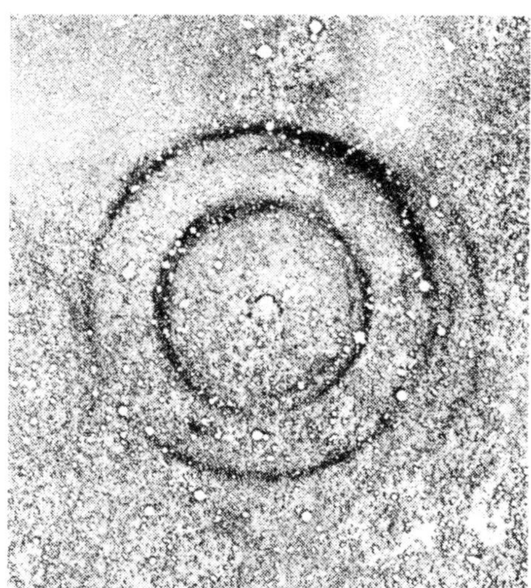

Figure 5. A similar photograph to Figures 3 and 4, dated 21 November 1989, 1001 days after the supernova first appeared. (Image courtesy of David Malin and the Anglo–Australian Observatory.)

proximity. It remains a difficult study, a situation that should improve as the supernova continues to fade. There is intense interest in the immediate environs of SN1987A, primarily because it can tell us something about the history of the star before its catastrophic demise.

According to the best theoretical accounts, the star that became Supernova 1987A had a complicated life story. It started massive and hot. As its hydrogen fuel became exhausted, and different nuclear processes took over, the star had to adjust itself, growing huge and cool. In that supergiant phase it slowly shed its outer layers – something approaching ten times the entire mass of the Sun – and consequently grew smaller and hotter once more. This second hot phase was accompanied by a bubble-blowing outflow like that of the stars in NGC 2044, so that the gas shed as a supergiant was ploughed into a small, hollow bubble around the star. Before the bubble attained great size . . . *bang!* Supernova. The material shed in the supergiant phase at least partially accounts for the echo features closest to the supernova.

THE ULTRAVIOLET FLASH

Theorists also tell us of an ultraviolet flash. If you were standing near a star that turned into a supernova, the first hint you would have of the inner catastrophe would be when a violent ripple progressing outwards through the star reached its surface. Should you ever find yourself in this predicament, ensure you have strong sunglasses and a liberal coating of suntan cream, for the manifestation is a burst of ultraviolet radiation lasting maybe as much as one hour, and many million times more intense than sunlight.

Of course, the UV flash has never been seen, because supernovae give us no advance warning, so nobody has ever been watching at the critical time. Anyway, it's all too far into the ultraviolet for our eyes. There is therefore considerable interest in verifying the UV flash by indirect methods.

Being first out, the UV flash leads the expanding light echo. We might instead be able to record extra ultraviolet radiation at the outer edge of the echo rings, but for one fact. Ultraviolet is very readily absorbed, so that it would almost certainly fail to penetrate the wall of gas close in to the supernova. The interest is therefore intense in studying that innermost echo.

The ploughed-up supergiant gas suffers a violent case of sunburn as it basks in the UV flash. Ultraviolet ionizes gas to create glowing nebulae like the famous one in Orion. Once ionized, the gas fades only slowly, taking perhaps a decade to unwind. This is a different type of light echo, but an echo all the same. In my initial analogy you can liken it to a cliff daubed with fluorescent paint. We are now watching the slow fading of the paint.

As the gas fades, the radiation it emits tells of the physical and chemical state and is amenable to study by spectroscopy. The ionized gas around SN1987A was discovered even before the outer echoes became visible, and has become much easier to study as the supernova faded.

INFRARED SPECTRA

Spectroscopy of the ejected material has, from the outset, been the optical astronomer's most valuable tool to understand the entire phenomenon. I teamed up with Peter Meikle and Jason Spyromilio from Imperial College in London to pursue infrared spectroscopy. Peter is the driving force behind the study; a quiet, competent physicist who contrasts so dramatically with Jason, young and flamboyant, that the three of us make quite a comedy act when together at the telescope. This appearance belies a deal of hard work, particularly by Peter, Jason, and students at Imperial College, to interpret the excellent data we have garnered.

Infrared spectroscopy of supernovae was a new field, and we were lucky to find that some of the most important clues to the entire supernova phenomenon lay there, awaiting our study. Like eighteenth-century explorers, we charted unmapped terrain and made exciting discoveries. As the nineteenth century progressed, the oceans and land masses of the world became known so that exploration lost much of its romance; in the same way the supernova settled down to become predictable as it began to fade from our view. In this endeavour, therefore, the frenetic pace eased.

Or so I thought, two years ago. But I had failed to consider the ultraviolet flash and its attendant echo so, just as the subject had become routine, we began to see a spectral feature whose intensity remained constant as SN1987A faded. With each observation this feature seemed to rise from the ashes of the star, and to dominate the spectrum more.

The feature was caused entirely by helium. And it, too, proved to be another vital clue to the supernova. I have not the space to enter into the rather complex physics of the situation. Suffice it to say that three months of our research went into understanding what was going on, and we emerged with clear evidence of two important pieces of information.

First, the supergiant phase didn't generate a spherical bubble of gas. Instead, the star chose to throw off material in a narrow spray on one side, or perhaps in two opposite directions. There is mounting evidence that this so-called bipolar pattern is not at all uncommon in many celestial situations, but astronomers argue about how it comes to pass. By finding an example in this star we have eliminated some of the possibilities. The orientation of the outflow parallels the elongation in the supernova's ejected material that has been recorded by several techniques. Somehow SN1987A knew its precursor had a favourite direction in space, and preferred that direction despite the star's total devastation.

Second, we were able to measure the mix of hydrogen and helium in the gas, finding much more of the latter than is normal in the universe. But this is exactly what was predicted by the theoreticians. In its supergiant phase, SN1987A threw off gas that had resulted from the conversion of hydrogen to helium. The proportion of the two had been predicted; our observations verified that prediction.

WILL IT EVER GO AWAY?

At this stage in the evolution of Supernova 1987A I could reasonably expect to have virtually finished with it. Instead of that, I am engrossed in two continuing studies of different versions of the echo. Both require considerable work over several years more. Ah well – it is this very unpredictability that gives research its appeal.

Can I even guess when SN1987A really will 'go away' and let me continue the research I was doing in 1986? I have learned not to be so rash. After all, there is a huge mass of gas billowing outwards at high speed, heading for an inevitable collision with the material shed by the supergiant. It should arrive some time in the 1990s. When it does, dramatic things will ensue: events never before witnessed. We will drop everything we are doing to study the happening in great detail: and adrenalin will flow faster yet.

EPILOGUE (WRITTEN BY PROFESSOR FRED WATSON IN JUNE 2012)

When you think of an echo, what do you imagine? A shouted 'hello', reflected from a distant rock face, perhaps? Or the dying reverberation of music in a great cathedral? You might be surprised to learn from this marvellous 1991 article by David Allen that astronomers are fond of another type of echo – but one that involves light instead of sound.

Imagine a cloud of dust in space being lit not, as often occurs, by the constant glow of starlight, but by the intense, searing flash of a nearby exploding star. The brilliant pulse of light might only last for a few weeks before the star fades back into obscurity. But because light travels at a finite speed through space (300,000 km per second), and because of the very great distances involved, the cloud of dust might be lit up months or years after the explosion occurred. This effect is called a light echo, and is analogous to the audible echo of a burst of sound like the shouted 'hello'.

The best-known example was caused by the exploding star, Supernova 1987A, and that is the topic of David Allen's article. This object, in the Large Magellanic Cloud (one of our nearest-neighbour galaxies), reached naked-eye visibility for a few weeks in May 1987. A couple of years later, long after its light had faded, two faint rings were seen around it in images taken by astrophotographer David Malin at the Anglo–Australian Telescope (Figure 6). Analysis by the two Davids and their colleagues showed that the rings matched the colour of the supernova when it was at its peak, proving that they were echoes of its light.

So much for the broad-brush picture. But what is perhaps more inspiring about David Allen's *1991 Yearbook of Astronomy* article is his description of the process of deduction that led him to discover the origin of the rings. He explains in vivid detail how he showed that they are caused by light scattered towards us by two thin sheets of dust in front of the supernova, a scenario that would be far from obvious to the casual observer. You can sense the excitement he felt as he made the calculations that revealed this geometry.

David's article goes on to explain how infrared spectroscopy suggested that the bubble of gas created by the supernova's progenitor was bipolar – elongated, rather than spherical. And he was right on the

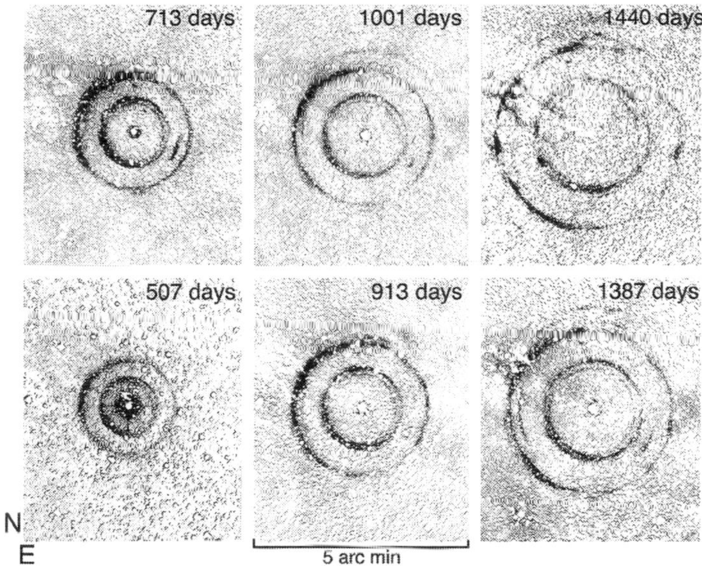

Figure 6. A six-image composite, made in the same way as Figures 3, 4 and 5, showing the development of the SN1987A light echo over a period of 1440 days (almost four years) beginning with the lower left image. The numbers in each panel show the number of days after 23 February 1987, the date when the supernova first appeared. (Image sequence courtesy of David Malin and the Anglo-Australian Observatory.)

money. Only a few years after the article was published, images from the then-new Hubble Space Telescope showed a bright ring of light, where the ejected material from the supernova was beginning to collide with material shed by the progenitor star, as predicted in David's final paragraph (Figure 7). The collision is still ongoing, but we now know much more. Thanks to detailed spectroscopic observations made in 2010 with the Very Large Telescope in Chile, the ring of material can be seen in 3-d as the 'waistline' of an hourglass-shaped bubble of debris. A bipolar nebula – exactly as David had surmised.

Sadly, David Allen did not live to see these exciting discoveries. He died on 26 July 1994 at the age of forty-seven, from a brain tumour. But I can well imagine the glee with which he would have welcomed our present-day knowledge of Supernova 1987A.

I think David would have been thrilled, too, with more recent

Figure 7. This image, taken in February 1994 (only five months before the death of David Allen) with the Hubble Space Telescope's Wide Field and Planetary Camera 2, shows the full system of three rings of glowing gas surrounding Supernova 1987A. (Image courtesy of P. Challis (Harvard-Smithsonian Center for Astrophysics).)

observations of light echoes. In 2002, for example, observations were made of an object that was rather more complex than the dust-sheets of Supernova 1987A. A distant (20,000 light years) and rather unstable star in our own galaxy, called V838 Monocerotis, brightened to 600,000 times the Sun's luminosity before fading again. The Hubble Space Telescope revealed that it was surrounded by a large and complex dust shell, and, as the sphere of light from the star expanded, different parts of the shell were illuminated in a complex bull's-eye pattern of light and shade (Figure 8). Once again, because the geometry of the light-echo is well understood, astronomers could use it to probe the complex structure of the V838 Monocerotis dust cloud in a technique very similar to the computer tomography we're familiar with in the medical world.

During May to December 2002, the V838 Monocerotis nebula appeared to expand from four to seven light years in diameter, a so-called 'super-luminal' (or faster-than-light) expansion. This is illusory,

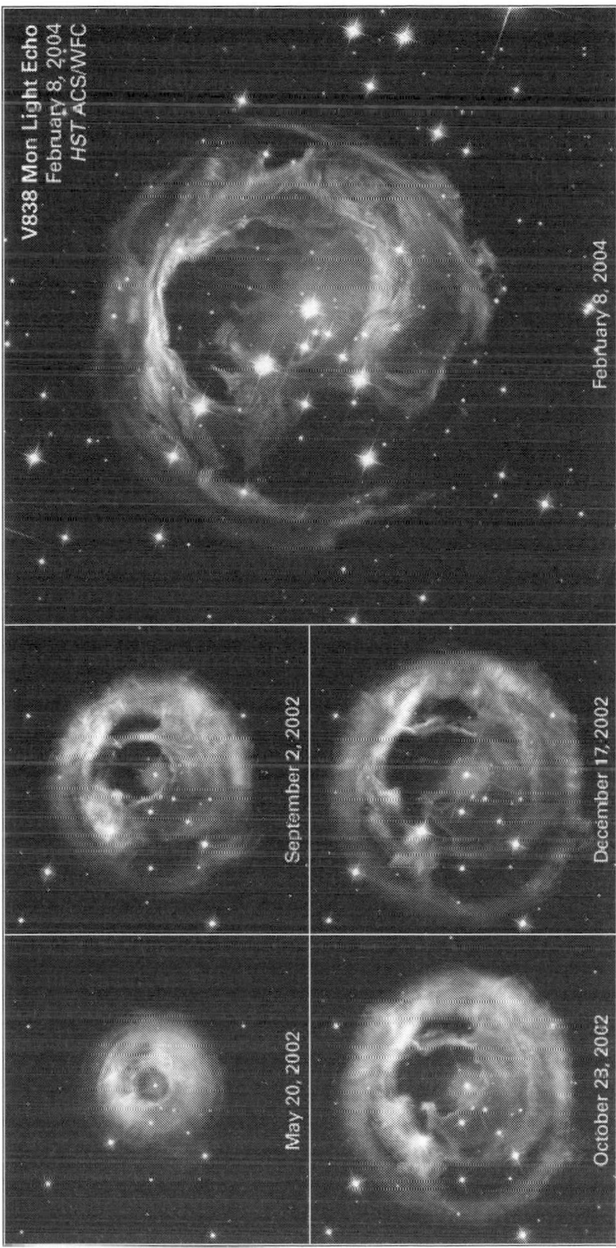

Figure 8. Sequence of Hubble Telescope images, showing the evolution of the light echo of V838 Monocerotis. The strong impression of an expanding shell of material conveyed by these images is illusory. As David Allen's article explains, what we are seeing is the effect of an expanding ellipsoid of illumination, which is concave, and not convex, towards the observer. (Image courtesy of HST ACS/WFC.)

of course, and results from the way the expanding light shell appears to illuminate its surroundings.

More intriguingly, light echoes can be used to study objects that were shining in the distant past. In 2008, for example, astronomers using the Subaru Telescope in Hawaii observed light from dust clouds illuminated by a supernova that had shone brilliantly for a few months in 1572, when it was observed by the great Danish astronomer, Tycho Brahe. The faint echoes were reflected from dust clouds a long way from the supernova, adding an extra path length of 436 light years to its distance of around 9000 light years. How extraordinary that modern spectroscopic analysis can be applied to exactly the same light that Tycho had observed four centuries earlier, revealing details he could never have imagined.

And finally, I'm sure David would have relished the mysterious blob we call 'Hanny's *Voorwerp*' (Dutch for 'Hanny's object'). It takes its name from a Dutch school teacher, Hanny van Arkel, who discovered it while she was participating as an amateur volunteer in the Galaxy Zoo programme. This admirable citizen science project mobilizes the public to classify galaxy images from the Sloan Digital Sky Survey.

Hanny's *Voorwerp* is huge (some 50,000 light years across), bright green, and close to a distant galaxy. Both the galaxy, known as IC 2497, and the *Voorwerp* itself, are about 650 million light years away. While there are competing theories as to its origin, perhaps the best guess is that we are seeing the emission of light from gas that has been excited by intense radiation from an energetic quasar in IC 2497's core. The quasar has now faded into obscurity, but 100,000 years or so after the event, its emission still manifests itself in Hanny's *Voorwerp*.

As echoes go, that's quite a long time.

The Lescarbault Legacy

RICHARD BAUM

Among those who tracked Mercury during its transit of the Sun on 8 May 1845 was one Edmond Modeste Lescarbault (1814–1894), a shy, somewhat diffident medical student with a passion for stargazing. By 1837, aged twenty-three, he felt confident enough to question the integrity of the celebrated Titius-Bode Law, an empirical relation between planetary distances that works well out to Uranus, and early in the century had guided astronomers to the large gap between Mars and Jupiter where four small bodies, the asteroids Ceres, Pallas, Juno and Vesta, were found. Now as he watched Mercury's progress across the solar background, an idea occurred to him. What if there are other planets closer to the Sun than Mercury? At intervals, he mused, they too must transit the Sun, and be well seen in its marginal zones, i.e., near the limb, betraying their existence as small black spots either entering or leaving the solar disk.

It was a first, if unwitting step towards a sequence of events that, in due course, launched astronomers on an extraordinary quest: the hunt for the planet of romance. A pursuit which, in its twists of circumstance, parallels the large body of stories about lost riches that mushroomed in the South-Western states of North America following the Spanish entrada; stories of searches for lost gold and silver mines and other treasure which have since become an integral part of regional folklore. Someone makes a discovery then loses it and is haunted for the rest of his or her life trying to find it. Such a strand folklorists classify as a 'motif', defined as 'the smallest element in a story having the power to persist in tradition'. This is the nature of 'Lescarbault's legacy'. It is an adventure, not for wealth or material gain, but in ideas – the intellectual equivalent of the myth-building process that stirs the imagination and causes the loquacious to fantasize and exaggerate, as did those who motivated the Conquistador Francisco Vasquez

Coronado (1510–1554) with richly adorned accounts of the fabulous but mythical Seven Cities of Cibola.

To exemplify the analogy in astronomical terms, think of the *canali* of Mars, and the many fantasies connected with early selenography. Something imperfectly glimpsed, something odd at the limit of vision that did not conform to expectation flickered briefly into consciousness during the long vigils of the night. Imagination stirred, and golden threads snaked through the tapestry of the mind as the appeal of the strange and the remote took hold. Yet rarely was the feature more than a shadow and a dream – such is the essence of Lescarbault's legacy.

Of course, his idea exhibited neither novelty nor originality. Still it provoked a major reaction, for planet hunting had yet to jade the sensibilities. As with Spain and the wealth of the New World, presumptions about interior planets had been in vogue for several decades. The German physicist and geodesist Johann Friedrich Benzenberg (1777–1846) had published a list of mysterious transits in his *Die Sternschnuppen sind Steine aus den Mondvulkanen* (The shooting stars are stones from volcanoes on the moon; Bonn, 1834), and on that premise had deployed the large refracting telescope of his private observatory at Bilk, near Düsseldorf, to search for intramercurial planets – bodies circling the Sun inside the orbit of Mercury. Whether Lescarbault knew of this, or that the German amateur Samuel Heinrich Schwabe (1789–1875), the pharmacist of Dessau acclaimed for his detection of the eleven-year sunspot cycle, had prefaced his long series of solar observations by searching for interior planets, is not known. Further proof that the idea was in the air came from the popular Scottish science writer Thomas Dick (1774–1857) who surmised, 'It is not improbable that a planet may exist . . . between the orbit of Mercury and the Sun' in his bestselling *Celestial Scenery* (1838).

Born at Chateaudun in 1814, Lescarbault obtained his medical diploma from the Faculty of Paris in 1848, and set up a practice in Orgères-en-Beauce, a small town in the arrondissement of Chateaudun, in the department of the Eure and the Loire, France, where he was to stay for almost a quarter of a century.

Meanwhile, the inventory of the planetary system continued to lengthen. More asteroids were known, while in 1846, Neptune materialized out of the hieroglyphics that crowded the papers on the desk of the French mathematician Urbain Jean Joseph Le Verrier (1811–1877), 'the man who discovered a planet with the point of his pen,' as Arago said. In 1845, though, all our would-be planet-finder could do was dream as he tracked Mercury in its passage across the Sun.

No doubt, the mounting number of asteroids fuelled his ambition and kept his dream alive. By 1858, his circumstances allowed more leisure and his search began. By then he had acquired an altazimuth refractor made by Cauche in 1838. Of 10 cm aperture and 1.46 m focal length, with a finder magnifying six times, it was mounted in a small revolving dome, and supported by a wooden pillar resting on three feet, the points of which rested on a frame, also with three feet and having screws to level the instrument (Figures 1 and 2).

To measure angular distance, he made a simple but effective device consisting of cross wires in the telescope and its finder, and a circular card about 15 cm in diameter graduated to half-degrees. He satisfied himself of its reliability and accuracy by test measures of lunar features using the French astronomer Jean-Dominique Cassini's lunar chart as control.

Position angles were taken and values transferred to a celestial globe. In this way he was able to determine the length of chord described by objects moving across the Sun. Values were tentative and rough, but at least they quantified the observations. To compensate for the lack of

Figure 1. Lescarbault's home and observatory c.1859. Figures unidentified. Photographer unknown. (Image from Richard Baum Collection.)

Figure 2. Lescarbault's observatory as it appeared in 1980. The dome replaced by slate and a small spire. Photographed by Richard Baum. (Image from Richard Baum Collection.)

standard time he regulated his timepiece by observing the meridian passage of the Sun with a small transit instrument.

Whenever leisure allowed, and an observational opportunity availed itself, with his watch regulated, and his equipment adjusted, he thus proceeded to examine the periphery of the solar disk in the hope of catching a small dark speck, at ingress or egress. After repeated searches, about 4 p.m. on 26 March 1859, his persistence was finally rewarded by the appearance of a small, well-defined, perfectly round black spot, less than a quarter the size of Mercury. It was very close to the edge of the Sun and advancing steadily across the yellow background. At that moment, his professional assistance was required, obliging him to descend into his surgery directly below the observatory. Having attended to his patient, he swiftly resumed his place and continued to log the movement of the spot, carefully noting times and position angles until it left the solar background at 5h 16m 55s p.m., Orgères time. As the spot was already on the Sun's face when he first spotted it, he estimated that it had taken about ninety minutes to traverse the solar background from limb to limb. Obviously, if real, the object was at a great distance from the observer.

Lescarbault felt certain of its planetary character, yet for some obscure reason refrained from an immediate announcement. His explanation seems rather diffuse to say the least. Nine months passed without a repetition, and in all likelihood, but for reading about Le Verrier's struggle with the theory of Mercury in *Cosmos*, a scientific magazine edited by the French Jesuit physicist L'Abbé Moigno (1804–1884), the mystery surrounding his observation might not have made headlines. Would history have taken a different route in consequence?

THE PROBLEM WITH MERCURY

'No planet has exacted more pain and trouble of astronomers than Mercury, and has rewarded them with so much anxiety and so many obstacles,' was the blunt truth Le Verrier proffered to the French Academy of Sciences even before he began to worry about Uranus.

If its orbit was circular and in the plane of the ecliptic, Mercury would transit the Sun once in each synodic period and the epoch of the transit could be calculated with comparative ease. The orbit is very eccentric, though, and inclined seven degrees to the ecliptic, hence, in order to predict transits, common multiples of the sidereal and synodic revolutions of Mercury and the earth must be found, and the relative positions of Mercury and the Sun determined from the tables.

Johannes Kepler (1571–1630), the German mathematician, astronomer and astrologer first attempted this very complex calculation in 1627. After completing the Rudolphine Tables, he predicted a transit would take place on 7 November 1631, allowing a possible error of about a day. La Hire's calculations for 1707 erred by almost that amount. Those for 1753 were out by several hours, a fact that induced Lalande to construct new tables. He erred by fifty-three minutes with the 1786 transit, and prepared new tables. These gave better results for 1789, 1799 and 1802 but were still unsatisfactory. Revised tables by Lindenau in 1813 were no more successful.

This was the state of things in 1843 when Le Verrier submitted his *A New Determination of the Orbit and the Perturbations of Mercury* to the French Academy of Sciences, but his tables for 1848 also failed. 'I was greatly surprised,' he remarked to the Academy on 2 July 1849, 'when, in working on the theory of Mercury, I discovered that the mean motion of that planet, deduced from the observations, was much smaller than the value derived from the comparison of ancient and modern observations . . . and my efforts to arrive at a theory free from this discrepancy have been fruitless.'

Though absolutely certain of his optical premise, the difficulty with the observations of passages by the ascending and descending nodes of the planet's orbit signalled he was in trouble. 'This is utterly absurd,' he exclaimed, 'one will eliminate the errors of the May transits, without introducing new ones into the November passages, only by modifying the values of the time-dependent elements of the orbit. The two corrections required must cancel out in the November [ascending node] passages while also accounting for the divergences observed in the May passages [descending node].'

In effect, the discrepancies, though small, meant the planet's elliptical orbit performed a complete yet inexplicable swing around the Sun once every three million years. This was anathema to him, the imperial mathematician, who, under the trident of Neptune, had fortified the sacred vessel of Newtonianism. To cite the distinguished American philosopher of science Norwood Russell Hanson (1924–1967): 'For Le Verrier the theory of Mercury was like a frayed garden hose conveying high-pressure steam. When he adjusted it here, it sprang a leak there; he hadn't enough hands, or ideas, to plug every hole at once.' Plainly, Newtonianism was under threat from a hidden foundation crack.

By 1859, Le Verrier did have a compromise solution. On 12 September that year, he presented a letter addressed to Hervé Auguste Étienne Albans Faye (1814–1902), Secretary of the French Academy of Sciences, in which he stated, 'There is, then, with the perihelion of Mercury, a progressive displacement reaching 38" per century, and this is not explained.' According to his calculations, the observations could be represented by the accepted elements of Mercury if the 38 arc second excess was added to the secular motion of the perihelion. To justify such a correction, there must be an increase of 10 per cent in the mass attributed to Venus from sixty years' meridian observations. If the increase in mass is admitted, 'we must conclude, either that the secular variation of the obliquity of the ecliptic, deduced from observation, is affected with errors by no means probable, or that the obliquity is changed by other causes wholly unknown to us. If, on the other hand, we regard the variation of the obliquity of the ecliptic, and the causes which produce it, as well established, we must believe that the excess of motion in the perihelion of Mercury is due to some unknown action.'

A planet smaller than Mercury, about half its distance from the Sun and travelling in a circular orbit slightly inclined to that of Mercury, seemed to be the most likely cause. Here he acknowledged a difficulty. The object must always be very near the Sun, and thus hard to locate. Observation in twilight would be difficult, for the planet would set almost with the Sun, conversely at sunrise. Why had it failed to appear during an eclipse? Was it hidden behind the Sun? 'Shades of Philolaus' Counter-Earth Antichthon!' was Hanson's incredulous response. However, the French mathematician Joseph Liouville (1809–1882) had dispensed with the 'hidden planet' notion in 1842 by demonstrating that it was physically impossible for a straight-line arrangement viz., earth-Sun-counter earth, to remain stable.

'All these difficulties disappear', Le Verrier argued, 'if we admit, in place of a single planet, a ring of small bodies circulating between Mercury and the Sun.' As these bodies must frequently transit the Sun, he cautioned astronomers to be vigilant of any unusual dark speck which might appear on the solar disk.

Hervé Auguste Étienne Albans Faye (1814–1902), Secretary of the French Academy of Sciences, endorsed the idea, and suggested that a total eclipse of the Sun offered the best observational opportunity. He also recommended the use of photography. Unaware of the searches by Benzenberg and Schwabe, he added that objects of the type specified had not hitherto been identified simply because no one had bothered to search for them. In the event, the archives abounded with reports of unidentified bodies crossing the Sun.

LESCARBAULT'S DEBUT

On 2 January 1860, Le Verrier announced '*Une nouvelle planète*' at the public meeting of the French Academy of Sciences. Parisian society was electrified. What had happened over the festive season to produce such a momentous declaration?

Lescarbault, it appears from what he tells us, was driven to report his observation by what he had read in *Cosmos.* Was there a deeper motive? Did he, for instance, realize that he stood at the confluence of two different but converging streams of thought, his own speculative whim and Le Verrier's solid hypothesis based on the grandiose conception of Isaac Newton? It was, in truth, the Neptune affair in reverse with empirical evidence preceding prediction. Is it possible, therefore, that, spurred on by feelings of rivalry, he finally decided to break silence? Whatever his motive, under the date of 22 December 1859, he penned a complete account of what he had seen and asked a local colleague, M. Vallée, the Honorary Inspector-General of Roads and Bridges, to deliver it into the august presence of Le Verrier at the Paris Observatory.

Le Verrier was at first non-plussed, unwilling to attach any credit to the claim. Who was this doctor, this obscure watcher of the skies who presumed so much? Did he seek to impose a fraud on him, the very person who had predicted what the allegation claimed to fulfil? Why had he waited nine months to make known his observation? His explanation was wholly insufficient to attract unqualified belief. Even so, he later admitted, 'the details were such as allowed me to place in it a certain confidence'. After all, if it was true, it would complete the unfinished business with Mercury, and restore the stability of the Newtonian edifice. It has to be remembered that Le Verrier saw the problem in terms of the Neptune affair; where Uranus's aberrant behaviour was rendered intelligible via the law of gravitation. Now confronted by a somewhat analogous situation with Mercury, he naturally pressed the same pattern of explanation into service. The difference, of course, was Lescarbault and his observation. For where the imperial mathematician was obliged to request optical proof for his prediction of Neptune, in the case of Mercury, if the allegation was genuine, visual proof already existed in advance of the publication of his theoretical findings! Curiosity eroded doubt and, finally, he resolved to ascertain the truth by confronting the doctor in person on his home ground.

With the son of M. Vallée as a witness to his intended interrogation, Le Verrier left Paris by rail on Friday, 30 December. As Orgères was 19 km from the nearest railhead, the last part of the journey was accomplished on foot through open country.

Arriving at Orgères, they soon located the home of Lescarbault, a single-storey barnlike structure with a high-pitched roof, distinguished by a tall turret at one end and two small extensions at the other. It was plain and strongly built with a tall arched door and large window of the same design, standing in extensive grounds, and was easily distinguished by the shuttered dome crowning the turret.

Le Verrier, secretly hopeful his prophecy had been fulfilled, but suspicious of the claimant's motive, walked briskly up to the house and announced his presence with an imperious rapping on the door. Lescarbault himself answered but the haughty Le Verrier declined to introduce himself. 'One should have seen M. Lescarbault,' reported L'Abbé Moigno, who had the story first-hand from Le Verrier, 'so small, so simple, so modest, and so timid, in order to understand the emotion with which he was seized.' A sharp contrast to the self-assurance of the Paris mathematician, whose height and blunt intonation, for which he was noted, had taken him by surprise.

'It is then you, sir,' Le Verrier said, fixing the doctor with a stern look, 'who pretend to have observed the intra-mercurial planet, and who has committed the grave offence of keeping your observation secret for nine months. I warn you,' he admonished, 'that I have come here with the intention of doing justice to your pretensions, and of demonstrating either that you have been dishonest or deceived. Tell me unequivocally, then, what you have seen.'

Shaken and taken aback at having his privacy so rudely disturbed, not to mention the superior tone of his visitor, the doctor stammered out a description of the phenomenon. 'You will then have determined the time of the first and last contact; and are you aware that the observation of the first contact is one of such extreme delicacy that professional astronomers often fail in observing it?' Le Verrier queried.

'Pardon me, sir,' replied Lescarbault, having by now regained his composure, 'I do not pretend to have seized the precise moment of contact. The round spot was upon the disk when I first perceived it. I measured carefully its distance from the margin, and, expecting that it would describe an equal distance, I counted the time which it took to describe this second distance, and I thus determined approximately the instant of its entry.'

'To count the time is easy to say, but where is your chronometer?' Le Verrier demanded, his disbelief becoming obvious.

'My chronometer is a watch with minutes, the faithful companion of my professional journeys.'

'What!' exploded Le Verrier, his suspicions fully aroused. 'With that old watch, showing only minutes, dare you talk of estimating seconds?'

'Pardon me,' the doctor meekly replied, 'I have also a pendulum which nearly beats seconds.'

'Show me this pendulum,' Le Verrier said. Lescarbault produced an ivory ball attached to a silken thread which was hung on a nail in the wall. 'I am anxious to see how skilfully you can thus reckon seconds.'

The doctor acquiesced. He fixed the upper end of the thread to a nail and plucked it. With the ivory ball at rest, he again drew it a little from the vertical and counted the number of oscillations corresponding to a minute as registered by his pocket watch. Le Verrier was unimpressed. 'This is not enough, it is one thing that your pendulum beats seconds, but it is another that you have the sentiment of the second beaten by your pendulum in order that you may count the seconds in observing.'

'Shall I venture to tell you,' the doctor replied, 'that my profession is to feel pulses and count their pulsations. My pendulum puts the second in my ears, and I have no difficulty in counting several successive seconds.'

'That is all very well for the chapter of time,' conceded a mollified Le Verrier, 'but in order to see so delicate a spot, you require a good telescope. Have you one?

'Yes, sir, I have succeeded, not without difficulty, privation, and suffering, to obtain for myself a telescope. After practising much economy I purchased from M. Cauche, an artist little known, though very clever, an object glass nearly ten centimetres in diameter. Knowing my enthusiasm and my poverty, he gave me the choice among several excellent ones; and as soon as I made the selection, I mounted it on a stand with all its parts; and I have recently indulged myself with a revolving platform, and a revolving roof [dome], which will soon be in action.'

Le Verrier's attitude softened. Suspicion soon flooded back, however, when he asked to see the original memorandum.

'Easy enough to ask', replied Lescarbault, fearful of its destruction, 'it was written on a small square of paper which I usually throw away! Yet I may still find it.' Filled with apprehension, he hunted around and happily came upon it – a laudanum-stained fragment of greasy paper acting as a bookmark in his copy of the *Connaissance des Temps.* Having carefully checked and compared it with the letter M. Vallée had brought to him, Le Verrier looked accusingly at Lescarbault: 'But, sir, you have falsified this observation; the time of emergence is four minutes too late.'

'It is,' came the reply; 'have the goodness to examine it more narrowly, and you will find that the four minutes is the error of my watch, regulated by sidereal time.'

'This is true; but how do you regulate your watch by sidereal time?

The small transit telescope was revealed, and that settled the matter.

Le Verrier then enquired how the contact points were measured. Again he pronounced satisfaction. His attitude by now had mellowed, and his tone was more conciliatory. Another shock waited however, when the doctor was taxed if he had attempted to calculate the planet's distance from the Sun. Here Lescarbault did confess incompetence. He was no mathematician and had not succeeded and, in part, this was the reason he had delayed an announcement of his discovery. Le Verrier pressed the issue, asking to see the rough drafts of the calculations. The doctor replied; 'My rough drafts! Paper is rather scarce with us. I am joiner as well as an astronomer. I calculate in my workshop, and I write upon the boards; and when I wish to use them in new calculations, I remove the old ones by planing!'

A quick visit to the workshop confirmed the truth of this statement, and his lack of mathematical abilities. With that, the questioning, which had lasted about one hour, came to an end.

Le Verrier was convinced an intramercurial planet had been seen, and, with grace and dignity, he congratulated Lescarbault. Before leaving for Paris he made extensive local enquiries about the private character of the doctor from local functionaries, only to discover he was highly regarded both as a person and doctor. With such high recommendations, Le Verrier returned home anxious both to obtain a reward for Lescarbault and to announce the addition of a new planet to the catalogue of the Solar System.

VULCAN!

There are two versions of this historic visit. One is the formal announcement Le Verrier made to the Academy of Sciences on 2 January 1860, the other the very dramatic and amusing version he gave amidst the sparkle and glitter of a New Year's Day party at the home of his father-in-law M. Choquet, the evening before, when the events were still fresh in his mind. Fortunately, L'Abbé Moigno was present among the guests and reproduced the informal version of the account, of which the foregoing is a paraphrase, more or less as it fell from Le Verrier's lips in the next issue of *Cosmos* (6 January 1860).

From the observations and on the supposition of a circular orbit, Le

Verrier calculated the planet completed one revolution in 19 days 17 hours at a mean distance of about 21 million km from the Sun in an orbit inclined 12° 10'. Transits would usually occur twice a year if the planet was in inferior conjunction about eighteen days before and after 3 April and 6 October (at the time of the ascending and descending nodes respectively). He concluded its mass was only a *seventeenth* part the mass of Mercury; too small to explain the whole of the anomaly he had detected. He further noted the planet would not stray more than eight degrees from the Sun and would be very faint, thus adding to the difficulty of its detection, which perhaps explained why it had not been found sooner. Here he was wrong as his critics were quick to demonstrate. As Faye predicted, it would be a strikingly conspicuous object during a total eclipse of the Sun. An earlier appellation suggested by Babinet was adopted and Vulcan officially became a constituent of the Solar System.

Midwinter Paris was taken by storm. Embellished and exaggerated in every syllable, the scientific melodrama of Orgères was the main topic of conversation. Garibaldi and the Italian troubles faded into the background, and Lescarbault was lionized and awarded the Legion of Honour. Colleagues and delegates from the scientific press proposed to invite him to a banquet at the Hotel de Louvre on 18 January. The medical fraternity in Chartres and Blois extended a similar offer. All were declined; Lescarbault pleading the simplicity of his lifestyle and 'retired habits', and his reluctance to forsake his patients.

No more was to be heard from him until three years before his death in 1894. Though minimal, his part was pivotal. If he had not declared his observation at the moment he did, it is interesting to contemplate how different things might have been.

After all, Le Verrier had many critics and enemies, and his 'hidden planet' hypothesis had not met with universal acceptance. Yet he was very popular, important and well connected, admired the world over – 'the Lion of every salon', it was said at the time. But his aloof and august manner engendered animosity and rivalry, and caused the staff at the Paris Observatory to go on strike, a matter that temporarily lost him his post as Director. Overall, although Lescarbault paled before this giant, and here one draws a parallel with the story of the Prince and the Pauper, his role in the affair is of greater consequence than it appears. Without his observation, Le Verrier had no need to travel to Orgères. No visit, no interrogation to amuse the populace and, importantly, no object to name, *ergo* no Vulcan. In other words, the encounter with Lescarbault enriches our perception of Le Verrier and Vulcan, and highlights the imaginative element that is such an important part of the drama.

Meanwhile, the archives were searched for evidence of historical transits. Swiss astronomer Johann Rudolf Wolf (1816–1893) and the English astronomer Richard Christopher Carrington (1826–1875) were successful and Le Verrier found several compatible with his hypothesis. Eventually, it was concluded there were at least three bodies inside the orbit of Mercury.

As the news spread, the global community made its plans. Transits were predicted for 29 March, and 2 and 7 April that year. Nothing was seen. The total solar eclipse on 18 July was eagerly anticipated, but again, Vulcan failed to appear. Euphoria chilled and tension heightened.

EXCITEMENT, DUST, AND DOUBT

Emmanuel Liais, a French astronomer in the service of the Brazilian Coast Survey, and a known rival of Le Verrier, launched a harsh critique denouncing Lescarbault as a fraud, and rejecting Le Verrier's hypothesis. On the question of the anomaly he said the excess motion was less than cited; that by admitting a possible error of about two degrees, in the obliquity of the ecliptic, and increasing the mass of Venus by one tenth, the whole of the 38 arc second excess could be explained.

Le Verrier's colleagues were furious, and a fierce debate raged. Still, the damage was done. Faith in the hypothesis sagged but ignominy was averted by the Manchester amateur astronomer W. Lummis. On 20 March 1862, he spotted a sharply defined black spot of a circular form moving across the face of the sun. Unfortunately, like Lescarbault, he, too, was distracted and was unable to complete the observation. French astromoners Valz and Radau determined elements for Vulcan from the sighting and arrived at values similar to those Le Verrier had computed from the Lescarbault report.

In December 1874, Le Verrier reviewed all the evidence. He reaffirmed his results, stating, 'The consequence is very clear. There is, without doubt, in the neighbourhood of Mercury, and between that planet and the sun, matter hitherto unknown. Does it consist of one, or several, small planets, or of asteroids, or even cosmic dust? Theory alone cannot decide this point.'

Three years later he died, still convinced of the reality of his prediction. Just before the end he again re-examined the problem and alerted the global network of observatories to be vigilant during the coming months as a re-appearance seemed likely. It was a useless gesture.

The climax came in 1878 at the total solar eclipse of 29 July. Public interest in Vulcan was at its height, fanned, no doubt, by Le Verrier's last prediction, and the eclipse offered a good opportunity to settle the vexed question once and for all. Totality would begin in north-eastern Asia. The Moon's 187 km-wide shadow would sweep across the Bering Strait into Alaska, strike Canada near its Pacific coastline, and then arc across the vast western United States. It would continue through Yellowstone and the majestic Wind River Range on to the central highland of Wyoming Territory. It was there astronomers gathered in large numbers at the isolated whistle-top of Separation, on the Union Pacific Railroad. James Craig Watson (1838–1880), a Canadian-born professor of astronomy and admirer of Le Verrier, along with others joined the US Naval Observatory party led by Simon Newcomb (1835–1909), which had set up camp alongside the railroad track (Figures 3 and 4).

Figure 3. Site of the US Naval Observatory eclipse camp at Separation, Wyoming, 29 July 1878 as it appeared in 2005. Separation ceased to exist when the track of the Union Pacific Railroad was moved further south soon after the eclipse took place. Photographed by Dr William Sheehan. (Image from Richard Baum Collection.)

Figure 4. Thomas A. Edison party of astronomers gathered at Rawlins, Wyoming, prior to the total eclipse of the Sun on 29 July 1878. Left to right: Professor George F. Barker, Robert M. Calbraith, Henry Morton, unidentified, Meyers, D. H. Talbor, M. F. Rae, Marshall Fox, James C. Watson, Mrs A. H. Watson, Mrs Henry Draper, Henry Draper, Thomas A. Edison and J. Norman Lockyer. (Image courtesy of United States Department of the Interior, National Park Service, Edison National Historic Site.)

Watson, a leading theoretical astronomer noted for his many discoveries of asteroids, staunchly supported the Vulcan hypothesis, and had hopes the planet would be found. What happened that day of wind and dust in the American West put Separation into world headlines and left posterity with an intriguing mystery. For during totality Watson claimed to have seen two disk-like objects in close proximity to the Sun, both of which he claimed were intramercurial planets.

Further south, near Denver, famed comet discoverer Lewis Swift (1820–1913) also had gone in search of Vulcan. He, too, described two suspicious objects near the Sun. And so the seeds of another dispute were sown.

These allegations were sharply contested by the German-American Director of the Litchfield Observatory, New York, Christian Heinrich Friedrich Peters (1813–1890), a notable planet-hunter, and a well-known critic of Le Verrier. He was dismissive of the Vulcan hypothesis and suggested Watson had mistaken the readings on his position circles and in consequence had misconstrued two well-known naked-eye stars in Cancer. He dismissed Swift as if with a wave of his hand, charging him with fabrication. In retrospect, his judgement, whatever its merit, is nevertheless unduly harsh, almost embittered. When evaluating the observations it is well to recall the conditions under which they were made. The period of totality is fleetingly brief and in 1878, it amounted to 2 minutes 40 seconds (Figure 5). The urgency, combined with the comparative unfamiliar nature of the event, induced hazards and distractions not usually encountered in routine work.

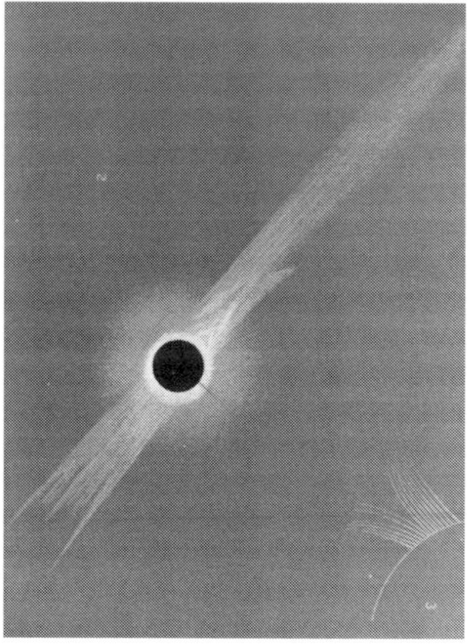

Figure 5. The outer corona at the total eclipse of the Sun on 29 July 1878. US Naval Observatory. (Image courtesy of Samuel Pierpont Lamgley. *The New Astronomy*, Boston (Ticknor and Company, 1888).)

EPILOGUE

Vigorous attempts to settle the problem at subsequent eclipses met with no success. A strange red star was whispered to have been seen by the French amateur entomologist and artist-astronomer Étienne Léopold Trouvelot during the 1883 eclipse. Two observers had a brief glimpse of a bright object between Venus and Mars at the eclipse of 22 January 1898.

Unidentified specks were detected on plates exposed by the Smithsonian expedition in May 1900, but confirmation was not forthcoming. The campaign continued at the solar eclipses of 1901, 1905 and 1908 when the relatively new technique of photography was applied. The conclusion reached by the American astronomer Charles Dillon Perrine (1867–1951) of the Lick Observatory who conducted the search, was that no planet with a diameter of around forty kilometres existed within the orbit of Mercury. Vulcan was never found because it did not exist.

Such sightings as were later reported failed to reignite interest. In 1899, with interest virtually gone and new ideas in circulation, Asaph Hall (1829–1907), the American known for his discovery of the two moons of Mars, noted that intramercurial objects 'are not mentioned [anymore] even by the astronomical writers of the *Atlantic Monthly*'.

It is, of course, true; transits were not infrequently reported but a new reality was abroad. If large numbers of planetary bodies did exist inside the orbit of Mercury, was it not remarkable that they avoided the scrutiny of competent observers? Yet the legend of the interior planet lives on as such things do. In recent years, investigators have scrutinized the immediate vicinity of the Sun for evidence of Vulcanoids, minor bodies inside the orbit of Mercury – a reminder of Le Verrier's inner asteroid ring – which, in perpetuating the search, emphasizes the analogy with the folklore of the South-West region of North America.

Vulcan, the planet of romance, or planet of fiction, to apply a more apt epithet, marked a downturn in Newtonian theory. The whole affair, so to speak, was created in 1859 at the desk of a mathematician in Paris out of a need to define something able to impress its influence on Mercury without imposing any sensible effect on other members of the solar family. Le Verrier's inability to achieve a positive result was not so much a failure as a brave attempt to save Newtonian theory with tools created from within itself. For, in effect, it signalled that our understanding of the physical world was less complete than imagined.

Albert Einstein (1879–1955) the German-born 'father of modern physics', established that fact fifty-six years later in 1915, at his desk in Berlin when he deftly removed the necessity for Vulcan with his General Theory of Relativity. One of the tests required by his prediction was that Mercury's perihelion should advance, quite without the intervention of another body, in the same direction and by the amount actually observed. Thus a slight adjustment of gravitational theory had at last banished Vulcan, and demonstrated that Einstein's mathematics are more precise than Newton's in dealing with bodies moving at high velocity in strong gravitational fields.

In retrospect, then, 1859 was a watershed year, one strangely auspicious in its indications. It is surrendered to Charles Darwin and *The Origin of Species*; to Bunsen and Kirchoff and the laws of spectrum analysis. It also marked the nadir of Newtonian supremacy. Vulcan told us of the future, and a concept of the physical world chaotic in comparison to the formal mathematical scheme of the *ancien regime.* Its drama echoes that of Coronado, and was best summed up at the solar eclipse of 1869 by C. H. F. Peters as 'a wild goose chase for Le Verrier's mythical birds'.

And what of Lescarbault, characterized as 'a kindly man, a bit of a dreamer who seems to have practised more astronomy than medicine'? The Pauper to Le Verrier the Prince. He vacated the scene in 1860 but re-surfaced one day in 1891 when, in his seventy-seventh year, this private, self-effacing little man reported another 'astronomical discovery' to the Academy of Sciences. He had seen 'a star of comparable brightness to Regulus, which I had never seen until today. It is beneath Theta in the Lion. I observed it only with the naked eye, on 10 and 11 January, in the early morning hours; and despite the weakening of my eyesight, I believe I saw it well, and was not the victim of an illusion.' Too late did he realize his mistake. Embarrassingly, he had delivered himself into the hands of his enemies. Camille Flammarion (1842–1925), a long-time critic, gleefully wasted no time in pointing out that he had not found a star, only the well-known body Saturn! It was a damaging blow to Lescarbault's esteem, and undermined the credibility of his 1859 observation.

Or did it? Today Le Verrier's ill-fated 'hidden planet' hypothesis is remembered only by historians of celestial mechanics, whereas Lescarbault is recalled for the mystery hidden in his report. He claims the spot had independent motion. Hence it could not have been a sunspot. What then? Was he deceived? Did he imagine what he believed he saw? We are now more aware of what goes on in the immediate vicinity of the Sun not to brush off such a claim as our predecessors did. Could it have been a Near Earth Asteroid, a meteoroid? Interplanetary space abounds with such material. Speculative interest was expressed in 2000 by the comet expert Jonathan Shanklin, Director of the Comet Section of the British Astronomical Association. Writing in the *Journal* of the Association, he noted how the Solar and Heliospheric Observatory (SOHO) had been very successful in discovering minor members of the Kreutz group of sungrazing comets. This led him to suggest that of the four objects reported by Watson and Swift at the 1878 eclipse, the one closest to the Sun seemed compatible with a small Kreutz sungrazer about five hours from perihelion: An intriguing supposition that leaves the field open to further conjecture.

Writers have been too ready to portray Lescarbault as a hapless incompetent. He built his own equipment, which even Le Verrier had to admit was effective in the face of the tasks its owner set. Lescarbault did have a concept of observational discipline, and he had the right instincts for how results should be processed. To a certain extent the Prince and the Pauper contrast applies to Flammarion and Lescarbault as well – Lescarbault had to scrimp to observe, whereas in comparison Flammarion led a materially charmed life. Lescarbault's handling of his Vulcan observation seems more responsible than Flammarion's 'plurality of worlds' observations, speculations and promotions. Flammarion's contempt was misplaced – all good, careful and disciplined observers end up making mistakes at some time in their career, because observing at the limits of one's equipment, skill and judgement is to balance on a perilously thin wire. The only way never to make an observational mistake is to never observe.

ACKNOWLEDGEMENTS

I would like to record my thanks to Professor W. J. Leatherbarrow, Director of the Lunar Section of the British Astronomical Association, and currently President of the Association (2011–13), to Nigel Longshaw, Randall Rosenfeld, Archivist of the Royal Astronomical Society of Canada, and Dr Jeremy Shears for critical comment during the preparation of the foregoing.

FURTHER READING

Baum, Richard and Sheehan, William, *In Search of Planet Vulcan. The Ghost in Newton's Clockwork Universe* (New York, Plenum, 1997)

Roseveare, N. T., *Mercury's Perihelion From Le Verrier to Einstein* (Oxford, Clarendon Press, 1982)

Touring the Subatomic Universe

FRED WATSON

Maybe we were just slow learners. It took us a while to suspect that there might be a pattern in what we were seeing. But eventually, after six nights of crystal-clear skies, we were certain of it . . .

The display would start early in the night, soon after the long twilight had ended, with the merest hint of a glow on the northern horizon. Then, around eight p.m. or so, we'd detect a thin greenish band, snaking from east to west through the far northern constellations that graced the unpolluted skies of Lyngenfjord. The Plough, Cassiopeia, Ursa Minor – star-patterns that were unfamiliar to those of us more accustomed to Crux Australis, Vela and Centaurus. As we watched, the green band would broaden, and we would begin to see delicately structured filaments of light stretching upwards, with more bands starting to fill the patches of dark sky.

And then, if we were lucky, the show would begin ramping up to its evening crescendo. Swirling curtains of light, twisted into impossible shapes, would sweep in waves across the sky, taking only seconds to form, brighten into prominence, and then fade again. Still more green bands would roam from east to west. And this silent display of nature's brilliance would be interrupted by our shouts of excitement, as we looked on in awe. 'Oh my God, look at that!' 'Behind you – turn around!' 'Overhead . . . LOOK UP!!'

By now, our early fumblings with tripods, ISO settings and exposure times had metamorphosed into a model of proficiency. With surprising deftness, we performed the nightly balancing act of getting enough light on to the image sensors of our cameras while not blurring out the fleeting structure in the aurorae. And all that while trying not to slip over in the powdery snow, or expose fingers and thumbs for too long to the frozen air. But there was still the almost insurmountable problem of knowing where best to point our cameras, as the sky exploded into curtains of green, tinged above with red, and below with a pale magenta.

At their height, these displays could bring a rare climax, with the formation – almost overhead – of an auroral corona (Figure 1). Like crystals precipitating out of a super-saturated chemical solution, green, finger-like rays would burst from the zenith before our astonished eyes. This extraordinary effect is a trick of perspective, caused by parallel columns of light forming along the near-vertical lines of the Earth's magnetic field. Except in the most extreme conditions of space weather, it is exclusively the province of those who venture to the planet's polar regions.

Figure 1. Rivalling the first-quarter Moon in intensity, an auroral corona bursts over the skies of Lyngenfjord in the Norwegian Arctic. (Image courtesy of Anne Spencer.)

But, this week, that meant us – for our vantage point in far-northern Norway was well within the Arctic Circle. On those frozen, magical nights at the beginning of 2012, members of the 'Fire in the Sky' study tour from down-under wouldn't have wanted to be anywhere else in the world (Figure 2).

Figure 2. In daylight, nature's magnificence is further revealed in the peaks and glaciers of the Lyngen Alps across the fjord. (Image courtesy of Fred Watson.)

PARTICLES FROM THE SUN

In earlier times, the Sami people who inhabited this frostbitten country regarded the aurora borealis with fearful awe. Legends about the origin of the eerie lights were intertwined with traditions concerning the souls of the departed, their fabled camp fires flickering mysteriously beyond the northern horizon. In Western European culture, too, the northern lights belonged in the realm of mystery. A woodcut of 1570, for example – now in the Crawford Collection of the Royal Observatory, Edinburgh – depicts the aurora as a row of heavenly candles above the clouds. No doubt the artist was at a loss to imagine how else such an unearthly apparition could be depicted.

It was only at the turn of the twentieth century that the first glimmer of understanding began to emerge as to the true cause of the aurora. But it was an idea so outlandish, and so unpalatable to the scientific establishment of the time, that it was rejected almost out of hand, particularly in Britain. It generated a human saga of epic proportions – and it had its origins not far from the small Norwegian town of Alta, where our study tour had begun its rendezvous with the dazzling aurorae of 2012.

Just to the west of Alta, in the spine of mountains that dominates

Norway's far-northern coastline, a high plateau called Haldde became the site of the world's first auroral observatory. It was set up by a visionary Oslo University scientist called Kristian Birkeland, who had a wild notion that the aurora borealis is caused by electric currents in space, and is somehow linked with the Earth's magnetic field. To this end, Birkeland and a handful of colleagues wintered over on two mountain peaks at Haldde during the bleak months of darkness from late 1899 to early 1900 (Figure 3).

Figure 3. Kristian Birkeland's book on his first expedition to Haldde made a political statement as well as reporting on his scientific observations of the aurora borealis. Controversially, the Norwegian flag was shown without its Swedish counterpart in the corner, which was mandatory in the days before Norway gained its independence. (Image courtesy of Fred Watson.)

Magnetometer measurements were correlated with appearances of the lights, and attempts were made to photograph the auroral displays from the peaks – Sukkertop and Talviktop – in order to estimate their height by triangulation. This was intended to reveal whether or not the aurora actually touched the ground, since many scientists still believed it was purely a meteorological phenomenon. In his first expeditions, Birkeland failed to make these measurements due to the insensitivity of his photographic equipment, but observations a few years later proved that aurorae do, indeed, populate the upper atmosphere at levels above ninety kilometres or so.

It was Birkeland's formulation of a radical theory in the wake of that early work that made him so unpopular with British scientists. He postulated that the Sun was a source of the subatomic particles that were then known as cathode rays (but are today called electrons), and that their interaction with the Earth's magnetic field near the poles caused them to excite the atoms of the upper atmosphere into a frenzy of luminescence. He later suggested that positive ions, too – protons – might also be emitted from the Sun, and play their part in the celestial light-show.

The idea that anything other than heat, light and gravity could come from the Sun was what so incensed Birkeland's critics. And Britain's Royal Society was in the van, believing that its scientific heritage endowed it with ownership of these phenomena. After all, they had been the province of Newton, Herschel and Maxwell – among others. But subatomic particles from the Sun? Proposed by a *Norwegian*? Rubbish.

As the twentieth century dawned and matured into the gentility of the Edwardian era, Birkeland's ideas were put to the test at the University of Oslo. In his Terrella ('little Earth'), he set up a magnetized model of the Earth, enclosed in a vacuum chamber, which could be bombarded with electrons. Sure enough, a recognizable simulation of the aurora was generated. Locally, at least, Birkeland gained credibility, and was encouraged to continue his relentless pursuit of the aurora's mechanism.

But with the drums of war beating loudly on the horizon, Birkeland also turned his attention to other inventions. He invested considerable time and effort in the development of a novel electric cannon. At the same time, more peaceful ambitions drove him to work on the fixation of atmospheric nitrogen for fertiliser by means of an electric furnace, which, in turn, led to the creation of the Norsk Hydro company – still a major industrial force today.

Birkeland would have been the first to admit that these diversions were undertaken merely to allow him to fund his obsession with pure research. As the First World War rumbled on to its dreadful course, his work took him to Egypt, where he embarked on observations of the zodiacal light to discover whether this, too, had an electromagnetic origin. Sadly, however, amidst growing paranoia about being persecuted by the British authorities, Birkeland began to sink into a mire of mental instability from which he never recovered. Eventually, in 1917, he died in Tokyo at the age of only forty-nine.

SPACE WEATHER

With Birkeland gone, there was no longer a champion for his ideas, and the notion of electric currents carried through space by a solar wind of particles dropped off the agenda in the face of a hostile scientific establishment. The origin of the polar aurorae was relegated to the too-hard basket. But half a century after his death, when the space age was still in its first-flush of youth, magnetic measurements made from Earth orbit revealed what ground-based observations had failed to detect. Yes, space is full of subatomic particles – negatively *and* positively charged. Astonishingly, on both counts, Birkeland had been right.

Even today, Birkeland's work is not well known outside his native country, although he is celebrated on the Norwegian 200-Kroner note – as well as in the small museum at Alta. Not quite so insightful as his auroral investigations, however, were Birkeland's researches into the zodiacal light. This is now known to originate in the flattened ring of dust particles surrounding the Sun and enveloping the inner planets. While the streams of electrons and protons from the Sun certainly buffet the dust particles, they do not play a role in exciting them to luminosity, as happens with atoms in the Earth's upper atmosphere. Rather, the zodiacal light is caused simply by the reflection and scattering of sunlight.

Aurorae, on the other hand, are the result of a highly complex interaction between the solar wind and the Earth's magnetic field. Our modern understanding includes such subtleties as the aurora's occurrence in a circular zone around each magnetic pole rather than a concentration of light at the pole itself – something that was not explained by Birkeland's theory. And our knowledge of the energies carried by the solar particles lets us understand why we see aurorae of different colours at different heights. The prominent green bands occur roughly between 100 and 200 km, and are caused by the excitation of atmospheric oxygen atoms. Often, there are extended pillars of red light above 200 km, which are again due to oxygen atoms, but now at a lower energy. In the highest energy displays, molecules of nitrogen are excited below 100 km, causing them to emit red, blue and violet light, giving a characteristic magenta fringe to the underside of bright aurorae.

Most readers of this yearbook will be aware that the northern and southern aurorae are the visible indicators of the strength and density of the solar wind – parameters we now include within the heading of 'space weather'. It's a great name, conveying graphically the cosmic buffeting of the Earth that space scientists have been aware of for several decades.

Of course, we are all taught at school that the space between the planets is a vacuum. That is essentially true, since the particle densities are extremely low (approximately five million particles per cubic metre near the Earth – roughly a billion billion times fewer than in the atmosphere at sea level). However, the rapid ejection of material from the Sun can energize this 'vacuum' to a very high degree.

Even when the Sun is in its quiescent state, the solar wind amounts to a million tonnes of subatomic material per second, travelling at speeds up to a million km per hour. It shapes the magnetosphere – the elongated bubble defining the region of the Earth's magnetic influence in space. Periodically, however, the Sun undergoes so-called coronal mass ejections (CMEs), which propel energetic particles and magnetic fields into space at up to 1200 km per second and generate powerful shock waves in the solar wind (Figure 4). Similar activity is caused by solar flares, which are violent explosions in the Sun's atmosphere. The underlying mechanism behind both CMEs and flares is now thought to be the sudden release of magnetic stress in the Sun – a kind of gigantic magnetic 'twang'.

The end product of this monumental dumping of energy into the inner Solar System is a geomagnetic storm, in which Earth's magnetosphere is seriously disturbed. Aurorae become visible nearer the equator than normal, and the intensity of the ground-level magnetic field increases dramatically – as was noted by Birkeland. Such phenomena tend to occur near the maximum of the Sun's eleven-year cycle of sunspot activity, and each event typically lasts for a day or two.

Because the magnetosphere is being bombarded by far more high-energy particles than usual, geomagnetic storms can damage electronic hardware in orbiting spacecraft, and are also hazardous for human space-travellers (and perhaps even airline passengers flying over the Earth's poles). They can also reduce the lifetimes of satellites in low-

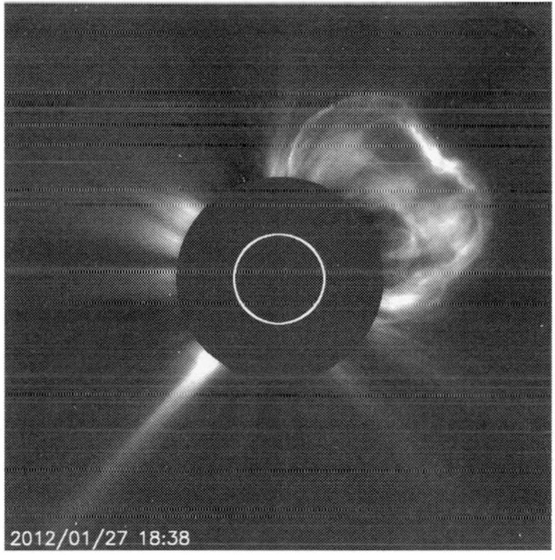

Figure 4. A coronal mass ejection (CME) triggered by a powerful solar flare is captured by the Solar and Heliospheric Observatory (SOHO) as it blasts off the Sun on 27 January 2012. The actual disk of the Sun, indicated by the white circle at centre, is hidden in this view by an instrument called a coronagraph. The coronagraph creates an artificial eclipse by blocking the too-bright light from the Sun's surface, allowing us to view the CME. (Image courtesy of SOHO/ESA&NASA.)

Earth orbit by increasing the atmospheric drag they experience.

It is the sweeping of strong magnetic fields over electricity transmission or telephone lines that causes power and telecommunications outages, however. Moving a wire through a magnetic field causes an electrical current to flow, and the same applies with a moving field and a stationary wire. Very large stray currents can thus be generated by geomagnetic storms, tripping overload equipment and bringing the distribution system to a halt.

An example of this occurred in March 1989, when a large geomagnetic storm caused a blackout in Quebec that affected nine million customers. Today, while our world is even more vulnerable because of the huge increase in the use of electronic systems in everyday life, we are also better informed about the occurrence of flares and CMEs by a flotilla of spacecraft monitoring the Sun in great detail. And, of course, the world wide web is the easiest way to keep up with the action – assuming the Sun doesn't switch that off, too . . .

ADVENTURES IN THE NORTH

The effects of subatomic particles interacting with a magnetic field are seen most vividly in the aurorae, which makes the polar regions, well . . . a magnet for tourists. That was what took me and my two small groups of Australian study-tourists to far-northern Scandinavia at the beginning of 2012, in celebration of the Sun's peak of activity. Perhaps I should explain that astronomy tourism has become a significant adjunct to my day-job at the Australian Astronomical Observatory (AAO), and I thoroughly relish the prospect of taking interested people to see the places where scientific ideas originate. These folk come from all walks of life – including science – but their common passion is a hunger for learning.

Fortunately, the AAO and the government department that operates it are wholly supportive of the venture. And Marnie, my tour organizer, has a real flair for putting together marvellous itineraries that also cater for the sometimes not-so-interested partners of the study tourists. Thus it was that after our rendezvous with the northern lights, we embarked on a tour of the other attractions of the region, both scientific and otherwise.

Alta and Lyngenfjord are within striking distance of Narvik, which boasts a spectacular railway line linking the famous port with Kiruna in northern Sweden. And Kiruna has a rocket range – the Esrange, operated by the Swedish Space Agency (Figure 5), whose restricted airspace will eventually be used by Richard Branson in connection with his Spaceport Europe. From Kiruna airport, Virgin Galactic will fly well-heeled passengers *through* the aurora borealis, as opposed to their simply observing it from below. This is perhaps the most intimate encounter possible between humans and subatomic particles . . .

Southern Scandinavia, too, has much of interest. Not far from Stockholm is Kvistaberg, where the historic University of Uppsala has custody of one of the largest Schmidt telescopes (wide-angle reflecting telescopes) in the Northern Hemisphere. With an aperture of one metre, the Kvistaberg Schmidt bears a distinct family resemblance to its larger siblings, the 1.2-metre Oschin Schmidt Telescope at Palomar Mountain, and the AAO's very own United Kingdom Schmidt Telescope at Siding Spring Observatory in New South Wales. The most striking difference is the steep polar angle of the telescope's equatorial

Figure 5. Launch tower at the Esrange Space Centre at Kiruna in northern Sweden. Until the mid-2000s, the tower was used to launch Skylark sounding rockets for upper atmosphere research. (Image courtesy of Fred Watson.)

mounting – equal to its latitude – of some sixty degrees, compared with the thirty degrees or so of the two larger telescopes. Although equipped with a CCD camera, the Kvistaberg Schmidt is little used today due to indifferent weather conditions compared with overseas sites.

Perhaps the jewel of southern Scandinavia, however, is the little island of Ven in the Öresund – the strait separating Sweden and Denmark (Figure 6). Here, in the late sixteenth century, the greatest astronomer of the pre-telescopic era plied his trade. The noble Tycho Brahe built two observatories, whose few broken remains make Ven a mecca for astronomy enthusiasts. In summertime the island is delightful, but a winter mantle of snow and surrounding ice floes with the occasional basking seal made our February visits just as magical.

Figure 6. The former Danish island of Hven is now Swedish territory and is known as Ven. Here, in the late sixteenth century, Tycho Brahe, the 'lord of the stars', made his high-precision measurements of the positions of stars and planets. In this wintertime view, Ven is surrounded by the ice floes of the Öresund. (Image courtesy of Fred Watson.)

And so on. There is little space here to mention our visit to Tartu in Estonia to see Joseph Fraunhofer's masterpiece: his 24-cm 'Great Dorpat Refractor' of 1824, which set the pattern for refracting (lens) telescopes throughout the nineteenth century. Nor our visit to Tuorla Observatory in Finland, where the tall tower housing its one-metre telescope of 1959 sprouts incongruously from a snow-laden pine forest (Figure 7). Nor an enchanting night-time excursion to Copenhagen's curious Round Tower of 1642, whose 209-metre long internal spiral ramp allowed horse-drawn carriages to access an astronomical observing platform high above the streets of the city. Oh, and then there was Iceland, with its volcanoes, hot springs and glaciers . . .

But I digress from the main theme of this article. Remarkable though they are, these places have little to do with subatomic particles being guided by magnetic fields. So where else might you find such things? And what could be their link with astronomy? And, most important of all, can you go and see them . . . ?

Figure 7. The one-metre telescope of the Tuorla Observatory in Finland is mounted on a high tower to raise it above the level of the surrounding forest. Tuorla is operated by the University of Turku, and is the largest astronomical research institute in the country. (Image courtesy of Fred Watson.)

PARTICLES IN COLLISION

One of astronomy's greatest contemporary puzzles is dark matter, the stuff that seems to hold galaxy clusters together by its gravitation, and likewise stops individual galaxies tearing themselves apart by their rapid rotation. How do we know dark matter is real? It reveals itself only by one thing – the effect of its gravitational attraction on matter that we can see. Other than that, we have no way of detecting it, at least for the time being. We can, however, sense this gravitational 'smoking gun' by a number of different methods, and they all give the same answer: the ordinary matter we see by the radiation it emits (for example, in stars and glowing gas clouds) or absorbs (for example, in dust clouds silhouetted against a bright background) only amounts to one sixth of what gravity tells us is there. Embarrassingly for astronomers, dark matter outweighs visible matter by five to one – and we don't yet know what it is.

It is the observations made by astronomers that have alerted us to the presence of dark matter, and they have also shown us where in the universe it is to be found. It turns out that dark matter and visible matter concentrate together. 'Beacons of light on hills of dark matter' is one eloquent description of clusters of galaxies, as revealed by studies that use a phenomenon known as gravitational lensing to map the distribution of their mass. Astronomical observations have also ruled out dark objects like black holes or cold orphan planets as candidates for dark matter, and have shown that it is more likely to be vast swathes of subatomic particles that have mass, but don't interact with the various particles that constitute normal matter.

By combining their various studies of dark matter, astronomers hope to learn enough about its behaviour for a clear leader to emerge from the various competing models. These have been built primarily by the theoretical physicists who study subatomic particles. But it is in the experimental research accompanying this that the quest for dark matter is taking perhaps its most exciting turn – if you'll pardon the pun – bringing hopes of a real breakthrough. It also leads us straight back to the underlying theme of this article, for those experiments are taking place at one of the most inspiring scientific institutions in the world. And it is an institution that makes visitors welcome.

Most readers of this yearbook will be familiar with the Large Hadron Collider (LHC), the giant atom-smasher straddling the Swiss-French border near Geneva, which is operated by CERN, the European Centre for Nuclear Research. This machine is the successor to an earlier facility called the Large Electron-Positron Collider (LEP), which occupied a 27 kilometre-long circular tunnel that had been excavated for it in the 1980s. When plans were made for a new machine, the same tunnel was to be used, so there was a lengthy period when physics took second place to engineering on the site. In December 2009, however, the LHC was fired up and, apart from one early mishap, it has performed as brilliantly as expected.

The LHC's role is to accelerate two streams of subatomic particles around their circular paths in opposite directions, achieving speeds close to the speed of light – and then collide them together. To date, protons and lead ions have been collided in separate experiments aimed at exploring different phenomena. And, by the way, while common sense suggests that the closing speed of the colliding particles should be nearly twice the speed of light, that's not the case. At such extreme velocities, relativity tramples on common sense. Adding up the particles' velocities under relativistic rules produces a sum that never exceeds the speed of light – although it is greater than each of the component velocities.

Making particles collide might sound simple, but the technology required to achieve it is little short of astonishing. As one of the largest scientific experiments in the world, the LHC bursts with engineering superlatives. Unfortunately for spectators – or perhaps fortunately – most of the high-energy action takes place deep underground, where the old LEP tunnel now houses twin vacuum tubes containing the particle beams. The plumbing alone is staggering. For example, rope-like skeins of microscopic copper tubes run for kilometres carrying supercooled liquid helium. The vacuum through which the particles travel is ten times lower than the vacuum at the surface of the Moon. And those particles are kept on track by superconducting magnets that are colder than space itself (Figure 8). Some folk have commented

Figure 8. One of the 'magnets' of the Large Hadron Collider (LHC), through which counter-rotating particle beams are carried at speeds close to the speed of light. The exposed plumbing gives only a hint of the astounding complexity of these units. It is the two small tubes near the centre that carry the beams themselves. (Image courtesy of Fred Watson.)

unfavourably on CERN's six billion Euro price tag for the LHC, but my personal view is that it's amazing they have been able to build it so cheaply.

You can probably tell that I'm rather impressed by the LHC and, to date, I've had the opportunity of making two visits there. The first time was with our 'Stargazer II' study tour in 2010, when we were hosted by Drs Klaus Bätzner and Quentin King, who graciously fielded all our questions about the machine and its various experiments. A highlight of this tour was lunch in one of the CERN cafeterias, where thousands of enthusiastic scientists and engineers spend their lunchtimes eating, drinking and, well, talking physics (Figure 9). The excitement there is palpable, and it remained in our eyes and ears as we boarded the TGV (high-speed train) to Paris later that afternoon. We touched almost 300 km per hour on that journey, a minuscule fraction of the almost 300,000 km per second reached by the protons circulating in the collider. Unlike the protons, however, we didn't collide.

My second visit was probably a consequence of the first, because, by then, everyone thought I was an expert on the LHC. Yes, well . . . I did

Figure 9. In the relaxed surroundings of the CERN cafeteria at the LHC site near Geneva, Australian physicists discuss prospects for finding the Higgs boson, while an Australian TV crew eavesdrops. (Image courtesy of Fred Watson.)

try to put them right. But in some ways, this second visit was even more exciting, because we were accompanied by a camera crew aiming to produce a down-to-Earth TV documentary about this most esoteric of human endeavours. This led us to parts of the facility that are normally out of bounds, such as the inside of some of the experimental control rooms. And, once again, it led us to the cafeteria, where the buzz was captured on camera with the help of a handful of Australian and Kiwi scientists, who were very happy to give us a lunchtime taste of their work.

In an echo of the TGV ride, it also led us to a ridge in the Jura Mountains behind Geneva, to a place called Col de la Faucille, which boasts a gut-wrenching roller-coaster ride intended to simulate an Olympic luge track. The idea was to give viewers a hint of what it might feel like to be a subatomic particle circulating around the LHC by mounting a camera on one of the cars, while we presenters took turns in expounding the details of impossible sideways accelerations around the ride's hairpin bends. This time, we only reached speeds of around 40 km per hour – but when your rear end is just a few centimetres above the track, I can tell you, it feels a lot like the speed of light.

HUNTING FOR DARK MATTER WITH SUSY

All the marvellous engineering notwithstanding, it is the LHC's potential for scientific discovery that excites physicists and astronomers. By smashing together particles such as protons, the LHC is effectively acting as a gigantic super-microscope, probing matter on the smallest scales by examining the cascade of subatomic debris released in the collisions. Its first job is to test something called the 'standard model', which is a hierarchy of twelve particles of matter and four so-called 'force particles' carrying three of the four fundamental forces of nature. Why four particles to carry three forces? Because one of them, the so-called weak nuclear force, is greedy, and needs two. And what happened to the other fundamental force? That one is gravity, and it is actually beyond this standard model at present. That is because we don't yet have a satisfactory theory of quantum gravity describing the way it acts on very small scales.

The standard model does, however, predict a particle that had not yet been found at the time of my visits, and that is the Higgs boson, a

1960s postulate of Professor Peter Higgs and his colleagues at Edinburgh University. The Higgs particle is thought to endow all the other standard model particles with the property of mass, so it's fundamental to the theory. And one of the first tasks of the LHC is to discover where it could be hiding. Not whereabouts in space, but whereabouts in the range of energies that characterizes the size of things in the subatomic world. By the end of 2011, the first hint of the Higgs had been announced. The evidence was spotted not just in one, but in two of the experiments being carried out at the LHC, a necessary condition for the detection.

But confirmation of the Higgs's existence and a measurement of its mass require much more work, and could take some time. What if it isn't confirmed, after all? Are we back to square one? In some ways, that would be even more interesting, because it means the standard model is incomplete or otherwise flawed, requiring the development of new physics to explain how mass is distributed among subatomic particles.

It is another area of new physics – an extension to the standard model called SUSY, for supersymmetry – that carries the hopes of astronomers as to the nature of dark matter. The idea behind SUSY is that each particle has a massive 'shadow' particle that has not yet been detected. Thus there might be an entire suite of undiscovered particles that, together, constitute a supersymmetric version of the standard model. The expected characteristics of at least one of these shadow particles would exactly fit the bill for dark matter. It would be massive, but would not interact with normal matter, except through gravity.

Finding SUSY at the LHC is turning out to be just as difficult as the hunt for the Higgs, with most of the simpler models already having been ruled out. But astronomers and physicists alike remain hopeful of a breakthrough in this area. The odds are that when the first announcement of the nature of dark matter is made, it will come not from a telescope, but from a particle collider.

Meanwhile, a recent incident at CERN provides a tantalizing glimpse of the reception that awaits the discoverers of any hint of new physics. In September 2011, an extraordinary announcement was made of exotic subatomic particles called neutrinos being clocked travelling ever so slightly *faster* than the speed of light. Once a candidate for dark matter, but now known not to exist in sufficiently large numbers, neutrinos barely interact with normal matter at all, making them very difficult to deal with. These particular ones were being fired from an accelerator smaller than the LHC, also at CERN, known as the Super Proton Synchrotron (SPS), through solid rock to a detector at Gran Sasso in Italy, 732 km away, in an experiment called OPERA (Oscillation Project with Emulsion-tRacking Apparatus). The journey time at the speed of light is 2.44 milliseconds, but these little chaps seemed to be arriving 60 nanoseconds too soon.

Although it's an idea beloved of science fiction writers, most folk know that faster-than-light travel is firmly prohibited by Einstein's Special Theory of Relativity. That's because accelerating an object to the speed of light requires infinite energy. This tried and tested rule is at the heart of our understanding of the universe, but it is just possible that the physical world actually consists of more than the three dimensions of space and one of time that we perceive – and might have hidden 'higher dimensions'. If that is the case, then objects like neutrinos might be able to take short cuts through these higher dimensions, arriving fractionally before they otherwise would, while still obeying the laws of special relativity. Thus, the result of the OPERA experiment was viewed as a tantalizing hint of the existence of such new physics, and attracted intense media interest.

The overwhelming view of the world's scientists, however, was that the faster-than-light effect would be shown to have an explanation entirely within the realm of normal physics. And, little more than six months after the original announcement, that's exactly what happened. A tiny measurement error, apparently caused by a loose connection, had resulted in an incorrect calculation of the neutrinos' flight times. Red-faced, the head of the experiment resigned his position – despite having been ultra-cautious when he'd first presented his astonishing result.

So – RIP new physics? Not necessarily. It remains possible that, some day, we might find similar evidence that does stand up to scrutiny, with implications for science that are truly staggering. The discovery of hidden dimensions would allow new thinking on a wide range of topics, including the origin of that other big cosmic mystery, dark energy, which is responsible for the accelerating expansion of the universe. At present, we have few clues as to its nature. New physics would feed directly into our understanding of reality at the most profound level.

For those of us cheering them on from the sidelines, this is a most exciting time to be watching the work of scientists at the LHC (Figure 10). There is unprecedented interest among ordinary people in what is

Figure 10. 'Walk through as if you owned the place.' The TV producer's exhortation to the author at the CERN Computer Centre did nothing to dispel the fear of immediate eviction as an imposter. (Image courtesy of Howard Sacre.)

happening there, and, as a result, the world of physics may even be starting to lose its image as the exclusive province of nerds. Books, articles, and even TV segments featuring insane roller-coaster rides, are all helping to satisfy the public's hunger for information.

The contrast between bustling high-tech experiments at the LHC and dancing aurorae silently illuminating the Arctic night could hardly be greater. But both have, at their heart, streams of subatomic particles being guided by magnetic fields. And both have much to tell us about the workings of the universe for those prepared to listen. Especially those who can do their listening on holiday – in a study tour.

ACKNOWLEDGMENTS

As always, it is a pleasure to acknowledge the support of the Director and staff of the Australian Astronomical Observatory, which is a division of the Department of Industry, Innovation, Science, Research and Tertiary Education. To my fellow-travellers in the study tours described here, I send a big thank-you for their enthusiasm and support. It's also a great pleasure to acknowledge Howard Sacre, of Nine Network Australia, for masterminding the documentary filming at the LHC. Likewise, I'm indebted to Paul Cass of AAO for ensuring that I didn't visit the Arctic without having read Lucy Jago's book on Birkeland. And, finally, none of these tours would have been possible without the consummate expertise and impeccable planning of my manager, Marnie Ogg.

FURTHER READING

CERN (2012. 'The Large Hadron Collider' (http://public.web.cern.ch/public/en/lhc/lhc-en.html, accessed June 2012).

Tim Herd, *Kaleidoscope Sky* (Abrams Books, 2007). [Stunningly illustrated work on atmospheric optical phenomena, with a useful chapter on the aurora.]

Lucy Jago, *The Northern Lights* (Hamish Hamilton, 2001). [Sympathetic account of the troubled life of Kristian Birkeland.]

Jason Palmer (2012). 'Higgs boson hints multiply in US Tevatron facility data' (http://www.bbc.co.uk/news/science-environment-17269647, accessed June 2012)

Jason Palmer (2012). 'Neutrinos clocked at light-speed in new Icarus test' (http://www.bbc.co.uk/news/science-environment-17364682, accessed June 2012)

Storms on Saturn

PAUL G. ABEL

INTRODUCTION

The atmospheres of gas giant planets like Jupiter and Saturn are violent and turbulent places. Although these two worlds are of a similar composition, even a cursory glance through a modest telescope reveals that the markings on the disk of Saturn are of a more subtle and elusive nature than those of the dramatic Jupiter. Bright, long-lived storms are a permanent feature of the Jovian atmosphere; the Great Red Spot (GRS) has been whirling in the planet's upper atmosphere for several hundred years and, similarly, the bright ovals of the south temperate region have lasted for decades. The dark belts and bright zones, too, are normally very well differentiated.

Saturn, however, is rather coy, and instead prefers to hide its details from all but the determined observer. The disk is divided into belts and zones, but in a more subtle way. Often it can be very difficult even with a large telescope to tell with any certainty where a belt ends and a brighter zone begins. Long-lived features like Jupiter's GRS seem to be absent from the Saturnian atmosphere, and the planet has gained the reputation of being a somewhat quieter version of Jupiter. In fact, Saturn's atmosphere is every bit as dynamic as Jupiter's, but all of those striking features are hidden beneath a layer of haze or smog, and it is this haze which hides all but the most prominent of atmospheric features from our telescopes.

GREAT WHITE SPOTS

In the past 136 years or so, there have been six recorded instances of huge storms occurring on Saturn. These so-called Great White Spot

(GWS) outbreaks are ten times larger than the regular storms which occur in Saturn's atmosphere, they are much rarer, and they seem to occur every thirty years or so. This has led some to suppose that these large storms occur with a change in season on the planet; remembering, of course, that Saturn takes 29.46 years to complete one orbit of the Sun, so a Saturnian season is typically 7–8 years in length. Such major storms on Saturn take the form of immense, brilliant white ovals, often forming in, and completely disrupting, a bright zone like the Equatorial Zone. The most notable storm of this type was (before 2010) the brilliant white spot recorded in 1933 by the comedian W. T. (Will) Hay, and since 2013 marks the eightieth anniversary of its discovery, the author decided to devote this *Yearbook* article to the huge storms of Saturn. (For more on Will Hay, see the article by Martin Mobberley elsewhere in this *Yearbook*.)

During the 2010–11 apparition, Saturn surprised everyone by producing the most violent, dramatic and long-lived storm ever observed on the planet. This was the first storm of any size to be detected in the Northern Hemisphere during the current Saturnian year (in the northern spring season) and it appeared roughly ten years earlier than might have been expected from previous GWS cycles; the next such outbreak had been expected to occur in about 2020. The storm produced the most powerful lightning ever detected, and the resulting storm swept through and disrupted much of the North Tropical Zone/North Temperate Zone. As with previous GWS outbreaks, this storm grew very quickly in both brightness and size; in just one week it extended from a length of about 3,000 kilometres to around 8,000 kilometres. At the time of writing (February 2012), it seems that the storm is still ongoing, albeit in a much reduced capacity. Are we seeing a new dramatic side to Saturn? It is hard to be sure, but certainly we should be on our guard!

Even with the highly successful *Cassini* spacecraft out at Saturn, there is still much work for the dedicated amateur observer to do. In the following sections, we shall look briefly at our current understanding of Saturn's atmosphere, examine the major historical storms of the past, including the great outbreak in 2010–11, and look at ways in which we can go on to detect and monitor new storms. Although we cannot be certain when a new GWS will appear, we can make sure that we are prepared and ready to study it.

ANATOMY OF SATURN

The atmosphere of Saturn is very different to that of the Earth. Saturn is a gas giant, and as the name implies, it is comprised chiefly of gases – about 88 per cent hydrogen and 11 per cent helium by mass, with traces of other substances such as methane, ammonia and water. Figure 1 shows a cutaway of the theoretical internal structure of the planet. Clearly we understand a reasonable amount about the atmosphere, but deep down in the hot murky depths, where we have yet to venture, we must assume – in the absence of evidence to the contrary – that our models of the planet's interior are reasonably accurate.

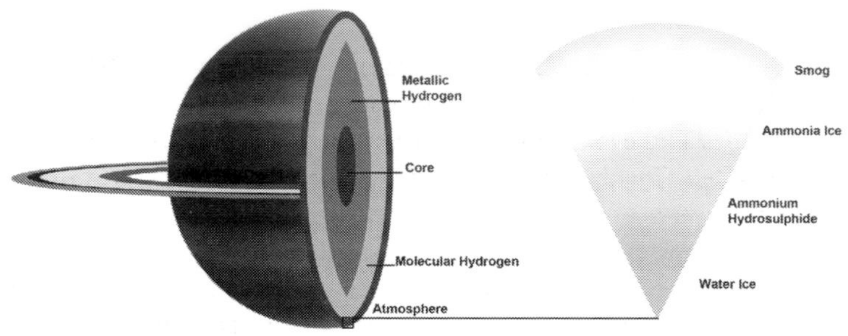

Figure 1. Cutaway of Saturn showing the three main cloud decks and the probable internal structure of the planet.

At the top of the atmosphere we have the main cloud decks. Like Jupiter's, the cloud decks are quite colourful; the author always finds the various belts and zones of Saturn to be lovely pastel shades of yellows, creams and browns. The colours are due to the various gases which are found there – the browns being compounds of hydrogen sulphide and the whites and creams being due to ammonia ice crystals.

Saturn's troposphere – the region of its atmosphere where the main 'cloud decks' are located – is thought to be differentiated into three separate layers, the composition changing as we move down through the troposphere. The boundary between the troposphere (lower atmosphere) and the stratosphere (upper atmosphere) is known as the tropopause. It is here (where the pressure is about 0.1 bar) that the high haze or smog layer is located. The temperatures in the troposphere range from about -180°C at the tropopause to about +80°C and the location of the clouds within this region is predicted based on the temperature at which vapour will condense into droplets. In other words, the point at which condensation occurs, on the atmospheric temperature profile, is where the clouds ought to be.

At the top of the visible cloud deck (about 100 kilometres below the tropopause, at a pressure of about 1 bar) we have the bright clouds of ammonia ice crystals. Up here the temperature is, on average, a chilly -140°C. As we move downwards to the next layer, the temperature and pressure increase and we come to the second cloud deck composed chiefly of ammonium hydrosulphide clouds. This layer can be found about 170 km below the tropopause, where the pressure is a few bars and is a relatively warmer -70°C. Beneath this, extending from about 250 km below the tropopause, we have the lowest cloud deck which consists mainly of water clouds at a temperature of around 0°C (close to the freezing point of water) at a pressure of about 10 bars.

As we descend ever deeper, the pressure increases and the molecular hydrogen that makes up the largest proportion of the atmosphere slowly changes to a liquid. Moving deeper still, something very peculiar happens. In the depths of the body of Saturn, the temperature and pressure are so high that the hydrogen is transformed into an extremely unusual substance called liquid metallic hydrogen. It is this vast swirling metallic substance that is thought to give rise to Saturn's considerable magnetic field. Beneath this layer there is a small, hot rocky core about ten times the mass of the Earth, located right at the centre of the planet.

Saturn, like Jupiter, gives out far more heat than it receives from the Sun. Certainly the so-called Kelvin-Helmholtz mechanism (the process whereby a planet cools, which, in turn, causes it to shrink, and so the resulting increased pressure heats up the core) accounts for much of it, though by no means all. Helium raining from higher levels to lower ones, generating friction (and therefore heat), is also thought to be a factor. This imbalance of internal heat radiated compared with the heat received from the Sun leads to a very dynamic atmosphere and is responsible for the very high wind speeds on Saturn – nearly 1,800 kilometres per hour at the equator.

One question the author is frequently asked when speaking about Saturn, is: 'How do we know it is not solid?' Certainly, if Saturn was composed of rock like the Earth, it would have a much greater mean density (mass per unit volume) and a much stronger gravitational field. Indeed, the mean density of Saturn at ~0.7 gm/cm^3 is less than that of water, so it is often said that if one could find an ocean large enough, Saturn would float! However, a further clue to the internal composition of Saturn can be gained simply by looking at it. Unlike the rocky terrestrial planets, Saturn is decidedly oblate in appearance and this is due to its rotation. Any rotating body tends to form an oblate spheroid (i.e., it is flattened at the poles and bulges at the equator) rather than a sphere, and the difference between the equatorial and polar diameters of a planet is called the equatorial bulge. Saturn, which has an equatorial diameter of 120,536 km and a polar diameter of 108,728 km, is the planet with the largest equatorial bulge in the Solar System (11,808 km). Saturn also rotates as a non-solid body in that there are slightly different rotational periods for the equator and the remainder of the planet: Saturn's equator rotates once in 10 hours 14 minutes 00 seconds (this is known as System I), while the rest of the planet (System II) rotates once in 10 hours 38 minutes 25 seconds. There is a further system of longitude known as System III, which refers to the rotational period of Saturn's interior. Latest *Cassini* estimates place System III to have a period of 10 hours 32 minutes 35 seconds. To the newcomer this may sound a little confusing, so included here is a Saturn drawing with Systems I and II indicated on it (Figure 2). The names of the various belts and zones are also shown.

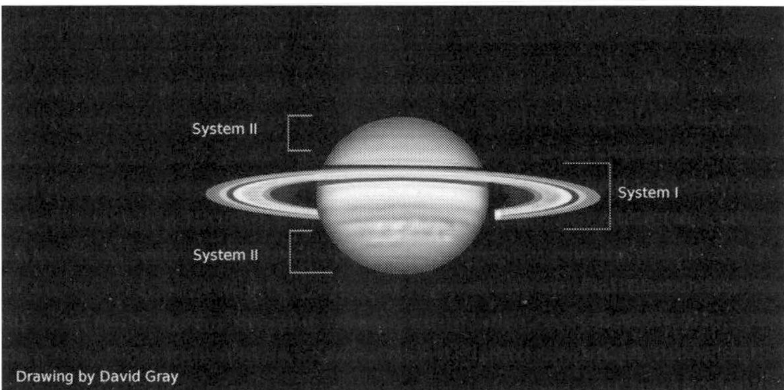

Figure 2. Saturn drawing showing the darker belts and lighter zones, with the locations of the System I and System II longitude systems indicated.

SATURN STORMS

Broadly speaking there are two types of Saturn storm. There are the small local storms which take the form of subtle white ovals, normally to be found within the various belts and zones of the planet. These may last anywhere from a few days to a number of months. They generally require a medium sized telescope (of, say, 150 mm aperture) or larger to see them. There may be a number of such ovals visible in any one apparition. There are also convective cloud systems with typical sizes ~2,000 kilometres, characterized by their brightness, irregular shape and rapid evolution in a matter of hours. Such features have been observed predominantly near planetographic latitude 38°N during the *Voyager* spacecraft flybys in 1980–81 and at latitude 38°S since 2002.

The truly large-scale storms – the Great White Spots – usually take the form of an immense white oval, often situated within the planet's Equatorial Zone (EZ), although the storm of 2010–11 occurred in the North Tropical Zone. These huge storms are quite rare, but when such an outbreak occurs they tend to be very brilliant and even a small telescope will show them. Great White Spots tend to be rather dynamic affairs, and over the course of a few months they usually spread out, making the zone in which they are located appear noticeably brighter; the belts near to the zone may also be affected.

While we cannot say with absolute certainty what the causes of the various storms are, it seems that the smaller ones are a function of the atmospheric dynamics common to planets like Jupiter and Saturn. It is thought that the GWS outbreaks are likely to be seasonal and seem to occur about once every Saturnian year (~30 Earth years). It is also interesting to note that all of the planet's six major GWS have occurred in the northern hemisphere of the planet – from the EZ(n) northwards.

The mechanism by which a GWS is formed is thought to be due to convection: the warmer gases lower down in Saturn's atmosphere push their way upwards to the colder cloud tops. Once at the top, the ammonia which existed as a vapour lower down is frozen out into ice crystals, and it is these bright ice crystals which give the storms their unique brilliant white appearance. Eventually, the powerful winds in the planet's atmosphere disperse the ice crystals, causing the ovals to become spread out longitudinally until they are dispersed along the zone, on occasions completely encircling the planet along an entire latitude band. When such a storm erupts they are probably the most striking features one can observe on Saturn.

GREAT STORMS OF THE PAST

In the early days of planetary astronomy, it was clearly important to build up some basic facts about the planets. As we have seen, the cloud markings on Saturn are very subtle and so the arrival of a brilliant GWS meant that estimates could be made of the rotational period of the planet. By timing how long it takes for a storm on the central meridian (an imaginary line joining the north and south poles through the centre of the planet's visible disk) to return to that position, it is possible to estimate the rotational period of Saturn.

Although the Italian astronomer Giovanni Cassini observed what he described as 'bright streaks' on the disk of Saturn, the first proper GWS outbreak to deliver useful information about the planet was discovered by the American astronomer Asaph Hall. On 7 December 1876, Hall was observing Saturn with the US Naval Observatory's 66-cm refracting telescope (the largest refracting telescope in the world at the time) when he noticed '. . . a well-defined spot, 2 or 3 arc seconds in diameter, situated just below the ring of Saturn'. Hall continued to observe the spot, although by 2 January 1877 he reported that the spot had become noticeably fainter and much less well defined. After this it seems he was hampered by poor weather and the planet's low altitude. As a result he was unable to observe it further. From his observations of the spot (which was at a planetographic latitude of 8° ± 3°N), Hall obtained an estimate for Saturn's rotational period of 10h 14m 23.8s, remarkably close to the currently accepted value!

There were further smaller bright spots recorded in 1891 and 1892, this time by Stanley Williams, who obtained a rotational period of 10h 14m 21.8s. The enthusiastic amateur astronomer and author of one of the early major popular books on astronomy (*Telescopic Work for Starlit Evenings*), William F. Denning, also observed these spots with his telescope in Liverpool. He obtained a rotational period of 10h 14m 23.8s.

Prior to the GWS outbreak of 2010–11, which is the longest lasting planet-encircling storm ever seen on Saturn, the previous record holder was the GWS of 1903, which lingered for 150 days. This outbreak (which occurred at a planetographic latitude of 36° ± 2°N) in the North Tropical Zone was discovered as a brilliant white, oval spot by American astronomer Edward Emerson Barnard on 15 June 1903. Subsequently, many observers recorded a large number of white or yellowish spots that developed at about the same latitude from July to December 1903.

The next major storm to occur on the planet was remarkable both in its brilliance and its discoverer: the Great White Spot of 1933, discovered by the celebrated stage and screen comedian Will Hay. Hay was a remarkable man in many ways. Not only was he a natural performer and a talented comic, but he was also a serious amateur astronomer, and he had established his own private observatory which housed an excellent 150 mm refractor.

On the night of 3 August 1933, at 22h 35m, Hay noticed a brilliant white spot in the EZ of Saturn (at a planetographic latitude of 2° ± 3°N). Hay must have been particularly observant since Saturn resided in the constellation of Sagittarius at the time and was inconveniently low down from the latitudes of the British Isles. Hay made an initial drawing (Figure 3a) of the GWS, and his drawing shows an unmistakable brilliant white oval, well defined on its following side and clearly the brightest feature on the disk. Another of Hay's drawings is shown in Figure 3b.

3a

3b

Figure 3. Two white spot drawings by W. T. Hay, (a) 3 August 1933 and (b) 9 August 1933.

Hay continued to watch Saturn over the following months, but as early as 12 August, he noted that the spot had lengthened appreciably. Hay calculated a rotational period of 10h 16m 17s for the spot, and watched as it followed the traditional pattern of dispersing around the EZ, eventually turning the whole of the EZ into a bright white region.

Almost thirty years later, in 1960, another GWS presented itself. In the early hours of 31 March 1960, the South African amateur astronomer J. H. Botham was observing Saturn. Normally Botham found the planet to have a somewhat bland appearance, but on this particular night he was struck by the appearance of a large white oval in the planet's North North Temperate Zone (at a planetographic latitude of 58° ± 1°N). Botham found that the spot took seventy minutes to cross the central meridian, but by 12 April, he noted that it took one hour twenty minutes. Clearly the GWS had lengthened in the same manner as its predecessors, and Botham was of the impression that the zone to either side of the spot was brighter. The spot was also discovered independently by the French astronomer Dr Audouin Dollfus using the fine 60-cm refractor at Pic du Midi on 27 April 1960. Other observers noted bright spots in the unusually active and bright NNTZ from April through to September 1960. Subsequently, on 28 July 1962, Patrick Moore recorded a bright oval in the EZ using George Hole's 61-centimetre reflector (Figure 4).

Then, on 25 September 1990, American amateur astronomer Stuart Wilber, using a 25-cm reflector at Las Cruces, New Mexico, discovered a bright white spot, again in the EZ. He informed Clyde Tombaugh (discoverer of Pluto back in 1930) and very soon both amateurs and professionals were observing the spot, which was observed to drift southwards in latitude during the first days from 12° N to 5° ± 1°N. By

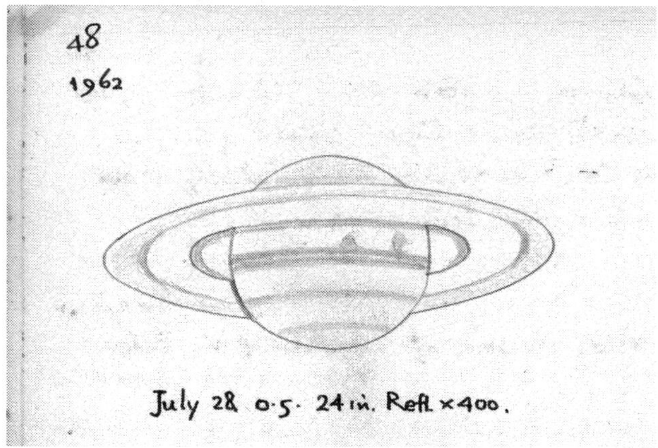

Figure 4. Drawing by Patrick Moore, made on 28 July 1962 with a 61-cm reflector and a power of x400, showing a large white spot in Saturn's Equatorial Zone.

10 October 1990, the whole of the EZ had been transformed into a brilliant white, as shown in David Gray's observation four days later (Figure 5a). This GWS outbreak was subsequently imaged by the Hubble Space Telescope (Figure 5b). Activity in the equatorial region

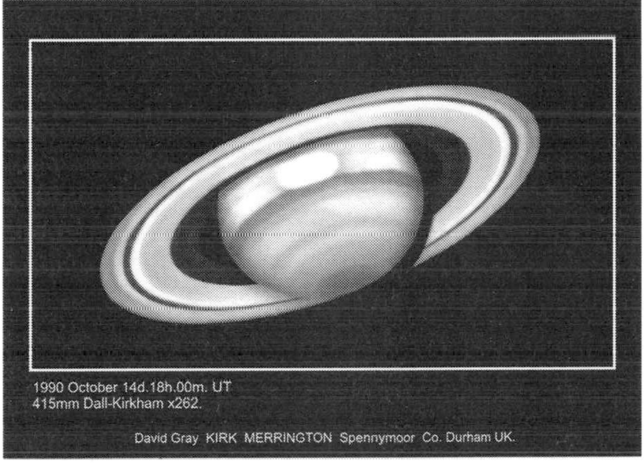

Figure 5a. White ovals in Saturn's Equatorial Zone as seen by David Gray at 1800 UT on 14 October 1990 with his 415mm Dall-Kirkham reflector, using a power of x262.

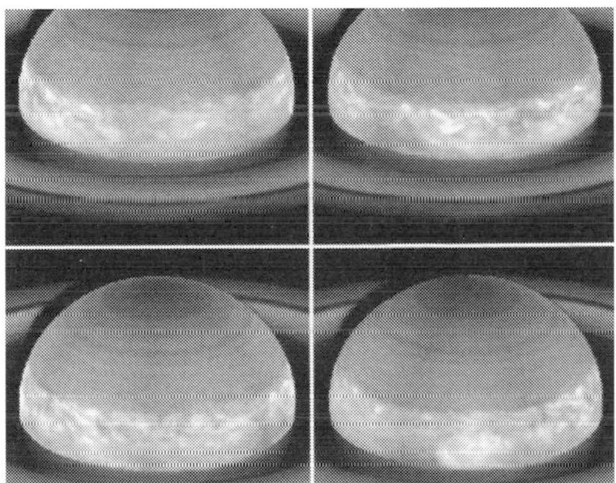

Figure 5b. Four frames from a movie of a GWS outbreak obtained by the Hubble Space Telescope's Wide-Field Planetary Camera on 17 November 1990. The movie covered one complete rotation of Saturn. The storm extended completely around the planet; in some places it appeared as great masses of clouds and in others as well-organized turbulence. (Images courtesy of NASA/Space Telescope Science Institute.)

was more or less continuous between 1991 and 1993 following the 1990 GWS event.

In July 1994, a further spot emerged, which was not the classical oval shape, and did not really fit the pattern of a typical GWS. (It endured through 1995, and was followed in 1996 and 1997 by other smaller-scale bright spots.) Looking rather like an arrow head, the Hubble Space Telescope imaged this unusual storm in the EZ of the planet on 1 December 1994 (Figure 6). Unfortunately, the planet was low down from the UK and not very well observed.

AN UNPRECEDENTED SATURNIAN STORM

The most dramatic, dynamic and unprecedented storm must be the GWS outbreak of 2010–11. (The author, for one, has never seen anything like it on Saturn!) The first hints that something unusual had occurred came when George Fischer announced that the radio and plasma wave instrument on the *Cassini* spacecraft had detected very strong electrical activity, probably due to powerful lightning bolts in a convective thunderstorm on Saturn, on 5 December 2010.

These electrical discharges were thought to be a consequence of seasonal changes in the northern hemisphere of the planet, which had passed through the vernal (spring) equinox in 2009. On the same day, 5 December 2010, *Cassini's* high-resolution cameras captured the first images of the storm, which rapidly became the largest ever observed on the planet from an interplanetary spacecraft.

The storm first appeared at a planetographic latitude of about 38° ± 1°N, although it subsequently drifted northwards to 41° ± 1°N and eventually wrapped itself around the entire planet to cover approximately 5 billion square kilometres. The biggest disturbance *Cassini* had previously witnessed on Saturn occurred in a latitude band in the southern hemisphere called 'Storm Alley' because of the prevalence of thunderstorms in this region. That storm lasted several months, from 2009 into 2010, and was actually a cluster of thunderstorms, each of which lasted up to five days or so and affected only the local weather. The recent great northern hemisphere disturbance is thought to be a single thunderstorm that raged continuously for more than 200 days and impacted almost one-fifth of the entire northern hemisphere!

Figure 6. A rare storm, which appeared as a white arrowhead-shaped feature near Saturn's equator, is shown in this composite image from the Hubble Space Telescope's Wide-Field Planetary Camera (WFPC2) on 1 December 1994. The east–west extent of this storm was about 13,000 kilometres. The image demonstrated that the storm's motion and size had changed little since its discovery in September 1994. The image is sharp enough to reveal that Saturn's prevailing winds shape a dark 'wedge' that eats into the western (left) side of the bright central cloud. The planet's strongest eastward winds are at the latitude of the wedge. To the north of this arrowhead-shaped feature, the winds decrease so that the storm centre is moving eastward, relative to the local flow. The clouds expanding north of the storm are swept westward by the winds at higher latitudes. The strong winds near the latitude of the dark wedge blow over the northern part of the storm, creating a secondary disturbance that generates the faint white clouds to the east (right) of the storm centre. The storm's white clouds are ammonia ice crystals that form when an upwelling of warmer gases pushes its way through Saturn's frigid cloud tops. (Image courtesy of NASA/Space Telescope Science Institute.)

The *Cassini* spacecraft's image mosaics and animations showed the GWS outbreak from its emergence as a tiny spot in a single image on 5 December 2010, through its subsequent growth into a storm so large that it completely encircled the planet by late January 2011 (Figures 7 and 8). *Cassini* subsequently acquired hundreds of images of the outbreak as part of the 'Saturn Storm Watch' campaign. These images,

Figure 7. This composite image from the *Cassini* spacecraft's wide-angle camera shows the tail of Saturn's huge northern storm on 12 January 2011. The head of the storm is beyond the horizon. The rings appear as a thin horizontal line in this view, which looks toward the northern, sunlit side of the rings from just above the ring plane. The shadow of the moon Enceladus is visible on the planet in the lower left of the image. (Image courtesy of NASA/JPL-Caltech/Space Science Institute.)

together with other high-quality images collected by *Cassini* since 2004, enabled scientists to trace back the subtle changes on the planet that preceded the storm's formation and have revealed insights into the storm's development, its wind speeds and the altitudes at which its changes occur. Although the storm's active convective phase ended in late June 2011, the turbulent clouds it created still lingered in the atmosphere a year after the storm's first appearance. Its 200-day active period also makes it the longest-lasting planet-encircling storm ever seen on Saturn. (The previous record holder was the outburst sighted in 1903 which lasted for 150 days.)

From the outset, the GWS outbreak was observed by amateurs. Indeed, on 5 December 2010, a small spot was detected in an image obtained by Japanese amateur astronomer T. Ikemura at planetographic

Figure 8. This image, captured by the *Cassini* spacecraft on 25 February 2011, was taken about 12 weeks after the GWS outbreak began, and the clouds by this time had formed a tail that wrapped around the planet. Some of the clouds moved south and got caught up in a current that flows to the east (to the right) relative to the storm head. This tail, which appeared as slightly blue clouds south and west (left) of the storm head, may be seen encountering the storm head in this view. The shadow cast by Saturn's rings has a strong seasonal effect, and it is possible that the occurrence of powerful storms in the northern hemisphere is related to the change of seasons after the planet's August 2009 equinox. (Image courtesy of NASA/JPL-Caltech/Space Science Institute.)

latitude 37.5°N. Subsequently, Chris Go of the Philippines managed to image the storm on 13 December, at which point it took the form of a small, very bright white oval-shaped region. By the time UK amateur Damian Peach imaged the storm on 26 December, it had clearly evolved; it had both brightened and disrupted the North Tropical Zone considerably. Another UK observer, Dr Richard McKim, produced an excellent drawing showing how the storm appeared to him on 4 January 2011 using his 410 mm reflector (Figure 9).

As usual the weather conditions conspired against the author, who did not manage to get a look at the region containing the storm until

Figure 9. Drawing of the 2010–11 storm by Dr Richard McKim in Upper Benefield as observed on 4 January 2011 at 0600 UT with his 410 mm Dall-Kirkham reflector, using powers of x256 and x410.

March! By then the storm had developed into a long trailing object, and its appearance resembled a comet. The author's drawing (Figure 10) shows how the storm appeared through a blue Wratten #80A filter. Even though it had been raging for months, it was still a brilliant feature. At the time of writing (February 2012) it appeared that there were still remnants of the storm in Saturn's North Tropical Zone.

Joerg Mosch has published drift charts of the NED (Northern Electrostatic Disturbance) because of the increase in radio and plasma

Figure 10. The author's drawing of the 2010–11 storm using a Wratten #80A filter, made on 28 March 2011 at 2353 UT with a 203 mm Newtonian reflector, and a power of x250.

interference on the ALPO Japan website (http://alpo-j.asahikawa-med.ac.jp/kk11/s110728r.htm). From his measurements (using over 280 videos), Mosch, using WINJUPOS 9.1.5, has determined that the storm had a drift rate of some 2.8° per day, and went around the whole of the planet once in just 160 days.

Analysis of this remarkable storm will continue for some time, but in their paper in the journal *Nature*, Agustin Sánchez-Lavega and colleagues[1] have suggested (using numerical simulations based on observations of the storm) that the winds of Saturn continue unperturbed down to at least the warmer, water, cloud deck. Interestingly, solar radiation cannot penetrate to this depth. They also suggest that the storm was a manifestation of seasonal changes now occurring on the planet (although it is admitted this is much earlier than would normally be expected, if a cycle of about one Saturnian year is adopted).

In a subsequent paper in the same issue of *Nature*, George Fischer and colleagues[2] present their findings in relation to the enormous lightning discharges that the *Cassini* spacecraft detected. They explain the spectacular visible white plume as the result of high-altitude clouds which moved up and past the ammonia cloud layer due to the strong vertical convection that such thunderstorms generate. The team reported a peak rate of up to ten lightning flashes a second, and suggests the storm is probably the result of seasonal changes. The storm was one of the most powerful and dramatic to occur on Saturn for many decades, perhaps the most powerful of the six Great White Spot outbreaks recorded to date on the planet (Figure 11).

In a detailed analysis of the long-term evolution of Saturn's sixth Great White Spot event in 2010–11, from ground-based observations, submitted to the journal *Icarus*, Agustin Sánchez-Lavega and colleagues[3] concluded that '. . . the GWS formed a planetary-scale disturbance that encircled the planet in 50 days, covering the planetographic latitude band between 24.6° and 44.8°N, or about 22,000 kilometres in meridional extent and 280,000 kilometres in full zonal circumference length. The head of the GWS was located at an averaged

[1] Sánchez-Lavega, A., *et al.*, Deep winds beneath Saturn's upper clouds form a seasonal long-lived planetary-scale storm, *Nature*, **475**, 71–74, 7 July 2011.

[2] Fischer, G., *et al.*, A giant thunderstorm on Saturn, *Nature*, **475**, 75–77, 7 July 2011.

[3] Sánchez-Lavega, A., *et al.*, Ground-based Observations of the Long-term Evolution and Death of Saturn's 2010 Great White Spot, *Icarus*, submitted in revised form: 8 May 2012.

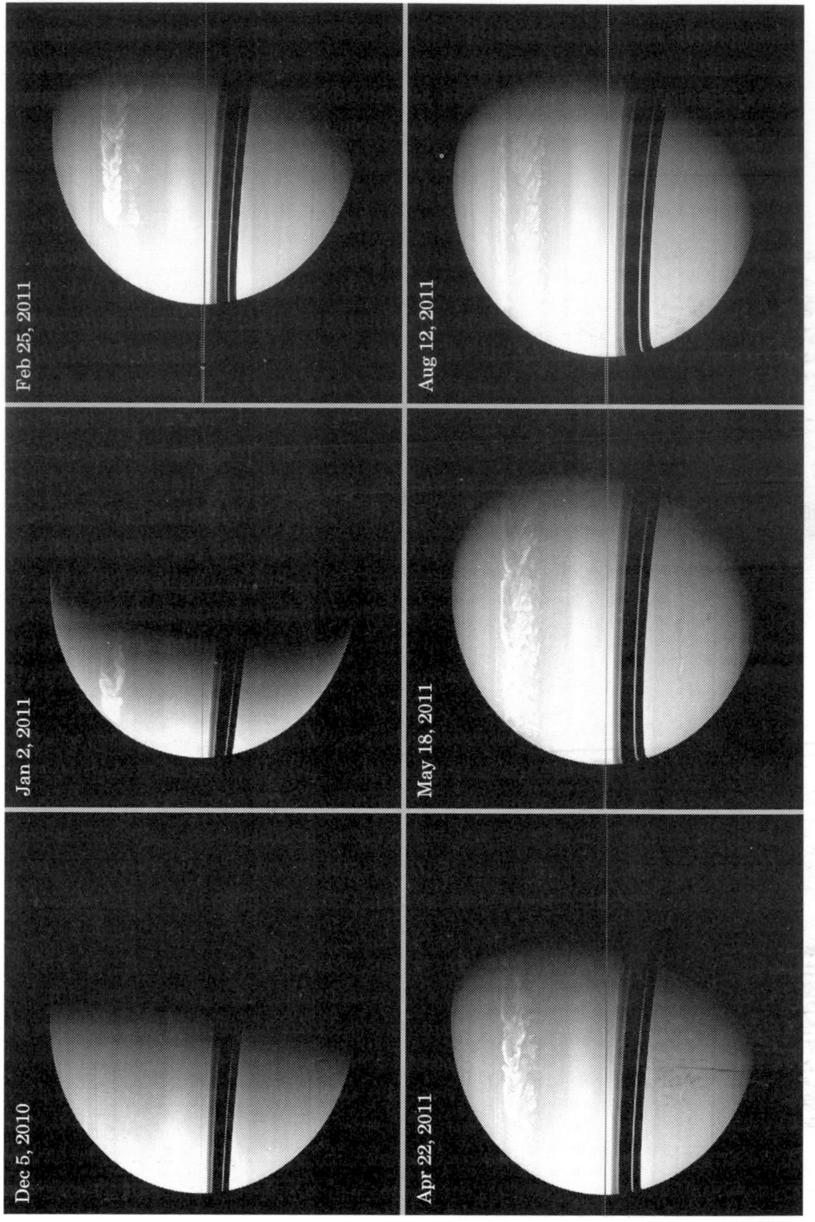
Dec 5, 2010
Jan 2, 2011
Feb 25, 2011
Apr 22, 2011
May 18, 2011
Aug 12, 2011

latitude of 40.8 ± 1°N in the peak of a westward jet and showed a mean linear drift in System III longitude of 2.793 degrees per day, equivalent to a mean zonal velocity of u = -27.9 metres per second, with maximum speed fluctuations around this mean of -5.3 to + 2.7 metres per second.'

CONTINUAL MONITORING

As we have seen, we can expect large GWS outbreaks to occur about every thirty years or so, but the 2010–11 storm has shown that powerful dynamic storms can suddenly manifest themselves on the planet at any time!

If you have a telescope of at least 150 mm aperture, then it is strongly recommended that you keep an eye (either Mark 1 or digital) on the planet. Remember the features of Saturn are subtle, so it takes some patience before the details become more obvious. Do not expect to see fine details when you have just stepped out of a brightly lit room to a telescope which has yet to reach equilibrium with the outdoor temperature!

Figure 11 (opposite). This series of images from the *Cassini* spacecraft chronicle the GWS outburst from its start in late 2010 through mid-2011, showing how the distinct head of the storm quickly grew large but eventually became engulfed by the storm's tail. In the earliest image of the storm, taken on 5 December 2010 (top left), the storm appears only as a small, white cloud on the terminator between the day side and night side of the planet. The next view, taken on 2 January 2011 (top middle), shows that the head quickly grew much larger and a tail began to trail a great distance eastward. Some of the clouds moved south and got caught up in a current that flows to the east (to the right) relative to the storm head. By 25 February (top right image) this tail (which appeared as slightly blue clouds south and now west (left) of the storm head), can be seen encountering the storm. The 22 April image (bottom left) is one of *Cassini's* last views of the storm when it still had a recognizable head; the tail is south of the head and is well established by this time. The 18 May view (bottom middle) shows only the storm's tail. The head still existed at this time, but it is beyond the horizon and out of the field of view here. Between the time of the 18 May image and the last image on 12 August (bottom right), the head of the storm was engulfed by the part of the storm's tail that spread eastward at the same latitude as the head. The head has now lost its distinct identity and is just part of the jumble of the storm. (All images courtesy of NASA/JPL-Caltech/Space Science Institute.)

Always use a suitable magnification for the seeing conditions; many people are often tempted to push the magnification as far as their telescope will allow, but this is unwise, because unless the air is very steady, all you will get is a large pudding-shaped blob in the eyepiece! Magnifying powers from x160 – x250 are quite suitable for Saturn. If you have one, try using a blue filter (e.g. Wratten #80A) since bright cloud markings and storms often stand out much better in this colour light than in just integrated (or unfiltered) light. Remember, though, if you do observe something out of the ordinary, always put the date, time (in UT) and the telescope details that go with your drawing or image, otherwise the observation is of little value!

You can report your findings to the Saturn Section of the British Astronomical Association (BAA – www.britastro.org) – the leading UK body for amateur observers. The BAA has its own bi-monthly *Journal* and the various observing sections turn amateur observations into good-quality science: it's well worth becoming a member. Other organizations, such as the Association of Lunar and Planetary Observers (ALPO – http://alpo-astronomy.org/), will also be glad to receive your observations. One of the things you can do is to time accurately when a spot crosses the central meridian of the planet, and thereby determine its longitude. From several such observations it is possible to determine how long it takes the spot to go around the planet. By plotting the changing longitude over time you can also determine the drift rate in longitude. Storms in Saturn's atmosphere are subject to the jet streams in which the storms are situated, so by measuring their drift rates one can determine the speed of the jet streams.

With most large professional telescopes employed in more fundamental studies, it seems likely that amateur astronomers will be among the first people to observe the next great Saturn storm, and they have always been instrumental in recording and monitoring them.

Current plans to continue the *Cassini* spacecraft's mission until 2017 will provide opportunities for *Cassini* to witness further changes in the planet's atmosphere as the seasons progress from northern spring to northern summer.

Johannes Hevelius (1611–1687): Instrument Maker, Lunar Cartographer and Surveyor of the Heavens

ALLAN CHAPMAN

We often think of the Renaissance as an Italian phenomenon, with great figures like Columbus, Leonardo, Michelangelo and Galileo initiating new realms of endeavour in the arts, the sciences and the world of ideas. Yet there was a movement no less exhilarating which took place in Northern Europe – in Great Britain, France, Germany and the Baltic countries – albeit slightly later. For who can deny the world-changing impact of 'northerners' like René Descartes, Martin Luther, William Shakespeare, Rembrandt van Rijn and Albrecht Dürer in philosophy, theology and the arts disciplines? And in the sciences, indeed in astronomy alone, where would modern civilization be without Copernicus, Tycho Brahe, Robert Hooke, and the early Royal Society – and Johannes Hevelius? Yet while most of the above figures might be classed as 'household names', Hevelius seems to have fallen through the net of wider cultural memory. But, writing at the very end of this his four hundredth anniversary year, I would like to reinstate him; for in his own day, Hevelius was universally considered to be one of the giants of European astronomy.

THE 'NORTHERN RENAISSANCE' IN ASTRONOMY

As a native of Dantzig, Poland, though of German-speaking Lutheran Protestant descent, Hevelius was very much aware of what had taken place in astronomy in the two hundred years before he was born, on 28 January 1611. There had been his fellow-Pole Nicholas Copernicus, whose *De Revolutionibus Orbium Coelestium* ('On the Revolution of the Heavenly Spheres') had put forward the first coherent model of a Sun-centred, or heliocentric, Solar System. And whilst a fatal stroke in Poland had coincided with the publication of his book in Nuremburg in 1543, we must not forget that his ideas had been circulating around scientific Europe for thirty years beforehand, and that Latin pamphlets announcing Copernicus's ideas by his disciple Joachim Rhaeticus had been widely distributed in 1540 and 1541, without a backlash of any kind. There had even been an untroubled second edition of *De Revolutionibus* printed in Basel, Switzerland, in 1566.

Yet Copernicus, like many of his scientific contemporaries, had been aware of major weaknesses in his theory. For if the Earth really was spinning in space, why were we not flung off? And if the Sun was at the centre of rotation, why did a dropped brick fall to Earth, rather than flying off to the Sun? Copernicus' theory, in fact, was deeply mathematical in its argument and presentation, and in it he suggested various geometrical analogies which might provide evidence for the Earth's motion, but no clear proof.

And then, a generation later, and also on the Baltic, there had been Tycho Brahe, the exotic Danish nobleman who had built the most sophisticated pre-telescopic observatory in the world, although he admitted by 1586 that he could detect no seasonal stellar changes which might suggest that the Earth was in motion. Yet Tycho's investigation into whether or not the Earth moved in space came not from abstruse philosophical argument or from ingenious geometrical theorems, but from direct observations and measurement, using a progressive 'family' of exquisite mathematical instruments, each succeeding generation of which improved upon the accuracy and engineering sophistication of its predecessor. Indeed, I would suggest that Tycho was the first research scientist to identify scientific investigation with a progressive instrument technology. For Tycho saw new and better physical data as essential, fundamental fodder for new theoretical investigation. For only a new and growing accumulation of accurate celestial measurements could break out of the tail-chasing loop of sometimes centuries-old observations being given new theoretical twists as a way of explaining astronomical problems.

Indeed, this new fascination with technology was everywhere to be seen in the world of 1611, when Hevelius was born: printing presses, windmills, foundries with great water-powered forge hammers, firearms, complex clocks, musical organs, glass factories, spectacles, brewing on an industrial scale, and scores of new inventions related to ships and navigation – all too obvious, indeed, to someone living in a great sea port like Dantzig. And on his travels in Holland in the early 1630s, Hevelius would have become very much aware of the great windmill-powered pumping machines being built to drain the polder and reclaim hundreds of square miles of land from the sea.

In short, machines were everywhere in early seventeenth-century Europe: in science, in industry, and even in personal gadgets such as chiming pocket-watches, musical boxes and wind-up toys. And as the son of a well-to-do merchant living in a great maritime city, the young Hevelius would have been familiar with them. Hevelius's northern 'Baltic' astronomical inheritance and the fact that he lived in a society in which technical devices played an increasingly important role in life give us a sense of what helped to make him the practical astronomer that he became.

JOHANNES HEVELIUS THE MAN

Though known to the world in the Latinized spelling of his name, Jan Hewelke was born the son of Abraham Hewelke (Höwelke or Heweliusza) and Kordula Hecker. His parents seem to have been of Germanic origin, from Bohemia (modern south-east Germany or the Czech Republic), and were German-speaking Protestant Lutherans, though living and trading peacefully in a predominantly Roman Catholic country. Abraham Hewelke was a master brewer and merchant, while his wife Kordula was also of rich merchant stock – indeed, part of what might be called the new international 'merchant aristocracy' of the age. People belonging to this class were hard-working, thrifty, and public-service-minded, and tended to put a high value upon education and a knowledge of how the wider world worked.

As a boy Jan, or Johannes, was sent to schools where he could master Polish, and, also important for a merchant and a future man of learning, the classical languages of Latin and Greek, the root languages of European civilization and education, whether biblical studies, law, Greek philosophy and science, or the bawdy Roman poets! In fact, he became a fluent and elegant Latin writer, and all of his great astronomical works were written in that language, which made them instantly accessible to an educated Dutchman, Frenchman, Irishman, Spaniard, Italian, or German, who could read them without the need for a translation. Just as they could, for instance, read the Latin books of Tycho, Copernicus, or even of eminent doctors such as William Harvey. For Latin in the seventeenth century was very much a living language, and taught in every good school from Lisbon to Stockholm. Hevelius also possessed a good knowledge of French, German and, probably, English. Sadly, in our own day, however, Hevelius's Latinity has led to most people being unable to read him, hence diminishing his historical importance, and his works have never been translated. (In February 2011, however, Professor Mark Birkinshaw of Bristol University gave a splendid paper on him to the Royal Astronomical Society: see 'Further Reading' below.)

At the age of sixteen, Jan was sent to the gymnasium, or senior grammar school or college, in Dantzig, where he fell under the influence of a man who was to shape his future fame, to make him become something more than just another rich merchant. This was Peter Krüger, a schoolmaster who was also a skilled mathematics teacher and practitioner of astronomy, as well as being an admirer of Copernicus and Tycho. Then in 1630, when he was nineteen, Hevelius left Dantzig to complete his education abroad. First, he went to the famous Dutch university of Leiden to study law, which was seen as a natural training for a man expected to play an active public role in the life of his native city. For in the years ahead, quite apart from his fame as an astronomer, Hevelius would not only become a highly successful brewer and merchant, but would serve as Master of the Dantzig Brewers' Guild (rather similar to the City of London Livery Companies), a City Councillor, and Mayor of Dantzig, with good connections to the Royal Family of Poland.

In addition to his legal studies at Leiden, Hevelius travelled to England and France, making friendships in Paris with the famous Roman Catholic priest-astronomers Pierre Gassendi and Marin Mersenne, and the Catholic layman Ismail Boulliau (Bulliadus in Latin); and while in Avignon, on France's southern border with Italy, he met the German Jesuit Athanasius Kircher. It was Kircher, indeed, who, in addition to being an astronomer and mathematician of international standing, invented the 'magic lantern' for projecting images on to a screen using lenses, which was the ancestor of the modern projector. Formidable personages for a 22-year-old to get to know, and with whom to forge enduring friendships.

His studies and travels complete, Hevelius returned to Dantzig in 1634 to begin his life in business and public affairs. And as beer was the staple drink of northern Europe, and Dantzig was a great trading centre, it was not difficult for a brewer to become rich, especially if, like Hevelius, you brewed the special Jopen Polish beer. In 1635, he married Katherina Rebeschke, the daughter of a merchant, who brought her own property – in terms of Dantzig houses – into the marriage.

Yet another aspect of Hevelius the man was to play a crucial role in his subsequent astronomical achievement; for Hevelius was also a talented artist. And in addition to being a fine draughtsman (he had probably received some of his training from Peter Krüger), he mastered, around 1635, the delicate art of copper engraving. This enabled him to do two things: first, to transfer his own beautiful drawings of his observatory and its instruments from paper on to engraved copper plates; and second, to engrave upon his large brass sextants and quadrants the exquisitely precise measuring scales with which he read off angles in the sky. For the engraved copper plates of his drawings could, of course, be inked, put into a printing press, and turned into the breathtakingly beautiful technical plates that illustrated all of his great astronomical publications. Many of his signed published drawings make the source clear in the abbreviation '*J. Hevelius inv. et sculp.*' ('Johannes Hevelius devised and engraved'). The names Stech, Saal, Visscher, and Boij also appear as artists and engravers on quite a few of his published plates, especially those depicting large-scale views of instruments, the observatory, or incidental decorations; although it was always Hevelius himself who did the fine detailed drawing and engraving of the technical parts. Some of the beautiful surviving portraits of Hevelius, along with co-artist names on his plates, would suggest that Hevelius had several artist friends (Figure 1). And without doubt, his great published works were all sumptuously illustrated, which makes it possible for us to reconstruct numerous details of his observatory and the instruments which it contained. They also indicate that Hevelius saw art and science as intimately related, as did most seventeenth-century people – for art in those days resided in the true representation of nature.

And perhaps leading naturally on from his skill as an artist and technical illustrator, Hevelius was also a gifted hands-on craftsman and deviser. For while he no doubt employed professional metal-founders to cast the large parts of his observatory instruments, the fine finishing appears to have been his own handiwork. Indeed, what immediately

Figure 1. Portrait of Johannes Hevelius. (From *Selenographia*, Dantzig, 1647.)

strikes one when reading or examining his books is the deep, hands-on practicality of the man. First and foremost he was an observer, rather than a theoretician; and while the physical theories which he sometimes touches upon, such as when trying to explain why comets appear to move in parabolic orbits, might be wrong by purely modern standards, his skill as an instrument *user* was accomplished in the highest degree.

OBSERVING THE MOON AND PLANETS

Knowing what we do today, it is hard to think ourselves back into the mindset of pre-telescopic astronomy, or to imagine the bafflement which the first telescopic images of astronomical bodies occasioned. For it was exactly eighteen months before Hevelius's birth, on 26 July 1609, that the Englishman Thomas Harriot made the first – and still surviving – telescopic drawing of the Moon. Four months later, Galileo, living near Venice, made his own first drawings, and unlike

Harriot, published them in March 1610. Their impact reverberated around Europe. So the ancient Greeks had been wrong, for the Moon, instead of being a mirror reflecting the Earth, or a tarnished silvery ball, was a *world* in its own right, with shadow-casting geographical features.

Johannes Hevelius was one of the first astronomers in Europe to examine the Moon through the ×50 and even more powerful telescopes that were becoming available from the improved optical technology of the 1640s. And as a skilled craftsman in his own right, Hevelius described and illustrated an advanced lens-grinding machine of his own devising in his great lunar treatise, *Selenographia* ('Account of the Moon') in 1647. For with that machine he figured the 1.5-inch-diameter object glasses for his 6- and 12-foot-focus lunar telescopes. And like a good scientist, Hevelius described the design and manufacture of *all* the instruments with which he made his discoveries, as a guide for subsequent researchers. (To make certain that the reader could fully understand the inner working parts of his instruments, he used the convention of depicting the parts of the dismantled instrument as lying on the floor before the assembled, working piece.)

But it is in Chapter 8 that he really gets down to the new and perplexing business of trying to interpret the lunar surface. He draws it in detail, names many features – some of his names still survive – and prints his drawings in meticulously detailed engraved copper plates. And in *Selenographia* he did two things which were of especial importance. First, he printed forty lunar phase plates, showing the shadow changes and emergence of topographical features during the course of a lunation, before bringing all the drawings together in the first ever finely detailed published maps of the full Moon. Second, he made a physical study of the libration changes – originally discovered by Galileo in 1632 (latitude libration) and in 1637 (longitude) – and drew them on opposite sides of the big full Moon plate. On the other hand, lacking any idea of gravitational theory – which lay forty years in the future – Hevelius had no coherent explanation for the libration, or for the changing terrestrial, lunar and solar distances which caused it.

Hevelius's map was not the first printed Moon map, for Galileo had published a few rudimentary woodcut drawings in 1610, while Michael van Langren (Langrenius) and Matthias Hirschgarter had independently issued fine but less detailed printed Moon plates in 1643; but Hevelius was streets ahead of these in terms of technique, rigour, size and detail, and in the whole scope of his presentation. Quite simply,

Hevelius's printed plates – each one a foot in diameter – were the foundation-stone of all subsequent lunar cartography (Figure 2). And as one finds in most of Hevelius's subsequent sumptuously illustrated books, *Selenographia* is full of visual symbols indicative of the new science. For in addition to the allegorical title pages, replete with optical symbols, the corners of each Moon plate depict winged cherubs or classical art *putti*, with banners carrying Latin biblical quotations about the heavens (for Hevelius was a devout man), or else engaged in surveying, drawing, or observing with telescopes. Hevelius himself drew the Moon map plate, but his artist friend Adolph Boij added the decorative *putti*.

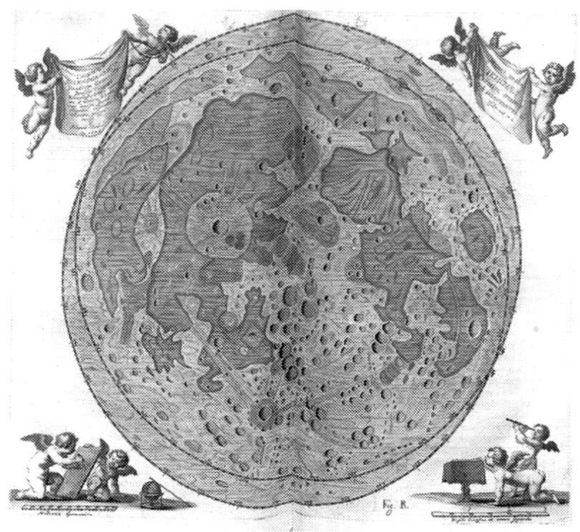

Figure 2. An example of one of Hevelius's printed plates of the full Moon showing librations on each side of the disk. The symbols at the corners depict winged cherubs or classical art *putti*. (From *Selenographia*, Dantzig, 1647.)

And in addition to the Moon, *Selenographia* contained several planetary drawings, his own, and reproductions of the drawings of contemporary astronomers, the errors in which bring home to us the limitation of his pre-1647 telescopes. Saturn is shown with the *ansae* or 'handles' mentioned by Galileo, which Christiaan Huygens, a mere decade later, would resolve into a ring. There is also a curiously gibbous Mars, and Jupiter's disk is covered with a series of lightning-like zigzags, which in the early 1660s, with better telescopes, Cassini and Hooke would resolve into the Jovian belt systems which we know today. Yet for the Moon his optical technology was fine, and quite capable of delineating light and dark contrast structures on a large variegated disk, while being as yet incapable of resolving faint features on a tiny shining sphere such as Jupiter. *Selenographia* also discusses Hevelius's studies of sunspots. Then around 1661, he modified his original 'helioscope' sunspot projector, later describing and illustrating it in detail in *Machina Coelestis, pars prior* ('Celestial Instruments, First Part') in 1673. Here, a telescope was fitted inside a form of universal joint (or 'scioptric ball'), to project the solar image on to a screen in a darkened chamber. Adjusting screws were used to track the moving image, while the sunspot positions were carefully recorded.

Selenographia, dedicated to King Ladislas IV of Poland, received international recognition from the astronomers of Europe. And among the eminent men of Europe to whom he sent complimentary copies, he included the Pope, who was impressed by this exquisite gift from a Polish Protestant! It not only set new standards in lunar cartography, but also emphasized Hevelius's concern with instrumentation and the necessity for precision and explicit description, and indeed turned astronomical representation into a new branch of the fine arts.

HEVELIUS'S TELESCOPES

The telescopes described in *Selenographia* were relatively feeble compared with those which Hevelius would build immediately thereafter. For, from around 1650 to 1690, the refracting telescope underwent rapid development which, at least in theory, tripled or quadrupled its magnifying power. This was made possible in part by responses from the glass-making industry to growing luxury demands in society at large: to the long-standing craft traditions of Italy, and especially in those Renaissance 'tiger economies' of Europe, most notably Holland and England. For here the rapidly expanding merchant classes were becoming opulent, especially from the new Atlantic trade with India, China and elsewhere, and were enjoying levels of personal economic and political liberty not possible in more rigid and hierarchic countries such as France and Spain. And then, as now, those with cash in their pockets wanted beautiful and interesting things with which to decorate their homes. And amongst these was a demand for bigger windows with large, clear panes, as well as for big mercury-backed looking-glass mirrors. The new skills developed by the luxury glass-blower, moreover, could be borrowed by the astronomer, as the Dutch brothers Christiaan and Constantijn Huygens found to their advantage. For if you went to the right glass-blowing firms, you could find mirror plate glasses over a foot square, which might contain three- or four-inch diameter areas of fairly pure transparency. And from these, good lens blanks could be cut out, and figured into fine single-element object glasses on lens-grinding lathes, similar to that described by Hevelius. (Some years later the wealthy Huygens brothers also described their own more advanced optical lathe for other astronomers to follow.)

These new larger, thicker blanks were to be figured into a series of long-focus telescope object glasses by craftsmen-scientists such as the Huygens brothers in Holland, Giuseppe Campani in Italy, Richard Reeves and Christopher Cocks in London, and Johannes Hevelius in Dantzig. Lenses with focal lengths that became longer as the technology for figuring very flat curves advanced: twenty, thirty, fifty, a hundred or more feet by the 1660s. Hevelius himself built the iconic long refractor, of 150 feet focus, described and illustrated in *Machina Coelestis* I (published in 1673), although the Royal Society in London still owns three beautiful lenses by Constantijn Huygens, the largest of which is nine inches in diameter, and of 210 feet focal length, though it is hard to know how it could have been used. (I have examined these Royal Society Huygens lenses myself, and can vouch for their transparency, while Professors A. A. Mills and M. L. Jones have published a detailed optical study of them: see 'Further Reading' below.)

But why did astronomers make these great long-focus glasses, which could only be used under perfect viewing conditions? Yes, they could help minimize chromatic aberration by spreading out the spurious colours produced by a simple lens across several inches of approximate focus, and hence make it easy to seek a quite sharp image in the yellow light, which could give a less distorting effect. But I would suggest that there was a more important reason. For the longer the focal length of the objective, the bigger the physical image at the focus, thereby making it easier to obtain fairly high magnifications at the eyepiece. The Moon, for instance, subtending an angle of a half-degree of arc, will produce a prime focus image 0.57 inches (14 mm) across at a focal length of 65.5 inches (1664 mm). And with a 100-foot (3,060 cm) focus lens, you can get an image the size of a dinner plate, so that even with simple eyepieces *theoretical* magnifications of around ×250 or more were feasible with decent definition. During the early 1660s, Robert Hooke in London used a 60-foot-focus telescope that gave him a ×173 magnification, with which he made a meticulous published drawing of the lunar crater Hipparchus.

And why was Hevelius, together with his colleagues across Europe, so concerned with high magnifications and yet not especially bothered by the necessarily dim images produced by these telescopes? I would suggest two reasons.

First, planetary astronomy, which we take for granted today, was new, exhilarating, and appeared full of potential 350 years ago. Instead of their being mere dots of light in the sky, post-1609 telescopes had shown the planets to be spherical worlds in their own right. Venus showed phases, and, some others believed (wrongly, as we now know), topographical features; Jupiter had four moons and perhaps surface markings; while Saturn had *ansae* or peculiar handles that seemed to change shape and which, in 1659, Huygens would correctly announce to be a ring, and Cassini in 1675 to be a divided ring. The Moon seemed a place of continents, mountains, odd-looking 'pits' (craters), and perhaps seas; while the Sun was spotted, and rotated on its axis in twenty-eight days. None of this had been imagined before 1609, and the obvious question was 'Are these worlds inhabited?' There was only one way to find out, and that was by building ever more powerful refracting telescopes with ever larger-focus object glasses producing ever higher magnifications.

Second, Galileo had drawn attention to the thirty-six stars that his telescope revealed in the Pleiades, the countless stars present in the Milky Way, and the non-naked-eye stars that one could see wherever one chose to point a telescope. And as each new, more powerful telescope revealed fresh shoals of hitherto invisible stars, not to mention peculiar glowing clouds (or *nebulae* in the Latin) in Orion or Andromeda, one was forced to ask the question 'How big is the universe?'

Very clearly, the closed cosmos of celestial spheres, bounded by a black sphere and fixed stars, which had been the scientific orthodoxy from Greek times to the telescope, was patently wrong. For the universe of 1660 seemed a much more weird place than that of 1600, posing as it did all sorts of questions about other worlds and infinite space, and the only way forward seemed to be through bigger and ever more powerful telescopes. This, I would argue, is what drove the development of the long-focus refractors.

In Chapter 20 of *Machina Coelestis* I, Hevelius gives a detailed description of his long telescopes, variously of 30, 40, 50, 60, 70, 140 and 150 feet focal length, while his 150-foot instrument with its 8-inch-diameter object glass – while not especially successful in practice – is generally regarded as the largest long-focus refractor ever to have been used for serious observation. And even taking into account the fact that the Dantzig (like the German) foot was equivalent to about 11 English inches, these Hevelian and other instruments were truly prodigious by any standard. (The foot, or Roman *pes*, and inch, or *pollex* (thumb-breadth), were in universal use across Europe, with slight variations, before the ideologues of post-Revolution France attempted to impose their new base 10 metric system after 1801.)

Managing these long, narrow tubes, on tall masts, using complex block and tackle adjustments, must have demanded enormous skill, patience and coordinated teamwork between the observer and the men working the ropes. Hevelius discusses tube design and mentions using bracing wires and other strengthening devices. It must have been impossible, however, to maintain optical collimation for hours on end, considering the moisture-absorbent thin wooden tubes and hemp ropes on a damp, still Baltic night. Things must have sagged out of adjustment, and demanded constant tweaking to remain in operation. Yet I suspect that, in a major seaport like Dantzig, Hevelius would have been able to find plenty of men – ex-sailors, building-site workers, and such – who had long since perfected their 'touch' in operating ropes, pulleys, yard-arms, beams, blocks and tackle, and who could work skilfully together and follow precise instructions to accomplish delicate manipulations.

COMETOGRAPHIA ('ACCOUNT OF COMETS'), 1668

Along with the rest of the universe, the telescope had brought into question the Greek ideas about comets. The definitive classical explanation, which had survived for nearly two thousand years, derived from Aristotle. He had regarded them as exhalations from the Earth which had caught fire in the air, and had discussed them as atmospheric phenomena in his *Meteorologia* ('On Things in the Air'), c.330 BC. But this explanation was first questioned by Hevelius's great 'northern Renaissance' hero, Tycho Brahe; for Tycho had made meticulous angular measurements of the positions of the brilliant comets of 1577 and 1588, using the then most accurate instruments in the world, and had been unable to detect any parallax whatsoever. Yet the Moon displayed a conspicuous parallax when measured against the fixed stars, which led Tycho to consider that comets must be well beyond the Moon, in astronomical space, and not merely in the air. Yet if they moved amongst the planets, what force drove them? And what shape could their orbits be? For as we saw, Tycho provides the first example of ancient astronomical explanations being undermined by fresh observations made with instruments of unprecedented accuracy. And while Tycho died eight years before Thomas Harriot first used a telescope to look at the Moon, the observations made with his precision-engineered and engraved large geometrical instruments detected discrepancies that were sometimes hard to reconcile with Greek theory. And what the telescope did was to continue, and vastly amplify, this process of observational challenge to ancient theory.

The comet of 1652 was the first such body to seriously engage Hevelius's attention, and over several years he had been working on a treatise dealing with the new post-Tycho-Brahe and post-telescope ideas on comets. This work was interrupted, however, by the comet of 1664, which led him to write a preliminary book, *Prodromus Cometicus* ('Cometary Prologue'), published in 1665, which (no doubt because he had received money from King Louis XIV of France) he strategically dedicated to Jean-Baptiste Colbert, the King's Finance Minister.

Of course, by 1668, when Hevelius published his great treatise *Cometographia* – now bearing the Latin dedication to Colbert's master, 'Ludovico XIV' – everybody agreed that comets were astronomical and not meteorological bodies. But what were they made of? Where did they come from? And what was the shape and cause of their orbits? These were some of the questions which he tried to address in his book.

Much of *Cometographia* was devoted to the comet of 1652, and his geometrical observations confirming it and the comet of 1664 to be space and not atmospheric bodies. Yet what was the shape of their orbits? For at this time, when many astronomers in Europe were slowly groping their way towards the notion of some *gravitating force*, to which Newton would give full and precise mathematical expression in 1687, the causes of astronomical motion, without the old crystalline spheres, were a puzzle. Was some invisible agency, such as magnetism, at work in moving the planets around the Sun, as Johannes Kepler had once speculated? Or could planetary motions derive from some sort of physical analogy based on Greek philosophical principles? The telescope, after all, showed that the planets were spherical bodies, so could their very *roundness* explain their natural tendency to encircle the Sun: round objects naturally tending to round paths?

At first, in the *Prodromus*, Hevelius had followed Kepler's thinking that, as short-lived, transitory, non-planetary bodies, comets must move not in curved but in straight paths, having no natural affinity with spheres.

By 1668, however, Hevelius had changed his mind. Realizing that he had not taken into consideration the Earth's own motion in space over the weeks he had been observing the 1664 comet, he now concluded that its orbit was curved, and concave to the Sun.

This new idea could also be elegantly dovetailed into the prevailing theoretical orthodoxy which argued that, as cometary material originated in effluvias and waste products arising from the surfaces of the outer planets (rather than the Earth), and the planets themselves were no doubt rotating on their own axes (as newly discovered markings on the surface of Jupiter and Mars suggested), then comets were likely to be thrown into space in precise geometrical curves, analogous to the ellipses in which Kepler had shown the planets to move. Similar in some ways, indeed, to the path described by a flying projectile such as an arrow (or better still, a present-day newly launched interplanetary space vehicle).

Hevelius was of the opinion that comets moved in parabolic orbits, although this was not an especially radical idea for the astronomers of 1668, with men like Robert Hooke and Christiaan Huygens already exploring such concepts. For if the planets moved in Keplerian elliptical orbits that derived from conic sections (or angled slices through the cone, following the c.150 BC Greek geometer Apollonius), then a parabola seemed a reasonable shape. In this aspect of his thinking, however, he was influenced by physical analogies and practical demonstrations, such as the paths described by terrestrial projectiles, as much as anything else.

Yet while classical theories of the physical nature of the heavens were being challenged, and demolished, one after the other in the post-telescopic age – explanations of the nature of the Sun, Moon and planets, of comets, of light and motion – the truth status of demonstrable Greek geometry remained untouched. For just as telescopic discoveries challenged physical theory, so the work of Tycho, Kepler, Jeremiah Horrocks, Borelli, Huygens, Hooke, Newton, Leibniz and others only enhanced the truth status of mathematics in all its branches. Infinite and immensely complex the 'new' universe might be, yet it clearly operated as a great clock and followed exact mathematical laws.

Hevelius's attempts to explain comets as material bodies was ingenious. Assuming comets to be the products of planetary effluvias, he suggested that the head, or nucleus, was a disk rather than a sphere, and that as it approached the Sun it began to dissolve away to produce the tail. At perihelion, the nucleus disk faced the Sun full on, enhancing its dissolution, while the disk's being edge-on to its line of motion provided minimal resistance, or braking effect, to the ether, hence explaining why comets accelerate at perihelion. (For like most astronomers of his day, Hevelius believed that space must contain an intangible substance or agent which conveyed light and solar heat, and which, perhaps, might dissolve comets.) To visualize Hevelius's idea of comets, imagine throwing a frisbee: thrown edge-on, they go fast, but if thrown face-on they fall down, due to air resistance. Think of the air as similar to how Hevelius envisaged the ether.

But perhaps the most enduringly significant parts of *Cometographia* were those chapters in which he compiled a list of 251 comets visible since biblical times down to 1665: a compilation which was later to provide useful source material for Edmond Halley's pivotal work on comets in 1705 (Figure 3).

MACHINA COELESTIS (1673), OR HEVELIUS'S INSTRUMENTS

I first came to examine Hevelius's books in 1973, when, as an Oxford doctoral student working on a history of astronomical instruments, I studied *Machina Coelestis pars prior* ('Celestial Instruments, First Part') in the Bodleian Library. Not only was this Hevelius's magnificent treatise on the instruments contained in his great Dantzig Observatory, but I was also delighted to find that I was working from Sir Christopher Wren's own copy. For Sir Christopher, himself an eminent astronomer, held a Professorship in astronomy at Gresham College, London, and the Savilian Chair at Oxford, before his subsequent fame as an architect (he had also been a student in my own college,

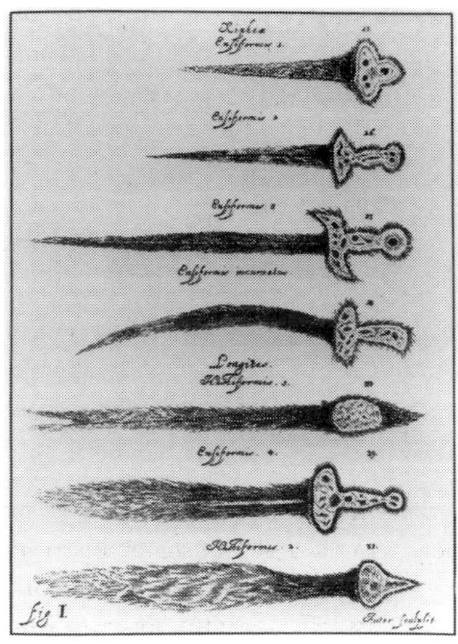

Figure 3. Some examples of Hevelius's illustrations of different cometary forms. (From *Cometographia*, Dantzig, 1668.)

Wadham). And like all of Hevelius's works, the *Machina* I is richly and sumptuously produced: printed on the finest-quality paper, still as white and supple as it was in 1673, larger than a modern *Encyclopaedia Britannica* volume, and two inches thick. (It is a myth that old books are by definition fragile. I routinely handle books four or five hundred years old that are still more robust than most 1960s paperbacks, and printed on much better-quality paper.)

Hevelius's 'Northern' – or Baltic – 'Renaissance' status, however, is abundantly confirmed in the *Machina*, for it is directly in the form and spirit of Tycho Brahe's *Astronomiae Instauratae Mechanica* ('Description of the Instruments used for the Reformation of Astronomy'), published in 1598. For in that work, Tycho described his Uraniborg, Denmark observatory before going on to delineate and illustrate each of the instruments which it contained, noting in detail their dimensions (in cubits, of about sixteen English inches), purpose, and sometimes how an instrument was superseded by another of improved design and accuracy.

Likewise, the *Machina Coelestis* I contains detailed accounts of over a dozen large graduated instruments, quadrants and sextants, in brass and wood, for measuring vertical (declination) and horizontal (azimuth, lateral, or right, ascension) angles. For when used in conjunction with each other, these instruments could measure the angles between the stars, or between the stars and the planets, to build up what would later be called Right Ascension and Declination co-ordinates from which tables and star maps could be built. These quadrants and sextants varied in size from three to nine feet in radius, and Hevelius began, probably in the 1630s, by completing a large altazimuth quadrant (to measure rotating angles in the vertical) which his old teacher Peter Krüger had begun.

A quadrant of 90° could be operated by a single observer, who would adjust the sighting arm – or alidade – to bring it in line with the object under observation, and then read off the angle on the scale. The sextant of sixty degrees was a two-observer instrument, however, and used to measure the angles between non-vertical, or laterally separated, pairs of objects in the sky. The assistant observer would sight one celestial object through a fixed sight on the sextant, corresponding to '0' degrees, while Hevelius himself would observe the other object through the sights on the centre-mounted alidade which could move across the arc, to read off the angular separation of that object from the other, on the sixty-degree brass scale. Although mounted in the altazimuth (as opposed to the equatorial), the sextant could be made to track objects across the sky for relatively short periods, by means of a screw which the assistant observer turned slowly, to push the sextant against a firm upright post in its mounting house.

While the basic design of Hevelius's quadrants and sextants was the same as Tycho's, where they differed was in their much more accurate naked-eye sights and graduated scales. All of the angle-measuring instruments described in the *Machina Coelestis* I used sophisticated naked-eye sights, in which the observers would use delicate micrometer screws to progressively narrow a pair of fine slits through which they were observing the star or planet being measured, to ensure enhanced accuracy. And then, to subdivide each degree, or part-degree, Hevelius equipped his instruments with those mathematical scales first described by Pierre Vernier in 1631. These Vernier scales enabled him to read to arc minutes, and perhaps arc seconds, to a level of precision which would not have been possible for Tycho in 1590. He even used a screw micrometer with a finger moving across a finely graduated circular dial to read off some subdivisions.

And in the majority of cases, Hevelius not only engraved these instrument scales with his own hands, using Euclidean geometrical techniques and striking off angles with beam compasses, but he also drew and engraved the pictures of his instruments, as his signature on some of the *Machina*'s plates depicting his scales and micrometers testifies. For Johannes Hevelius was merchant, brewer, scholar, inventor, craftsman, astronomer and artist all in one.

In one magnificent plate, Hevelius has his entire observatory depicted, consisting as it did of a large wooden platform built across the roofs of several houses that he owned. On it are depicted his quadrants and other angle-measuring instruments – surrounded by their moveable sheds or houses protecting both instrument and observer from the severe Baltic winter weather – as well as some of his smaller telescopes. The great 150-foot refractor, suspended from its own hoisting mast, was set up on a special site outside Dantzig, as the engraving of the instrument, with the distant background of city church spires, makes clear (Figure 4). And just as Tycho had named his observatory Uraniborg ('Castle of the Heavens'), so Hevelius referred to his as Stellaburgum or 'Citadel of the Stars'.

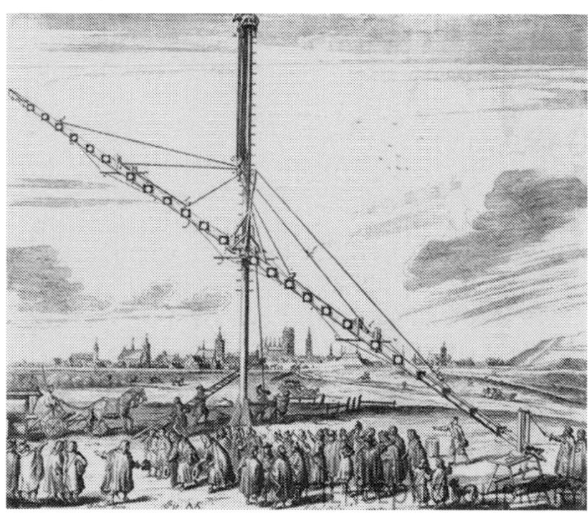

Figure 4. Hevelius's 150-foot telescope (German feet) or 140-foot (English), with 8-inch (approximate) diameter object glass. (From *Machina Coelestis* I, Dantzig, 1673.)

Then in 1679, Hevelius published Volume II of his masterpiece, entitled *Machina Coelestis pars posterior* ('Second Part'), which comprised the great catalogue of observations which he had built up with these instruments. It was a truly monumental piece of work, and one of the great achievements of 'Grand Amateur' astronomy.

When one examines the meticulously detailed pictures of the scales of his instruments in *Machina Coelestis* I, one gets an impression of their exquisite accuracy; without any doubt, the most accurately engraved astronomical scales in Europe in 1673, and capable of delineating angles to within ten arc seconds or so. But it was these scales, and their accuracy, that pitched Hevelius into an acrimonious dispute with Robert Hooke and, to a lesser degree, with John Flamsteed, in London: an unpleasant anomaly in his otherwise long, friendly relationship with England.

TELESCOPIC SIGHTS AND CONTROVERSY WITH ROBERT HOOKE

When Robert Hooke read his copy of *Machina Coelestis* I, and reported on Hevelius's work to the Royal Society (Hevelius had been elected an honoured overseas Fellow of the Royal Society in March 1664), he pounced upon what he considered a fundamental weakness in Hevelius's observatory and observing technique. As indicated above, all of Hevelius's angle-measuring instruments had open, naked-eye slit sights, like those of Tycho and most other preceding European astronomers, instead of those new telescopic sights which Hooke was actively promoting. Of course, Hevelius was fully familiar with Hooke's telescopic sights, but in the context of the still-incomplete theory of telescopic image formation, he argued that even when using cross-hairs in a Keplerian eyepiece it was still possible to slightly displace an apparent star-image if the observer moved his eye around in the field of view. Hence one could never be certain that the magnified star image was in the exact geometrical centre of the field, whereas with naked-eye sights one knew exactly where a star was in relation to the angular graduations on the instrument's degree scale.

Of course, Hevelius, as a pioneer of 'big telescope' astronomy for planetary study, had no objection to using telescopes as such: but he objected to using them for the making of accurate angular measurements. Hooke had first openly criticized Hevelius's use of naked-eye sights after receiving his copy of *Cometographia* in 1668 – Hevelius regularly distributed copies of his sumptuous books as gifts to learned friends across Europe – in the correspondence which had subsequently taken place between the two men. Feeling that he had proven his case that telescopic sights were superior to naked-eye ones, Hooke was somewhat incensed to find them ignored when he received *Machina Coelestis* I. But one must admit that Hooke rather overreacted when, in 1674, he published his *Animadversions on the First Part of Hevelius his Machina Coelestis* through the Royal Society.

The use of telescopes on the sighting arms of quadrants and sextants, though unsuccessfully experimented with since 1630, had first been made workable by the Yorkshire astronomer William Gascoigne in about 1640, though it was not until the early Royal Society began examining the deceased Gascoigne's and his friends Jeremiah Horrocks's and William Crabtree's surviving papers and artefacts in the early 1660s that their potential really emerged. And their greatest advocate became Hooke, who argued that if a telescope could enhance the acuity of the observer's eye forty-fold, then they could improve the accuracy of angular measurements forty-fold as well. Not only did Hooke believe that Hevelius's caveats about centring the star images in the telescope field could easily be overcome by using cross-hairs, but he also went on to conduct a brilliantly simple but cogent experiment. Hooke prepared a board on to which rows of distinct and carefully measured squares had been painted. Setting up the board in good light, he moved away from the board until he reached a point where he could no longer discern the individual squares. Then, using a geometrical surveying technique (for each square formed a long, thin triangle with his eye) and knowing the size of each square, and the measured separation point at which they all coalesced into one, he calculated that his eye could not resolve angles that were less than one arc minute. Therefore, and irrespective of the geometrical accuracy and elegance of his instruments, Hevelius's observations could be of no better than one-minute accuracy (which was equivalent to those of Tycho a hundred years before). Ophthalmically speaking, Hooke was exactly right, for even the standard modern physiological textbooks give the normal eye resolution as around one arc minute. It is unfortunate that he got so worked up about it, and relations with Hevelius deteriorated. Happily, things would improve somewhat in 1679 when the Royal

Society sent the amiable young genius Edmond Halley out to Dantzig to smooth the waters, as we shall soon see.

CATHERINA ELISABETHA KOOPMAN HEVELIUS, THE FIRST WOMAN ASTRONOMER

Hevelius's first wife Katharina died in 1662, and the following year the 52-year-old brewer-astronomer married another merchant's daughter. Catherina Elisabetha Koopman was not yet seventeen when she became the second Madame or Lady Hevelius, but by all accounts it was a happy and successful marriage. Elisabetha seems to have been highly intelligent, practical, and especially well-educated for a seventeenth-century woman. She had, or else acquired, a good knowledge of astronomy and geometry, and presumably of Latin, for that was the language used in all Hevelius's published books and much of his foreign correspondence, and it was in Latin that, on 1 November 1687, Elisabetha wrote to Thomas Gale, Secretary of the Royal Society, telling him of Johannes' death. She probably also knew German, the root language of Baltic Protestantism, and some French. They produced three daughters, all of whom lived to maturity: a considerable achievement, considering the high mortalities of that age.

Catherina Elisabetha, however, was destined to become the first women ever to be depicted in art as actually *doing science*, for on two of the fine-art plates in *Machina Coelestis* I, she is shown in the act of observing, as she operated the 'guide star' end of a sextant while her husband measured an angular separation (Figure 5). And after her husband's death, she played an active role in seeing his unpublished work through the press, in a way that would be recapitulated by Margaret Flamsteed after the death of her Astronomer Royal husband John after 1719.

But we would like to know more about Catherina Elisabetha from surviving Polish records.

HEVELIUS AND ENGLAND

We have already seen how Hevelius's exchanges with Robert Hooke were not of the friendliest character during the telescopic sights

Figure 5. Johannes and Elisabetha Hevelius observing an azimuth angle with the large brass sextant with naked-eye sights. This is the first depiction of a woman actually doing science. (From *Machina Coelestis* I, Dantzig, 1673.)

controversy (though it was Hooke who was the aggressive party), yet we must not forget that the incident would never have arisen in the first place had not the Dantzig brewer already been famous across Europe, with a long and friendly association with England. Hevelius first visited England on his Grand Tour as a law student in 1631, and was amongst the first of those overseas scientists of the very top rank to be invited to become a Fellow of the new Royal Society, in March 1664: less than four years after the Society's foundation, and only ten months after its Royal Charter of incorporation from King Charles II (May 1663). And he routinely sent gift copies of his books to his London friends – including Hooke. The Royal Society Library copies of *Machina Coelestis* and *Cometographia*, for example, contain fulsome Latin dedications in Hevelius's elegant handwriting.

But English astronomers must be eternally grateful to Hevelius for publishing one of the first great achievements in British astronomy: Jeremiah Horrocks's account of his own, and his friend William Crabtree's, first ever observed transit of Venus across the Sun's disk on 24 November 1639. Jeremiah Horrocks's own manuscript book, completed in 1640, which described and analysed many of the aspects of the Venus transit, had found its way to London following Horrocks's sudden death in January 1641 and Crabtree's in 1644. A copy of the manuscript seems to have come into the hands of Christiaan Huygens (who was also elected an overseas Fellow of the Royal Society) when he was visiting London in 1661, and was taken back to Holland. It appears to have been Huygens who sent on a handwritten copy of Horrocks's account to Hevelius in Dantzig.

It says much for Hevelius's generosity of spirit that he had Horrocks's *Venus in sole visa* ('Venus seen on the Sun') set up into the same fine and elegant typeface as was the hallmark of all his own publications. For on 3 May 1661, Hevelius had himself observed a transit of Mercury across the solar disk (his friend Pierre Gassendi had observed the very first known Mercury transit in Paris in 1631), and, like Gassendi, Hevelius had been amazed at the smallness of Mercury when seen on the Sun. So when in 1662 Hevelius published his own observation, *Mercurius in sole visus*, he accompanied it with the first ever printed treatise of Jeremiah Horrocks, and made it clear that he was proud to do so. Hevelius also had Horrocks's drawing of three Venus positions on the solar disk engraved to accompany Horrocks's printed text. So the achievement of a twenty-year-old English astronomer who had lived and worked in the small Lancashire village of Much Hoole, near Preston, and of his friend Crabtree, the Salford cloth merchant, was broadcast across the European world of learning in 1662 under the generous auspices of a Dantzig brewer astronomer. Indeed, we should not underestimate the significance of Hevelius's giving the first international recognition to two of those three north-country astronomers, all amateurs: Jeremiah Horrocks and his friends William Crabtree, and William Gascoigne of Middleton, Leeds.

In 1679, the Royal Society sent out Edmond Halley to Dantzig, as a sort of scientific diplomat, to examine independently Hevelius's naked-eye-sight instruments and, hopefully, heal the rift between him and Robert Hooke: two internationally eminent Fellows of the Royal Society. In fact, Halley was the ideal man for the mission, for not only was the 23-year-old widely travelled and already renowned across Europe as the cataloguer of the Southern Hemisphere stars, but he was also famously charming and congenial, and while himself a convinced user of telescopic sights, was known to be fair-minded and impartial.

He was also the son of a rich City of London merchant with a shrewd business head on his shoulders, and would have instinctively related to Hevelius's world and circumstances.

As soon as Halley arrived in Dantzig, on 26 May 1679, he and Hevelius began to observe together; and they did so over the next two months, until Halley left to return to London in July. Assisted by Madame Hevelius and several astronomer friends who worked with Hevelius, they made hundreds of observations of stellar and star-Moon angles. They used Hevelius's quadrant, large brass sextant, and the two-foot-radius telescopic sight quadrant which Halley brought with him, and with which he had observed the angles for his St Helena Southern Hemisphere star catalogue a couple of years before. The astronomers took turns with each other's instruments, Halley showing Hevelius and the Dantzig astronomers how to use his telescopic quadrant. And whilst Hevelius was still not convinced that telescopic-sight instruments were more accurate than naked-eye ones, Halley had to admit to the excellence of Hevelius's large brass instruments (and probably the 68-year-old astronomer's exceptional eyesight) which could often be relied upon to ten arc *seconds*.

So in a way, both sides had been correct. For while Hevelius misunderstood Robert Hooke's optical geometry of telescopic sights, and did not fully appreciate the stability of star images on the cross-hairs even if the observer's eye position changed slightly, his own instruments, combined with his visual acuity, had shown themselves to be capable of exceptionally high accuracy, exceeding the 'normal' limitations as defined by Hooke.

By 1713, however, by which time Halley was Savilian Professor of Geometry, a malicious tale was obviously circulating in Oxford that while in Dantzig, twenty-four years before, the then 23-year-old Halley had had an affair with the 33-year-old Madame Hevelius. However, its apocryphal status was emphasized by the Oxford diarist Thomas Hearne, who added: 'I am apt to think it false.' And while it is true that once back in England, Halley fulfilled a commission for Catherina Elisabetha of obtaining a sumptuous silk dress in the latest fashion, he was nonetheless concerned with being reimbursed for the £6 8- 4^{d} that he had been obliged to lay out for it. (In 1679, this sum would have been six months' wages for an English labourer.) On the other hand, Halley and Johannes and Elisabetha Hevelius and his new Dantzig astronomer friends corresponded in a very amiable fashion, Halley obtaining lenses and books for Johannes himself, as well as expensive clothes for his wife.

THE END OF THE DANTZIG OBSERVATORY

On 26 September 1679, some two months after Halley's return to London (and shortly before he left for France and Italy), tragedy struck Hevelius's great Dantzig observatory. For it was totally destroyed in a fire which Hevelius believed was due to the mischief of a servant. Many books, much correspondence, and other portable scientifically valuable items were fortunately saved, but all the instruments and a great deal else went up in the flames. What is amazing, however, is that the 68-year-old astronomer soon began to construct a second observatory, though it was never as grand as the original, and the shock of the loss naturally disheartened the great astronomer. He left an account of the disaster in his *Annus Climactericus* ('Year of Great Changes'), published in 1685. Edmond Halley, believing Hevelius to have perished in the blaze, sent a letter of condolence to Dantzig. Amazingly, it would not be until he reached Rome, in October 1681, that he discovered that Hevelius was indeed alive and in good health; and he immediately sent off a letter in French to Madame Hevelius and his friends in Dantzig expressing his joy at the news.

Yet while we, just like most seventeenth-century people, think of Hevelius as a great observational astronomer, we must not forget that in his own day he was also renowned as a former mayor and eminent civic dignitary of a great European commercial city. In 1660, for instance, he had been raised up to become a titular member of the Polish nobility, with a coat of arms incorporating the Polish Royal Crown, and an exemption from paying taxes. Various members of the Polish Royal Family regularly visited his observatory from 1660 onwards, and King Jan Sobieski III had made numerous visits between 1677 and 1683. When he created a number of new constellations from small unnamed clusters of stars between the main asterisms, Hevelius christened several in honour of his Royal patron. *Scutum Sobieski* ('Shield of Sobieski'), between Aquila and Sagittarius, for instance, was named in honour of King Jan III, who not only provided resources to help with the building of Hevelius's post-fire observatory, but was also an internationally renowned military hero who had rescued Vienna, and potentially much of the rest of Europe, from siege and imminent Turkish invasion in 1683 – hence the 'Sobieski Shield'.

Following Hevelius's death in 1687, his widow played a major part in seeing his remaining work through the press, most notably his *Uranographia* ('Account of the Heavens') and star maps, and *Prodromus Astronomiae* ('Astronomical Prologue'), both in 1690 (Figures 6 and 7). These posthumous works are often found bound up together. Yet by that time, the international centre of innovation for precision astronomical measurement had shifted to the Royal Observatory, Greenwich, under the directorship of the telescopic-sight astronomers Revd John Flamsteed, Edmond Halley, and their Astronomer Royal successors down to the late nineteenth century.

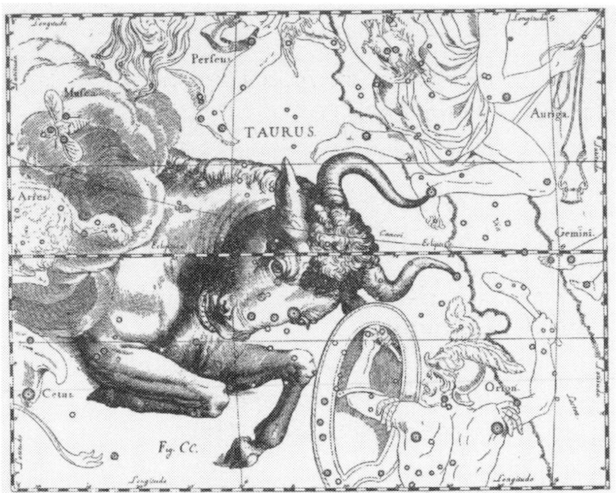

Figure 6. Hevelius's illustration of the Zodiacal constellation Taurus, the Bull. (From *Prodromus Astronomiae*, Dantzig, 1690.)

Johannes Hevelius was the last great naked-eye observer: the natural heir of the Tycho Brahe tradition and style of astronomy, and a Grand Amateur brewer-merchant astronomer, who enjoyed the friendship and patronage of the Polish and French kings, and was raised to noble status, and addressed by Halley as 'Lord Hevelius'. He was not by inclination a theoretician or a pure mathematician, but an artist, designer, craftsman, and masterly observer – with a Europe-wide reputation. Post-Soviet liberated modern Poland now rightly sees him as a national cultural icon,

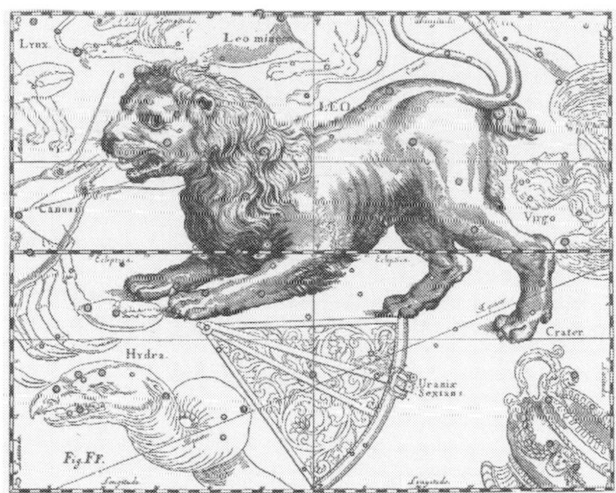

Figures 7. Hevelius's illustration of the Zodiacal constellation Leo, the Lion. (From *Prodromus Astronomiae*, Dantzig, 1690.)

and near the site of his old observatory the Polish people have erected a life-size statue in honour of his achievement. And he needs to be more widely known on a world level, for Hevelius was undoubtedly one of the founders of modern astronomy.

FURTHER READING

J. Hevelius, *Machina Coelestis, pars prior*, I (Dantzig, 1673).
J. Hevelius, *Cometographia* (Dantzig, 1668).
J. Hevelius, *Prodromus Cometi* (Dantzig, 1665).
J. Hevelius, *Selenographia* (Dantzig, 1647).
J. Hevelius, *Mercurius in sole visus* (Dantzig, 1662).

Original copies of Hevelius's works are lodged in several major British libraries, as well as in Europe and the US. (I have worked on those in the Bodleian and Christ Church Libraries, Oxford, and in the Royal Society Library.) Additionally, sixty-seven of Hevelius's Latin letters to the Royal Society, and one by Elisabetha Hevelius, are deposited in the Royal Society Library (Letters EL/H2).

Mark Birkinshaw, 'Johannes Hevelius, the Prussian Lynx at 400', paper given to the R.A.S. meeting on 11 February 2011, reported in *The Observatory*, Vol. 131, No. 1223 (August 2011), pp. 207–10.

Allan Chapman, *Dividing the Circle. The development of critical angular measurement in astronomy, 1500–1850* (Ellis Horwood and Simon and Schuster, Chichester, 1990; 2nd edn. Wiley-Praxis, 1995).

Allan Chapman, 'Christiaan Huygens (1629–1695), astronomer and mechanician', *Endeavour* 19, No. 4 (1995), 140–5.

Alan Cook, *Edmond Halley. Charting the Heavens and the Seas* (OUP, 1998): Chapter 4.

Thomas Hearne, *Remains and Collections*, Vol. 4, 13 Nov. 1713 (Oxford Historical Society, 1884), for story of Halley and Madame Hevelius.

R. Hooke, *Animadversions on the First Part of Hevelius, his Machina Coelestis* (London, 1674).

H. C. King, *The History of the Telescope* (London, 1955).

E. F. MacPike, *Hevelius, Flamsteed and Halley* (London, 1937).

Alan A. Mills and M. L. Jones, 'Three lenses by Constantijn Huygens in the possession of the Royal Society', *Annals of Science* 46 (1989), 173–82.

John D. North, 'Johannes Hevelius', *Dictionary of Scientific Biography* (Scribners, New York, 1970–80).

Charles Leeson Prince, *The Illustrated Account Given by Hevelius in his 'Machina Celestis' . . . of his Telescopes* (Lewes, 1882).

V. P. Sheglov (ed.), *J. Hevelius, The Star Atlas* (Tashkent, 'FAN' Press, 1968, 1970).

Ivan Volkoff, E. Franzgrote, and A. D. Larsen, *Johannes Hevelius and his Catalogue of Stars* (Brigham Young Univ. Press, 1971).

Ewen A. Whitaker, *Mapping and Naming the Moon. A History of Lunar Cartography and Nomenclature* (CUP, 1999).

Mary G. Winkler and Albert van Helden, 'Johannes Hevelius and the visual language of astronomy', in *Renaissance and Revolution. Humanists, Scholars, Craftsmen, and Natural Philosophers in early modern Europe*, ed. Judith V. Field and Frank A. J. L. James (CUP, 1993), pp. 97–116.

Donald K. Yeomans, *Comets. A Chronological History of Observation, Science, Myth, and Folklore* (Wiley Science, Chichester, N.Y., etc., 1991).

There have been other Hevelius studies in Polish and German.

Part Three

Miscellaneous

Some Interesting Variable Stars

JOHN ISLES

All variable stars are of potential interest, and hundreds of them can be observed with the slightest optical aid – even with a pair of binoculars. The stars in the list that follows include many that are popular with amateur observers, as well as some less well-known objects that are, nevertheless, suitable for study visually. The periods and ranges of many variables are not constant from one cycle to another, and some are completely irregular.

Finder charts are given after the list for those stars marked with an asterisk. These charts are adapted with permission from those issued by the Variable Star Section of the British Astronomical Association. Apart from the eclipsing variables and others in which the light changes are purely a geometrical effect, variable stars can be divided broadly into two classes: the pulsating stars, and the eruptive or cataclysmic variables.

Mira (Omicron Ceti) is the best-known member of the long-period subclass of pulsating red-giant stars. The chart is suitable for use in estimating the magnitude of Mira when it reaches naked-eye brightness – typically from about a month before the predicted date of maximum until two or three months after maximum. Predictions for Mira and other stars of its class follow the section of finder charts.

The semi-regular variables are less predictable, and generally have smaller ranges. V Canum Venaticorum is one of the more reliable ones, with steady oscillations in a six-month cycle. Z Ursae Majoris, easily found with binoculars near Delta, has a large range, and often shows double maxima owing to the presence of multiple periodicities in its light changes. The chart for Z is also suitable for observing another semi-regular star, RY Ursae Majoris. These semi-regular stars are mostly red giants or supergiants.

The RV Tauri stars are of earlier spectral class than the semi-regulars, and in a full cycle of variation they often show deep minima and double maxima that are separated by a secondary minimum. U Monocerotis is one of the brightest RV Tauri stars.

Among eruptive variable stars is the carbon-rich supergiant R Coronae Borealis. Its unpredictable eruptions cause it not to brighten, but to fade. This happens when one of the sooty clouds that the star throws out from time to time happens to come in our direction and blots out most of the star's light from our view. Much of the time R Coronae is bright enough to be seen in binoculars, and the chart can be used to estimate its magnitude. During the deepest minima, however, the star needs a telescope of 25 cm or larger aperture to be detected.

CH Cygni is a symbiotic star – that is, a close binary comprising a red giant and a hot dwarf star that interact physically, giving rise to outbursts. The system also shows semi-regular oscillations, and sudden fades and rises that may be connected with eclipses.

Observers can follow the changes of these variable stars by using the comparison stars whose magnitudes are given below each chart. Observations of variable stars by amateurs are of scientific value, provided they are collected and made available for analysis. This is done by several organizations, including the British Astronomical Association (see the list of astronomical societies at the end of this volume), the American Association of Variable Star Observers (49 Bay State Road, Cambridge, Massachusetts 02138, USA), and the Royal Astronomical Society of New Zealand (PO Box 3181, Wellington, New Zealand).

Star	RA		Declination		Range	Type	Period	Spectrum
	h	m	°	′			(days)	
R Andromedae	00	24.0	+38	35	5.8–14.9	Mira	409	S
W Andromedae	02	17.6	+44	18	6.7–14.6	Mira	396	S
U Antliae	10	35.2	−39	34	5–6	Irregular	—	C
Theta Apodis	14	05.3	−76	48	5–7	Semi-regular	119	M
R Aquarii	23	43.8	−15	17	5.8–12.4	Symbiotic	387	M+Pec
T Aquarii	20	49.9	−05	09	7.2–14.2	Mira	202	M
R Aquilae	19	06.4	+08	14	5.5–12.0	Mira	284	M
V Aquilae	19	04.4	−05	41	6.6–8.4	Semi-regular	353	C
Eta Aquilae	19	52.5	+01	00	3.5–4.4	Cepheid	7.2	F–G
U Arae	17	53.6	−51	41	7.7–14.1	Mira	225	M
R Arietis	02	16.1	+25	03	7.4–13.7	Mira	187	M
U Arietis	03	11.0	+14	48	7.2–15.2	Mira	371	M

Some Interesting Variable Stars

Star	RA		Declination		Range	Type	Period	Spectrum
	h	m	°	′			(days)	
R Aurigae	05	17.3	+53	35	6.7–13.9	Mira	458	M
Epsilon Aurigae	05	02.0	+43	49	2.9–3.8	Algol	9892	F+B
R Boötis	14	37.2	+26	44	6.2–13.1	Mira	223	M
X Camelopardalis	04	45.7	+75	06	7.4–14.2	Mira	144	K–M
R Cancri	08	16.6	+11	44	6.1–11.8	Mira	362	M
X Cancri	08	55.4	+17	14	5.6–7.5	Semi-regular	195?	C
R Canis Majoris	07	19.5	–16	24	5.7–6.3	Algol	1.1	F
VY Canis Majoris	07	23.0	–25	46	6.5–9.6	Unique	—	M
S Canis Minoris	07	32.7	+08	19	6.6–13.2	Mira	333	M
R Canum Ven.	13	49.0	+39	33	6.5–12.9	Mira	329	M
*V Canum Ven.	13	19.5	+45	32	6.5–8.6	Semi-regular	192	M
R Carinae	09	32.2	–62	47	3.9–10.5	Mira	309	M
S Carinae	10	09.4	–61	33	4.5–9.9	Mira	149	K–M
l Carinae	09	45.2	–62	30	3.3–4.2	Cepheid	35.5	F–K
Eta Carinae	10	45.1	–59	41	-0.8–7.9	Irregular	—	Pec
R Cassiopeiae	23	58.4	+51	24	4.7–13.5	Mira	430	M
S Cassiopeiae	01	19.7	+72	37	7.9–16.1	Mira	612	S
W Cassiopeiae	00	54.9	+58	34	7.8–12.5	Mira	406	C
Gamma Cas.	00	56.7	+60	43	1.6–3.0	Gamma Cas.	—	B
Rho Cassiopeiae	23	54.4	+57	30	4.1–6.2	Semi-regular	—	F–K
R Centauri	14	16.6	–59	55	5.3–11.8	Mira	546	M
S Centauri	12	24.6	–49	26	7–8	Semi-regular	65	C
T Centauri	13	41.8	–33	36	5.5–9.0	Semi-regular	90	K–M
S Cephei	21	35.2	+78	37	7.4–12.9	Mira	487	C
T Cephei	21	09.5	+68	29	5.2–11.3	Mira	388	M
Delta Cephei	22	29.2	+58	25	3.5–4.4	Cepheid	5.4	F–G
Mu Cephei	21	43.5	+58	47	3.4–5.1	Semi-regular	730	M
U Ceti	02	33.7	–13	09	6.8–13.4	Mira	235	M
W Ceti	00	02.1	–14	41	7.1–14.8	Mira	351	S
*Omicron Ceti	02	19.3	–02	59	2.0–10.1	Mira	332	M
R Chamaeleontis	08	21.8	–76	21	7.5–14.2	Mira	335	M
T Columbae	05	19.3	–33	42	6.6–12.7	Mira	226	M
R Comae Ber.	12	04.3	+18	47	7.1–14.6	Mira	363	M
*R Coronae Bor.	15	48.6	+28	09	5.7–14.8	R Coronae Bor.	—	C
S Coronae Bor.	15	21.4	+31	22	5.8–14.1	Mira	360	M
T Coronae Bor.	15	59.6	+25	55	2.0–10.8	Recurrent nova	—	M+Pec
V Coronae Bor.	15	49.5	+39	34	6.9–12.6	Mira	358	C
W Coronae Bor.	16	15.4	+37	48	7.8–14.3	Mira	238	M
R Corvi	12	19.6	–19	15	6.7–14.4	Mira	317	M
R Crucis	12	23.6	–61	38	6.4–7.2	Cepheid	5.8	F–G

Star	RA		Declination		Range	Type	Period	Spectrum
	h	m	°	′			(days)	
R Cygni	19	36.8	+50	12	6.1–14.4	Mira	426	S
U Cygni	20	19.6	+47	54	5.9–12.1	Mira	463	C
W Cygni	21	36.0	+45	22	5.0–7.6	Semi-regular	131	M
RT Cygni	19	43.6	+48	47	6.0–13.1	Mira	190	M
SS Cygni	21	42.7	+43	35	7.7–12.4	Dwarf nova	50±	K+Pec
*CH Cygni	19	24.5	+50	14	5.6–9.0	Symbiotic	—	M+B
Chi Cygni	19	50.6	+32	55	3.3–14.2	Mira	408	S
R Delphini	20	14.9	+09	05	7.6–13.8	Mira	285	M
U Delphini	20	45.5	+18	05	5.6–7.5	Semi-regular	110?	M
EU Delphini	20	37.9	+18	16	5.8–6.9	Semi-regular	60	M
Beta Doradûs	05	33.6	–62	29	3.5–4.1	Cepheid	9.8	F–G
R Draconis	16	32.7	+66	45	6.7–13.2	Mira	246	M
T Eridani	03	55.2	–24	02	7.2–13.2	Mira	252	M
R Fornacis	02	29.3	–26	06	7.5–13.0	Mira	389	C
R Geminorum	07	07.4	+22	42	6.0–14.0	Mira	370	S
U Geminorum	07	55.1	+22	00	8.2–14.9	Dwarf nova	105±	Pec+M
Zeta Geminorum	07	04.1	+20	34	3.6–4.2	Cepheid	10.2	F–G
Eta Geminorum	06	14.9	+22	30	3.2–3.9	Semi-regular	233	M
S Gruis	22	26.1	–48	26	6.0–15.0	Mira	402	M
S Herculis	16	51.9	+14	56	6.4–13.8	Mira	307	M
U Herculis	16	25.8	+18	54	6.4–13.4	Mira	406	M
Alpha Herculis	17	14.6	+14	23	2.7–4.0	Semi-regular	—	M
68, u Herculis	17	17.3	+33	06	4.7–5.4	Algol	2.1	B+B
R Horologii	02	53.9	–49	53	4.7–14.3	Mira	408	M
U Horologii	03	52.8	–45	50	6–14	Mira	348	M
R Hydrae	13	29.7	–23	17	3.5–10.9	Mira	389	M
U Hydrae	10	37.6	–13	23	4.3–6.5	Semi-regular	450?	C
VW Hydri	04	09.1	–71	18	8.4–14.4	Dwarf nova	27±	Pec
R Leonis	09	47.6	+11	26	4.4–11.3	Mira	310	M
R Leonis Minoris	09	45.6	+34	31	6.3–13.2	Mira	372	M
R Leporis	04	59.6	–14	48	5.5–11.7	Mira	427	C
Y Librae	15	11.7	–06	01	7.6–14.7	Mira	276	M
RS Librae	15	24.3	–22	55	7.0–13.0	Mira	218	M
Delta Librae	15	01.0	–08	31	4.9–5.9	Algol	2.3	A
R Lyncis	07	01.3	+55	20	7.2–14.3	Mira	379	S
R Lyrae	18	55.3	+43	57	3.9–5.0	Semi-regular	46?	M
RR Lyrae	19	25.5	+42	47	7.1–8.1	RR Lyrae	0.6	A–F
Beta Lyrae	18	50.1	+33	22	3.3–4.4	Eclipsing	12.9	B
U Microscopii	20	29.2	–40	25	7.0–14.4	Mira	334	M
*U Monocerotis	07	30.8	–09	47	5.9–7.8	RV Tauri	91	F–K

Some Interesting Variable Stars

Star	RA		Declination		Range	Type	Period	Spectrum
	h	m	°	′			(days)	
V Monocerotis	06	22.7	−02	12	6.0–13.9	Mira	340	M
R Normae	15	36.0	−49	30	6.5–13.9	Mira	508	M
T Normae	15	44.1	−54	59	6.2–13.6	Mira	241	M
R Octantis	05	26.1	−86	23	6.3–13.2	Mira	405	M
S Octantis	18	08.7	−86	48	7.2–14.0	Mira	259	M
V Ophiuchi	16	26.7	−12	26	7.3–11.6	Mira	297	C
X Ophiuchi	18	38.3	+08	50	5.9–9.2	Mira	329	M
RS Ophiuchi	17	50.2	−06	43	4.3–12.5	Recurrent nova	—	OB+M
U Orionis	05	55.8	+20	10	4.8–13.0	Mira	368	M
W Orionis	05	05.4	+01	11	5.9–7.7	Semi-regular	212	C
Alpha Orionis	05	55.2	+07	24	0.0–1.3	Semi-regular	2335	M
S Pavonis	19	55.2	−59	12	6.6–10.4	Semi-regular	381	M
Kappa Pavonis	18	56.9	−67	14	3.9–4.8	W Virginis	9.1	G
R Pegasi	23	06.8	+10	33	6.9–13.8	Mira	378	M
X Persei	03	55.4	+31	03	6.0–7.0	Gamma Cas.	—	O9.5
Beta Persei	03	08.2	+40	57	2.1–3.4	Algol	2.9	B
Zeta Phoenicis	01	08.4	−55	15	3.9–4.4	Algol	1.7	B+B
R Pictoris	04	46.2	−49	15	6.4–10.1	Semi-regular	171	M
RS Puppis	08	13.1	−34	35	6.5–7.7	Cepheid	41.4	F–G
L^2 Puppis	07	13.5	−44	39	2.6–6.2	Semi-regular	141	M
T Pyxidis	09	04.7	−32	23	6.5–15.3	Recurrent nova	7000±	Pec
U Sagittae	19	18.8	+19	37	6.5–9.3	Algol	3.4	B+G
WZ Sagittae	20	07.6	+17	42	7.0–15.5	Dwarf nova	1900±	A
R Sagittarii	19	16.7	−19	18	6.7–12.8	Mira	270	M
RR Sagittarii	19	55.9	−29	11	5.4–14.0	Mira	336	M
RT Sagittarii	20	17.7	−39	07	6.0–14.1	Mira	306	M
RU Sagittarii	19	58.7	−41	51	6.0–13.8	Mira	240	M
RY Sagittarii	19	16.5	−33	31	5.8–14.0	R Coronae Bor.	—	G
RR Scorpii	16	56.6	−30	35	5.0–12.4	Mira	281	M
RS Scorpii	16	55.6	−45	06	6.2–13.0	Mira	320	M
RT Scorpii	17	03.5	−36	55	7.0–15.2	Mira	449	S
Delta Scorpii	16	00.3	−22	37	1.6–2.3	Irregular	—	B
S Sculptoris	00	15.4	−32	03	5.5–13.6	Mira	363	M
R Scuti	18	47.5	−05	42	4.2–8.6	RV Tauri	146	G–K
R Serpentis	15	50.7	+15	08	5.2–14.4	Mira	356	M
S Serpentis	15	21.7	+14	19	7.0–14.1	Mira	372	M
T Tauri	04	22.0	+19	32	9.3–13.5	T Tauri	—	F–K
SU Tauri	05	49.1	+19	04	9.1–16.9	R Coronae Bor.	—	G
Lambda Tauri	04	00.7	+12	29	3.4–3.9	Algol	4.0	B+A
R Trianguli	02	37.0	+34	16	5.4–12.6	Mira	267	M

Star	RA		Declination		Range	Type	Period	Spectrum
	h	m	°	′			(days)	
R Ursae Majoris	10	44.6	+68	47	6.5–13.7	Mira	302	M
T Ursae Majoris	12	36.4	+59	29	6.6–13.5	Mira	257	M
*Z Ursae Majoris	11	56.5	+57	52	6.2–9.4	Semi-regular	196	M
*RY Ursae Majoris	12	20.5	+61	19	6.7–8.3	Semi-regular	310?	M
U Ursae Minoris	14	17.3	+66	48	7.1–13.0	Mira	331	M
R Virginis	12	38.5	+06	59	6.1–12.1	Mira	146	M
S Virginis	13	33.0	–07	12	6.3–13.2	Mira	375	M
SS Virginis	12	25.3	+00	48	6.0–9.6	Semi-regular	364	C
R Vulpeculae	21	04.4	+23	49	7.0–14.3	Mira	137	M
Z Vulpeculae	19	21.7	+25	34	7.3–8.9	Algol	2.5	B+A

V CANUM VENATICORUM 13h 19.5m +45° 32′ (2000)

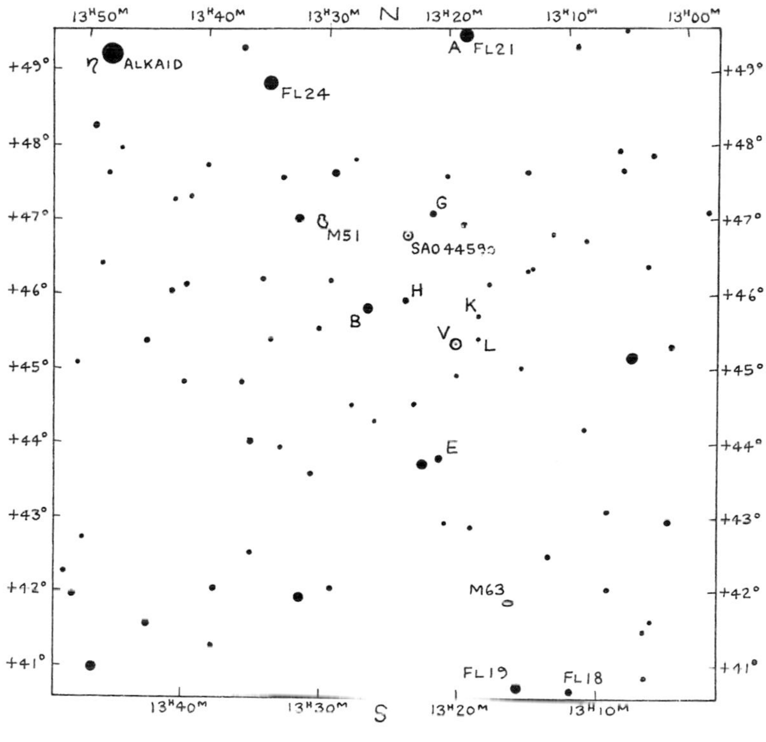

A 5.1	H 7.8
B 5.9	K 8.4
E 6.5	L 8.6
G 7.1	

○ (MIRA) CETI 02h 19.3m −02° 59′ (2000)

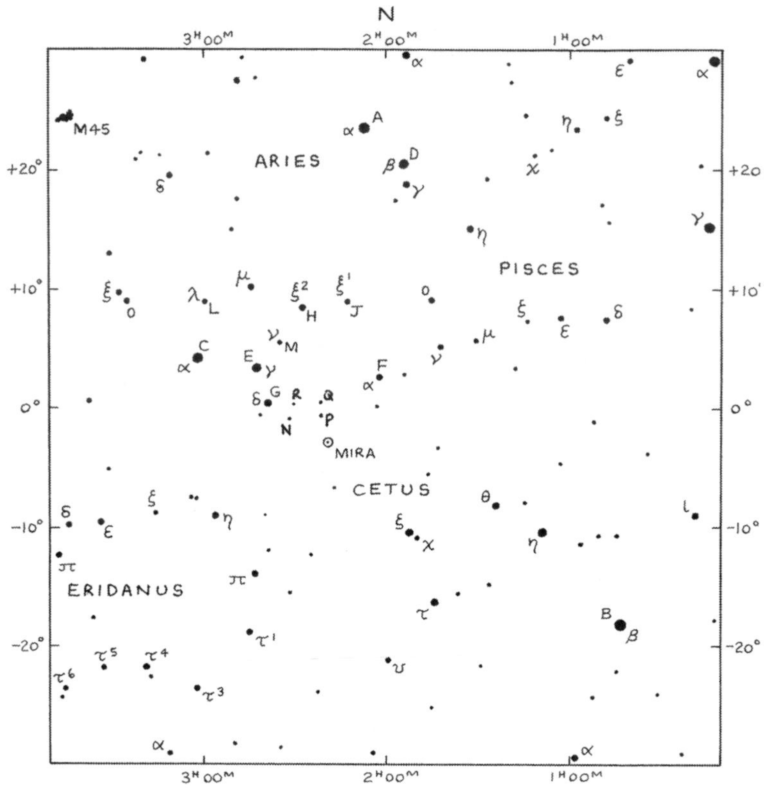

A 2.2	J 4.4
B 2.4	L 4.9
C 2.7	M 5.1
D 3.0	N 5.4
E 3.6	P 5.5
F 3.8	Q 5.7
G 4.1	R 6.1
H 4.3	

R CORONAE BOREALIS 15h 48.6m +28° 09′ (2000)

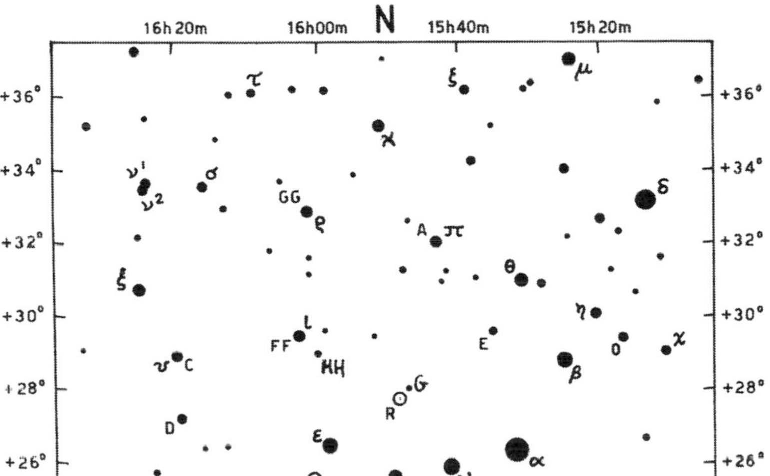

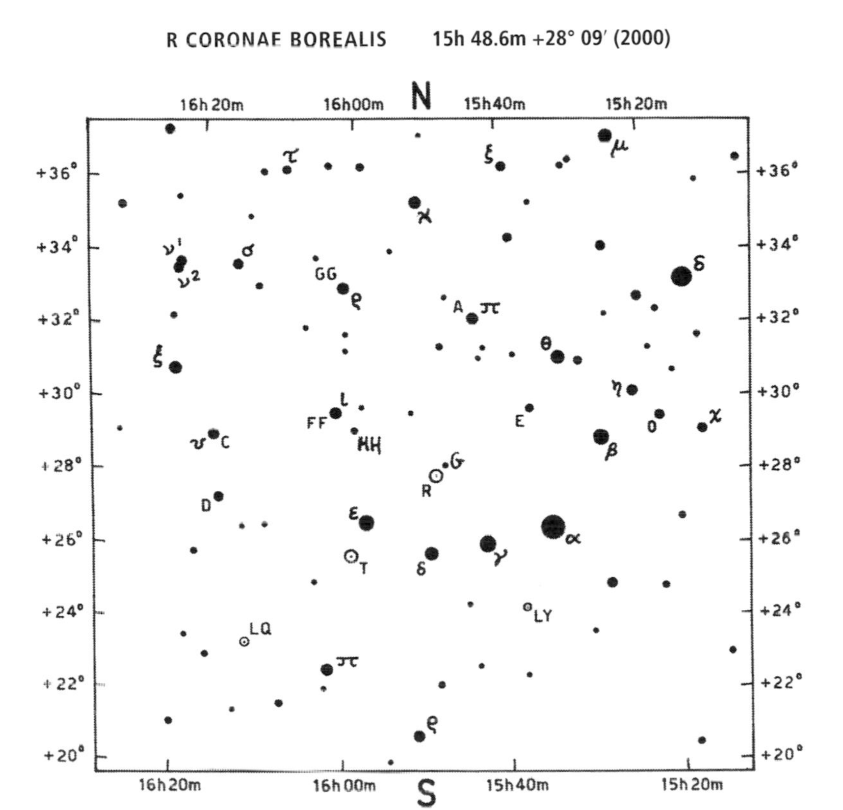

FF 5.0 C 5.8
GG 5.4 D 6.2
A 5.6 E 6.5
HH 7.1
G 7.4

CH CYGNI 19h 24.5m +50° 14′ (2000)

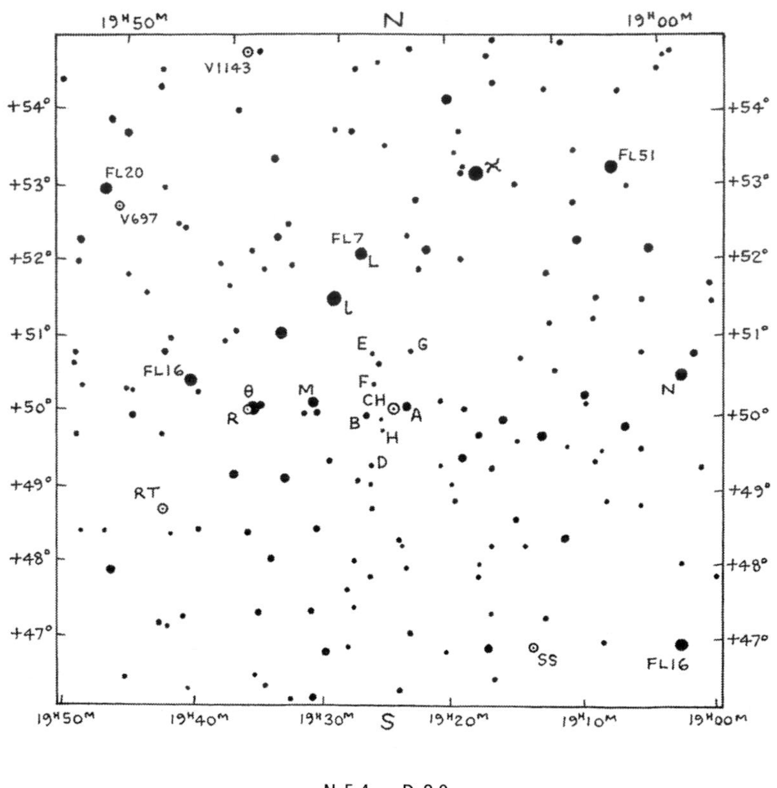

N 5.4	D 8.0
M 5.5	E 8.1
L 5.8	F 8.5
A 6.5	G 8.5
B 7.4	H 9.2

U MONOCEROTIS 07h 30.8m −09° 47′ (2000)

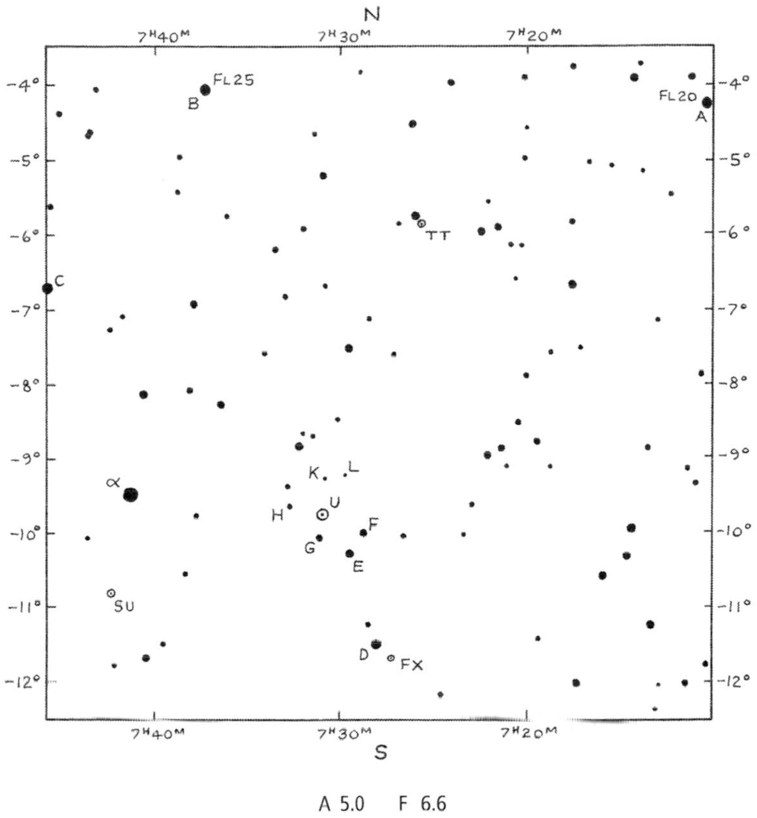

A	5.0	F	6.6
B	5.2	G	7.0
C	5.7	H	7.5
D	5.9	K	7.8
E	6.0	L	8.0

RY URSAE MAJORIS 12h 20.5m +61° 19′ (2000)
Z URSAE MAJORIS 11h 56.5m +57° 52′ (2000)

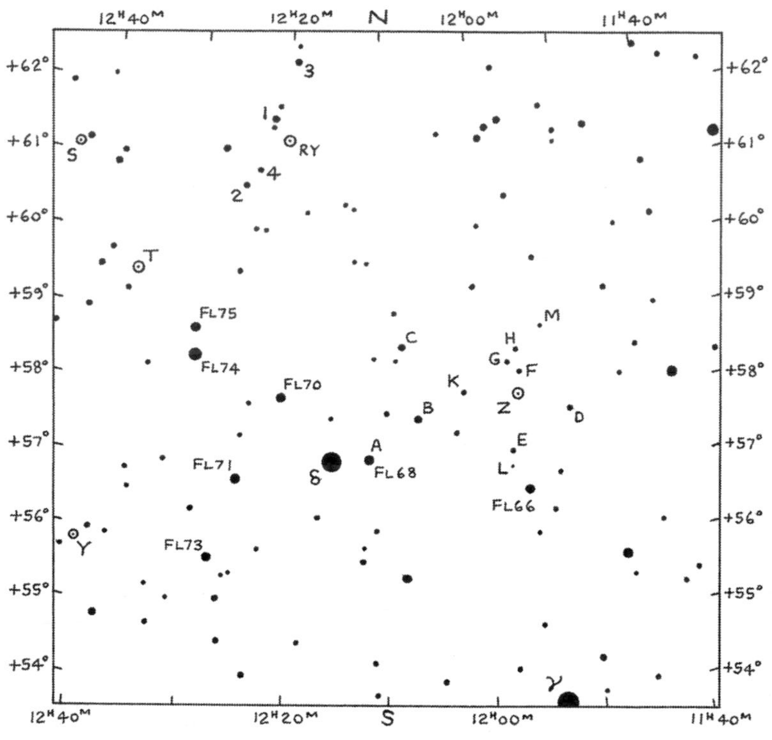

A	6.5	F	8.6	M	9.1
B	7.2	G	8.7	1	6.9
C	7.6	H	8.8	2	7.4
D	8.0	K	8.9	3	7.7
E	8.3	L	9.0	4	7.8

Mira Stars: Maxima, 2013

JOHN ISLES

Below are the predicted dates of maxima for Mira stars that reach magnitude 7.5 or brighter at an average maximum. Individual maxima can in some cases be brighter or fainter than average by a magnitude or more, and all dates are only approximate. The positions, extreme ranges and mean periods of these stars can be found in the preceding list of interesting variable stars.

Star	Mean magnitude at maximum	Dates of maxima
R Andromedae	6.9	6 Jan
W Andromedae	7.4	22 Nov
R Aquarii	6.5	7 Mar
R Aquilae	6.1	18 May
R Boötis	7.2	29 Mar, 7 Nov
R Cancri	6.8	25 July
S Canis Minoris	7.5	6 Aug
R Carinae	4.6	23 May
S Carinae	5.7	22 Feb, 21 July, 17 Dec
R Cassiopeiae	7.0	13 Oct
R Centauri	5.8	17 Feb (secondary max.), 18 Dec (primary max.)
T Cephei	6.0	22 Mar
U Ceti	7.5	8 June
Omicron Ceti	3.4	6 July
T Columbae	7.5	30 June
S Coronae Borealis	7.3	3 Sept
V Coronae Borealis	7.5	13 Nov
R Corvi	7.5	9 June
R Cygni	7.5	18 Aug

Star	Mean magnitude at maximum	Dates of maxima
U Cygni	7.2	5 Dec
RT Cygni	7.3	19 Mar, 25 Sept
Chi Cygni	5.2	5 May
R Geminorum	7.1	24 Jan
U Herculis	7.5	25 Feb
R Horologii	6.0	22 June
R Hydrae	4.5	27 Dec
R Leonis	5.8	30 Jan, 5 Dec
R Leonis Minoris	7.1	21 Feb
R Leporis	6.8	1 Sept
RS Librae	7.5	18 Feb, 23 Sept
V Monocerotis	7.0	19 Jan, 25 Dec
T Normae	7.4	11 Aug
V Ophiuchi	7.5	13 Mar
X Ophiuchi	6.8	16 Aug
U Orionis	6.3	25 Mar
R Sagittarii	7.3	2 Aug
RR Sagittarii	6.8	22 July
RT Sagittarii	7.0	1 Sept
RU Sagittarii	7.2	8 Feb, 6 Oct
RR Scorpii	5.9	28 Jan, 3 Nov
RS Scorpii	7.0	4 Aug
S Sculptoris	6.7	24 Dec
R Serpentis	6.9	11 Aug
R Trianguli	6.2	11 Apr
R Ursae Majoris	7.5	12 May
R Virginis	6.9	2 Mar, 26 July, 18 Dec
S Virginis	7.0	26 Feb

Some Interesting Double Stars

BOB ARGYLE

The positions, angles and separations given below correspond to epoch 2013.0.

No.	RA	Declin-ation	Star	Magni-tudes	Separa-tion	PA	Cata-logue	Comments
	h m	° ′			arcsec	°		
1	00 31.5	−62 58	β Tuc	4.4,4.8	27.1	169	LCL 119	Both stars again difficult doubles.
2	00 49.1	+57 49	η Cas	3.4,7.5	13.3	323	Σ60	Easy. Creamy, bluish. P = 480 years.
3	00 55.0	+23 38	36 And	6.0,6.4	1.1	327	Σ73	P = 168 years. Both yellow. Slowly opening.
4	01 13.7	+07 35	ζ Psc	5.6,6.5	23.1	63	Σ100	Yellow, reddish-white.
5	01 39.8	−56 12	p Eri	5.8,5.8	11.7	188	Δ5	Period – 484 years.
6	01 53.5	+19 18	γ Ari	4.8,4.8	7.5	1	Σ180	Very easy. Both white.
7	02 02.0	+02 46	α Psc	4.2,5.1	1.8	262	Σ202	Binary, period = 933 years.
8	02 03.9	+42 20	γ And	2.3,5.0	9.6	63	Σ205	Yellow, blue. Relatively fixed.
			γ2 And	5.1,6.3	0.1	94	OΣ38	BC now beyond range of amateur instruments.
9	02 29.1	+67 24	ι Cas AB	4.9,6.9	2.6	229	Σ262	AB is long-period binary. P = 620 years.
			ι Cas AC	4.9,8.4	7.2	118		
10	02 33.8	−28 14	ω For	5.0,7.7	10.8	245	HJ 3506	Common proper motion.
11	02 43.3	+03 14	γ Cet	3.5,7.3	2.3	298	Σ299	Not too easy.
12	02 58.3	−40 18	θ Eri	3.4,4.5	8.3	90	PZ 2	Both white.

No.	RA h m	Declin-ation ° ′	Star	Magni-tudes	Separa-tion arcsec	PA °	Cata-logue	Comments
13	02 59.2	+21 20	ε Ari	5.2,5.5	1.4	210	Σ333	Closing very slowly. P = 1216 years? Both white.
14	03 00.9	+52 21	Σ331 Per	5.3,6.7	12.0	85	–	Fixed.
15	03 12.1	–28 59	α For	4.0,7.0	5.3	300	HJ 3555	P = 269 years. B variable?
16	03 48.6	–37 37	*f* Eri	4.8,5.3	8.2	215	Δ16	Pale yellow. Fixed.
17	03 54.3	–02 57	32 Eri	4.8,6.1	6.9	348	Σ470	Fixed.
18	04 32.0	+53 55	1 Cam	5.7,6.8	10.3	308	Σ550	Fixed.
19	04 50.9	–53 28	ι Pic	5.6,6.4	12.4	58	Δ18	Good object for small apertures. Fixed.
20	05 13.2	–12 56	κ Lep	4.5,7.4	2.0	357	Σ661	Visible in 7.5 cm. Slowly closing.
21	05 14.5	–08 12	β Ori	0.1,6.8	9.5	204	Σ668	Companion once thought to be close double.
22	05 21.8	–24 46	41 Lep	5.4,6.6	3.4	93	HJ 3752	Deep yellow pair in a rich field.
23	05 24.5	–02 24	η Ori	3.8,4.8	1.8	77	DA 5	Slow-moving binary.
24	05 35.1	+09 56	λ Ori	3.6,5.5	4.3	44	Σ738	Fixed.
25	05 35.3	–05 23	θ Ori AB	6.7,7.9	8.6	32	Σ748	Trapezium in M42.
			θ Ori CD	5.1,6.7	13.4	61		
26	05 40.7	–01 57	ζ Ori	1.9,4.0	2.2	166	Σ774	Can be split in 7.5 cm. Long-period binary.
27	06 14.9	+22 30	η Gem	var,6.5	1.6	253	β1008	Well seen with 20 cm. Primary orange.
28	06 46.2	+59 27	12 Lyn AB	5.4,6.0,	1.9	68	Σ948	AB is binary, P = 908 years.
			12 Lyn AC	5.4,7.3	8.7	309	–	
29	07 08.7	–70 30	γ Vol	3.9,5.8	14.1	298	Δ42	Very slow binary.
30	07 16.6	–23 19	h3945 CMa	4.8,6.8	26.8	51	–	Contrasting colours. Yellow and blue.
31	07 20.1	+21 59	δ Gem	3.5,8.2	5.5	228	Σ1066	Not too easy. Yellow, pale blue.

Some Interesting Double Stars

No.	RA	Declin-ation	Star	Magni-tudes	Separa-tion	PA	Cata-logue	Comments
	h m	° ′			arcsec	°		
32	07 34.6	+31 53	α Gem	1.9,2.9	4.9	56	Σ1110	Widening. Easy with 7.5 cm.
33	07 38.8	−26 48	κ Pup	4.5,4.7	9.8	318	H III 27	Both white.
34	08 12.2	+17 39	ζ Cnc AB	5.6,6.0	1.1	27	Σ1196	Period (AB) = 59.6 years. Near maximum separation.
			ζ Cnc AB-C	5.0,6.2	5.9	67	Σ1196	Period (AB–C = 1115 years.
35	08 46.8	+06 25	ε Hyd	3.3,6.8	2.9	307	Σ1273	PA slowly increasing. A is a very close pair.
36	09 18.8	+36 48	38 Lyn	3.9,6.6	2.6	226	Σ1334	Almost fixed.
37	09 47.1	−65 04	υ Car	3.1,6.1	5.0	129	RMK 11	Fixed. Fine in small telescopes.
38	10 20.0	+19 50	γ Leo	2.2,3.5	4.6	126	Σ1424	Binary, period = 510 years. Both orange.
39	10 32.0	−45 04	s Vel	6.2,6.5	13.5	218	PZ 3	Fixed.
40	10 46.8	−49 26	μ Vel	2.7,6.4	2.6	56	R 155	P = 138 years. Near widest separation.
41	10 55.6	+24 45	54 Leo	4.5,6.3	6.6	111	Σ1487	Slowly widening. Pale yellow and white.
42	11 18.2	+31 32	ξ UMa	4.3,4.8	1.7	190	Σ1523	Binary, 59.9 years. Needs 7.5 cm.
43	11 23.9	+10 32	ι Leo	4.0,6.7	2.1	98	Σ1536	Binary, period = 186 years.
44	11 32.3	−29 16	N Hya	5.8,5.9	9.4	210	H III 96	Both yellow. Long-period binary.
45	12 14.0	−45 43	D Cen	5.6,6.8	2.8	243	RMK 14	Orange and white. Closing.
46	12 26.6	−63 06	α Cru	1.4,1.9	4.0	114	Δ252	Glorious pair. Third star in a low power field.
47	12 41.5	−48 58	γ Cen	2.9,2.9	0.1	225	HJ 4539	Period = 84 years. Nearing periastron. Both yellow.

No.	RA	Declin-ation	Star	Magni-tudes	Separa-tion	PA	Cata-logue	Comments
	h m	° ′			arcsec	°		
48	12 41.7	–01 27	γ Vir	3.5,3.5	2.0	11	Σ1670	Now widening quickly. Beautiful pair for 10 cm.
49	12 46.3	–68 06	β Mus	3.7,4.0	0.9	56	R 207	Both white. Closing slowly. P = 194 years.
50	12 54.6	–57 11	μ Cru	4.3,5.3	34.9	17	Δ126	Fixed. Both white.
51	12 56.0	+38 19	α CVn	2.9,5.5	19.3	229	Σ1692	Easy. Yellow, bluish.
52	13 22.6	–60 59	J Cen	4.6,6.5	60.0	343	Δ133	Fixed. A is a close pair.
53	13 24.0	+54 56	ζ UMa	2.3,4.0	14.4	152	Σ1744	Very easy. Naked-eye pair with Alcor.
54	13 51.8	–33 00	3 Cen	4.5,6.0	7.7	102	H III 101	Both white. Closing slowly.
55	14 39.6	–60 50	α Cen	0.0,1.2	4.9	265	RHD 1	Finest pair in the sky. P = 80 years. Closing.
56	14 41.1	+13 44	ζ Boo	4.5,4.6	0.5	292	Σ1865	Both white. Closing – highly inclined orbit.
57	14 45.0	+27 04	ε Boo	2.5,4.9	2.9	344	Σ1877	Yellow, blue. Fine pair.
58	14 46.0	–25 27	54 Hya	5.1,7.1	8.3	122	H III 97	Closing slowly.
59	14 49.3	–14 09	μ Lib	5.8,6.7	1.8	6	β106	Becoming wider. Fine in 7.5 cm.
60	14 51.4	+19 06	ξ Boo	4.7,7.0	5.8	305	Σ1888	Fine contrast. Easy. P = 151.6 years.
61	15 03.8	+47 39	44 Boo	5.3,6.2	1.3	64	Σ1909	Period = 210 years. Closing more quickly.
62	15 05.1	–47 03	π Lup	4.6,4.7	1.7	66	HJ 4728	Widening.
63	15 18.5	–47 53	μ Lup AB	5.1,5.2	1.1	300	HJ 4753	AB closing. Under-observed.
			μ Lup AC	4.4,7.2	22.7	127	Δ180	AC almost fixed.
64	15 23.4	–59 19	γ Cir	5.1,5.5	0.8	358	HJ 4757	Closing. Needs 20 cm. Long-period binary.
65	15 34.8	+10 33	δ Ser	4.2,5.2	4.0	172	Σ1954	Long-period binary.

No.	RA	Declination	Star	Magnitudes	Separation	PA	Catalogue	Comments
	h m	° ′			arcsec	°		
66	15 35.1	−41 10	γ Lup	3.5,3.6	0.8	277	HJ 4786	Binary. Period = 190 years. Needs 20 cm.
67	15 56.9	−33 58	γ Lup	5.3,5.8	10.2	49	PZ 4	Fixed.
68	16 14.7	+33 52	σ CrB	5.6,6.6	7.2	238	Σ2032	Long-period binary. Both white.
69	16 29.4	−26 26	α Sco	1.2,5.4	2.6	277	GNT 1	Red, green. Difficult from mid-northern latitudes.
70	16 30.9	+01 59	λ Oph	4.2,5.2	1.4	39	Σ2055	P = 129 years. Fairly difficult in small apertures.
71	16 41.3	+31 36	ζ Her	2.9,5.5	1.2	153	Σ2084	Period = 34.5 years. Now widening. Needs 20 cm.
72	17 05.3	+54 28	μ Dra	5.7,5.7	2.5	5	Σ2130	Period 812 years.
73	17 14.6	+14 24	α Her	var,5.4	4.6	103	Σ2140	Red, green. Long-period binary.
74	17 15.3	−26 35	36 Oph	5.1,5.1	5.0	142	SHJ 243	Period = 471 years.
75	17 23.7	+37 08	ρ Her	4.6,5.6	4.1	319	Σ2161	Slowly widening.
76	17 26.9	−45 51	HJ 4949 AB	5.6,6.5	2.1	251	HJ 4949	Beautiful coarse triple. All white.
			Δ 216 AC	7.1	105.0	310		
77	18 01.5	+21 36	95 Her	5.0,5.1	6.5	257	Σ2264	Colours thought variable in C19.
78	18 05.5	+02 30	70 Oph	4.2,6.0	6.1	128	Σ2272	Opening. Easy in 7.5 cm. P = 88.4 years.
79	18 06.8	−43 25	HJ 5014 CrA	5.7,5.7	1.7	1	–	Period = 450 years. Needs 10 cm.
80	18 25.4	−20 33	21 Sgr	5.0,7.4	1.7	279	JC 6	Slowly closing binary, orange and green.
81	18 35.9	+16 58	OΣ358 Her	6.8,7.0	1.5	148	–	Period = 380 years.
82	18 44.3	+39 40	ε1 Lyr	5.0,6.1	2.4	347	Σ2382	Quadruple system with epsilon2. Both pairs visible in 7.5 cm.
83	18 44.3	+39 40	ε2 Lyr	5.2,5.5	2.4	77	Σ2383	

No.	RA	Declin-ation	Star	Magni-tudes	Separa-tion	PA	Cata-logue	Comments
	h m	° ′			arcsec	°		
84	18 56.2	+04 12	θ Ser	4.5,5.4	22.4	104	Σ2417	Fixed. Very easy
85	19 06.4	–37 04	γ CrA	4.8,5.1	1.4	358	HJ 5084	Beautiful pair. Period = 122 years.
86	19 30.7	+27 58	β Cyg AB	3.1,5.1	34.3	54	Σ I 43	Glorious. Yellow, blue-greenish.
			β Cyg Aa	3.1,5.2	0.4	89	MCA 55	Aa. Very difficult. Period = 214 years
87	19 45.0	+45 08	δ Cyg	2.9,6.3	2.7	219	Σ2579	Slowly widening. Period = 780 years.
88	19 48.2	+70 16	ε Dra	3.8,7.4	3.0	19	Σ2603	Slow binary.
89	19 54.6	–08 14	57 Aql	5.7,6.4	36.0	170	Σ2594	Easy pair. Contrasting colours.
90	20 46.7	+16 07	γ Del	4.5,5.5	9.0	265	Σ2727	Easy. Yellowish. Long-period binary.
91	20 59.1	+04 18	ε Equ AB	6.0,6.3	0.4	283	Σ2737	Fine triple. AB a test for 30 cm. P = 101.5 years
			ε Equ AC	6.0,7.1	10.3	66		
92	21 06.9	+38 45	61 Cyg	5.2,6.0	31.4	152	Σ2758	Nearby binary. Both orange. Period = 678 years.
93	21 19.9	–53 27	θ Ind	4.5,7.0	7.0	271	HJ 5258	Pale yellow and reddish. Long-period binary.
94	21 44.1	+28 45	μ Cyg	4.8,6.1	1.6	319	Σ2822	Period = 789 years.
95	22 03.8	+64 37	ξ Cep	4.4,6.5	8.4	274	Σ2863	White and blue. Long-period binary.
96	22 14.3	–21 04	41 Aqr	5.6,6.7	5.1	113	H N 56	Yellowish and purple?
97	22 26.6	–16 45	53 Aqr	6.4,6.6	1.3	52	SHJ 345	Long-period binary; periastron in 2023.
98	22 28.8	–00 01	ζ Aqr	4.3,4.5	2.2	167	Σ2909	Period = 487 years. Slowly widening.
99	23 19.1	–13 28	94 Aqr	5.3,7.0	12.3	351	Σ2988	Yellow and orange. Probable binary.
100	23 59.5	+33 43	Σ3050 And	6.6,6.6	2.3	338	–	Period = 717 years. Visible in 7.5 cm.

Some Interesting Nebulae, Clusters and Galaxies

Object	RA		Declination		Remarks
	h	m	°	′	
M31 Andromedae	00	40.7	+41	05	Andromeda Galaxy, visible to naked eye.
H VIII 78 Cassiopeiae	00	41.3	+61	36	Fine cluster, between Gamma and Kappa Cassiopeiae.
M33 Trianguli	01	31.8	+30	28	Spiral. Difficult with small apertures.
H VI 33–4 Persei, C14	02	18.3	+56	59	Double cluster; Sword-handle.
Δ142 Doradûs	05	39.1	–69	09	Looped nebula round 30 Doradus. Naked eye. In Large Magellanic Cloud.
M1 Tauri	05	32.3	+22	00	Crab Nebula, near Zeta Tauri.
M42 Orionis	05	33.4	–05	24	Orion Nebula. Contains the famous Trapezium, Theta Orionis.
M35 Geminorum	06	06.5	+24	21	Open cluster near Eta Geminorum.
H VII 2 Monocerotis, C50	06	30.7	+04	53	Open cluster, just visible to naked eye.
M41 Canis Majoris	06	45.5	–20	42	Open cluster, just visible to naked eye.
M47 Puppis	07	34.3	–14	22	Mag. 5.2. Loose cluster.
H IV 64 Puppis	07	39.6	–18	05	Bright planetary in rich neighbourhood.
M46 Puppis	07	39.5	–14	42	Open cluster.
M44 Cancri	08	38	+20	07	Praesepe. Open cluster near Delta Cancri. Visible to naked eye.
M97 Ursae Majoris	11	12.6	+55	13	Owl Nebula, diameter 3′. Planetary.
Kappa Crucis, C94	12	50.7	–60	05	'Jewel Box'; open cluster, with stars of contrasting colours.
M3 Can. Ven.	13	40.6	+28	34	Bright globular.
Omega Centauri, C80	13	23.7	–47	03	Finest of all globulars. Easy with naked eye.
M80 Scorpii	16	14.9	–22	53	Globular, between Antares and Beta Scorpii.
M4 Scorpii	16	21.5	–26	26	Open cluster close to Antares.

Object	RA		Declination		Remarks
	h	m	°	′	
M13 Herculis	16	40	+36	31	Globular. Just visible to naked eye.
M92 Herculis	16	16.1	+43	11	Globular. Between Iota and Eta Herculis.
M6 Scorpii	17	36.8	–32	11	Open cluster; naked eye.
M7 Scorpii	17	50.6	–34	48	Very bright open cluster; naked eye.
M23 Sagittarii	17	54.8	–19	01	Open cluster nearly 50′ in diameter.
H IV 37 Draconis, C6	17	58.6	+66	38	Bright planetary.
M8 Sagittarii	18	01.4	–24	23	Lagoon Nebula. Gaseous. Just visible with naked eye.
NGC 6572 Ophiuchi	18	10.9	+06	50	Bright planetary, between Beta Ophiuchi and Zeta Aquilae.
M17 Sagittarii	18	18.8	–16	12	Omega Nebula. Gaseous. Large and bright.
M11 Scuti	18	49.0	–06	19	Wild Duck. Bright open cluster.
M57 Lyrae	18	52.6	+32	59	Ring Nebula. Brightest of planetaries.
M27 Vulpeculae	19	58.1	+22	37	Dumb-bell Nebula, near Gamma Sagittae.
H IV 1 Aquarii, C55	21	02.1	–11	31	Bright planetary, near Nu Aquarii.
M15 Pegasi	21	28.3	+12	01	Bright globular, near Epsilon Pegasi.
M39 Cygni	21	31.0	+48	17	Open cluster between Deneb and Alpha Lacertae. Well seen with low powers.

(M = Messier number; NGC = New General Catalogue number; C = Caldwell number.)

Our Contributors

Dr Stephen Webb is the author of several books, including *Where Is Everybody?* (winner of the Contact in Context award, and shortlisted for the 2003 Aventis Prize for science books) and the recently published *New Eyes on the Universe: Twelve Cosmic Mysteries and the Tools We Need To Solve Them* (which gives an overview of the many observatories that are planned or are already in construction – ALMA is just one of many amazing instruments that will soon be observing the universe!) He blogs on these and other topics in science at http://stephenwebb.info/.

Dr David M. Harland gained his BSc in astronomy in 1977 and a doctorate in computational science. Subsequently, he has taught computer science, worked in industry and managed academic research. In 1995 he 'retired' and has since published many books on space themes.

Nick James B.Sc., C.Eng. is the papers secretary of the British Astronomical Association and Assistant Director of its Comet Section. Professionally he is an engineer working in the space industry responsible for a team developing space communication and tracking systems. He has had a long-term interest in astronomical imaging, starting with film and progressing more recently to CCDs and digital cameras. He is joint author, with Gerald North, of *Observing Comets*, published by Springer in 2003.

Martin Mobberley is one of the UK's most active imagers of comets, planets, asteroids, variable stars, novae and supernovae and served as President of the British Astronomical Association from 1997 to 1999. In 2000, he was awarded the Association's Walter Goodacre Award. He is the sole author of seven popular astronomy books published by Springer as well as three children's 'Space Exploration' books published by Top That Publishing. In addition he has authored hundreds of articles in *Astronomy Now* and numerous other astronomical publications.

Dr David A. Allen's extraordinary life had an inauspicious start due to a misdiagnosed hip complaint. He spent seven years of his childhood with his body and legs encased in plaster, always claiming to have been 'born at the age of nine'. Educated in Cambridge, where he began his pioneering work on infrared astronomy, David soon gravitated to the Anglo–Australian Observatory. There, his outstanding scientific research was matched by a prolific output in popular astronomy, including many articles for the *Yearbook of Astronomy*. David died in July 1994 at the age of 47.

Richard Myer Baum is a former Director of the Mercury and Venus Section of the British Astronomical Association, an amateur astronomer and an independent scholar. He is author of *The Planets: Some Myths and Realities* (1973) (with W. Sheehan), *In Search of Planet Vulcan: The Ghost in Newton's Clockwork Universe* (1997) and *The Haunted Observatory* (2007). He has contributed to the *Journal of the British Astronomical Association*, *Journal for the History of Astronomy*, *Sky and Telescope* and many other publications including *The Dictionary of Nineteenth-Century British Scientists* (2004) and *The Biographical Encyclopedia of Astronomers* (2007).

Professor Fred Watson is Astronomer-in-Charge of the Australian Astronomical Observatory at Coonabarabran in north-western New South Wales, and one of Australia's best-known science communicators. He is a regular contributor to the *Yearbook of Astronomy*, and his recent books include *Universe* (for which he was Chief Consultant), *Stargazer – The Life and Times of the Telescope* and *Why is Uranus Upside Down – and Other Questions About the Universe*. In 2006 he was awarded the Australian Government Eureka Prize for Promoting Understanding of Science. In 2010 he was appointed a Member of the Order of Australia. Visit Fred's website at http://fredwatson.com.au/.

Paul G. Abel is a co-presenter on the BBC *Sky at Night* and works in the Department of Physics and Astronomy at the University of Leicester. Initially he trained as a mathematician and works in the area of theoretical physics concerned with Hawking Radiation – the radiation emitted by black holes. His other research interest is concerned with Quantum Field theories in curved space-time. He also teaches mathematics in the Physics Department and is an active amateur astronomer with a keen interest in the Moon and planets, who makes all of his observations visually, using drawings to capture the details. He is Assistant Director of the Saturn Section of the British Astronomical Association (BAA) and a Fellow of the Royal Astronomical Society. He frequently writes for various popular astronomy magazines.

Dr Allan Chapman, of Wadham College, Oxford, is probably Britain's leading authority on the history of astronomy. He has published many research papers and several books, as well as numerous popular accounts. He is a frequent and welcome contributor to the *Yearbook*.

Astronomical Societies in the British Isles

Association for Astronomy Education

Secretary: Teresa Grafton, The Association for Astronomy Education, c/o The Royal Astronomical Society, Burlington House, Piccadilly, London W1V 0NL.

Astronomical Society of Edinburgh

Secretary: Graham Rule, 105/19 Causewayside, Edinburgh EH9 1QG.

Website: www.roe.ac.uk/asewww/; *Email:* asewww@roe.ac.uk

Meetings: City Observatory, Calton Hill, Edinburgh. 1st Friday each month, 8 p.m.

Astronomical Society of Glasgow

Secretary: Mr David Degan, 5 Hillside Avenue, Alexandria, Dunbartonshire G83 0BB.

Website: www.astronomicalsocietyofglasgow.org.uk

Meetings: Royal College, University of Strathclyde, Montrose Street, Glasgow. 3rd Thursday each month, Sept.–Apr., 7.30 p.m.

Astronomical Society of Haringey

Secretary: Jerry Workman, 91 Greenslade Road, Barking, Essex IG11 9XF.

Meetings: Palm Court, Alexandra Palace, 3rd Wednesday each month, 8 p.m.

Astronomy Ireland

Secretary: Tony Ryan, PO Box 2888, Dublin 1, Eire.

Website: www.astronomy.ie; *Email:* info@astronomy.ie

Meetings: 2nd Monday of each month. Telescope meetings every clear Saturday.

British Astronomical Association

Assistant Secretary: Burlington House, Piccadilly, London W1V 9AG.

Meetings: Lecture Hall of Scientific Societies, Civil Service Commission Building, 23 Savile Row, London W1. Last Wednesday each month (Oct.–June), 5 p.m. and some Saturday afternoons.

Federation of Astronomical Societies

Secretary: Clive Down, 10 Glan-y-Llyn, North Cornelly, Bridgend, County Borough CF33 4EF.

Email: clivedown@btinternet.com

Junior Astronomical Society of Ireland

Secretary: K. Nolan, 5 St Patrick's Crescent, Rathcoole, Co. Dublin.

Meetings: The Royal Dublin Society, Ballsbridge, Dublin 4. Monthly.

Society for Popular Astronomy

Secretary: Guy Fennimore, 36 Fairway, Keyworth, Nottingham NG12 5DU.

Website: www.popastro.com; *Email:* SPAstronomy@aol.com

Meetings: Last Saturday in Jan., Apr., July, Oct., 2.30 p.m. in London.

Webb Deep-Sky Society

Membership Secretary/Treasurer: Steve Rayner, 11 Four Acres, Weston, Portland, Dorset DT5 2JG.

Email: stephen.rayner@tesco,net
Website: www.webbdeepsky.com

Aberdeen and District Astronomical Society
Secretary: Ian C. Giddings, 95 Brentfield Circle, Ellon, Aberdeenshire AB41 9DB.
Meetings: Robert Gordon's Institute of Technology, St Andrew's Street, Aberdeen. Fridays, 7.30 p.m.

Abingdon Astronomical Society (was **Fitzharry's Astronomical Society**)
Secretary: Chris Holt, 9 Rutherford Close, Abingdon, Oxon OX14 2AT.
Website: www.abingdonastro.org.uk; *Email:* info@abingdonastro.co.uk
Meetings: All Saints' Methodist Church Hall, Dorchester Crescent, Abingdon, Oxon. 2nd Monday Sept.–June, 8 p.m. and additional beginners' meetings and observing evenings as advertised.

Altrincham and District Astronomical Society
Secretary: Derek McComiskey, 33 Tottenham Drive, Manchester M23 9WH.
Meetings: Timperley Village Club. 1st Friday Sept.–June, 8 p.m.

Andover Astronomical Society
Secretary: Mrs S. Fisher, Staddlestones, Aughton, Kingston, Marlborough, Wiltshire SN8 3SA.
Meetings: Grately Village Hall. 3rd Thursday each month, 7.30 p.m.

Astra Astronomy Section
Secretary: c/o Duncan Lunan, Flat 65, Dalraida House, 56 Blythswood Court, Anderston, Glasgow G2 7PE.
Meetings: Airdrie Arts Centre, Anderson Street, Airdrie. Weekly.

Astrodome Mobile School Planetarium
Contact: Peter J. Golding, 53 City Way, Rochester, Kent ME1 2AX.
Website: www.astrodome.clara.co.uk; *Email:* astrodome@clara.co.uk

Aylesbury Astronomical Society
Secretary: Alan Smith, 182 Marley Fields, Leighton Buzzard, Bedfordshire LU7 8WN.
Meetings: 1st Monday in month at 8 p.m., venue in Aylesbury area. Details from Secretary.

Bassetlaw Astronomical Society
Secretary: Andrew Patton, 58 Holding, Worksop, Notts S81 0TD.
Meetings: Rhodesia Village Hall, Rhodesia, Worksop, Notts. 2nd and 4th Tuesdays of month at 7.45 p.m.

Batley & Spenborough Astronomical Society
Secretary: Robert Morton, 22 Links Avenue, Cleckheaton, West Yorks BD19 4EG.
Meetings: Milner K. Ford Observatory, Wilton Park, Batley. Every Thursday, 8 p.m.

Bedford Astronomical Society
Secretary: Mrs L. Harrington, 24 Swallowfield, Wyboston, Bedfordshire MK44 3AE.
Website: www.observer1.freeserve.co.uk/bashome.html
Meetings: Bedford School, Burnaby Rd, Bedford. Last Wednesday each month.

Bingham & Brooks Space Organization
Secretary: N. Bingham, 15 Hickmore's Lane, Lindfield, West Sussex.

Birmingham Astronomical Society
Contact: P. Bolas, 4 Moat Bank, Bretby, Burton-on-Trent DE15 0QJ.
Website: www.birmingham-astronomical.co.uk; *Email:* pbolas@aol.com
Meetings: Room 146, Aston University. Last Tuesday of month. Sept.–June (except Dec., moved to 1st week in Jan.).

Blackburn Leisure Astronomy Section
Secretary: Mr H. Murphy, 20 Princess Way, Beverley, East Yorkshire HU17 8PD.
Meetings: Blackburn Leisure Welfare. Mondays, 8 p.m.

Blackpool & District Astronomical Society
Secretary: Terry Devon, 30 Victory Road, Blackpool, Lancashire FY1 3JT.
Website: www.blackpoolastronomy.org.uk; *Email:* info@blackpoolastronomy.org.uk
Meetings: St Kentigern's Social Centre, Blackpool. 1st Wednesday of the month, 7.45 p.m.

Bolton Astronomical Society
Secretary: Peter Miskiw, 9 Hedley Street, Bolton, Lancashire BL1 3LE.
Meetings: Ladybridge Community Centre, Bolton. 1st and 3rd Tuesdays Sept.–May, 7.30 p.m.

Border Astronomy Society
Secretary: David Pettitt, 14 Sharp Grove, Carlisle, Cumbria CA2 5QR.
Website: www.members.aol.com/P3pub/page8.html
Email: davidpettitt@supanet.com
Meetings: The Observatory, Trinity School, Carlisle. Alternate Thursdays, 7.30 p.m., Sept.–May.

Boston Astronomers
Secretary: Mrs Lorraine Money, 18 College Park, Horncastle, Lincolnshire LN9 6RE.
Meetings: Blackfriars Arts Centre, Boston. 2nd Monday each month, 7.30 p.m.

Bradford Astronomical Society
Contact: Mrs J. Hilary Knaggs, 6 Meadow View, Wyke, Bradford BD12 9LA.
Website: www.bradford-astro.freeserve.co.uk/index.htm
Meetings: Eccleshill Library, Bradford. Alternate Mondays, 7.30 p.m.

Braintree, Halstead & District Astronomical Society
Secretary: Mr J. R. Green, 70 Dorothy Sayers Drive, Witham, Essex CM8 2LU.
Meetings: BT Social Club Hall, Witham Telephone Exchange. 3rd Thursday each month, 8 p.m.

Breckland Astronomical Society (was **Great Ellingham and District Astronomy Club**)
Contact: Martin Wolton, Willowbeck House, Pulham St Mary, Norfolk IP21 4QS.
Meetings: Great Ellingham Recreation Centre, Watton Road (B1077), Great Ellingham, 2nd Friday each month, 7.15 p.m.

Bridgend Astronomical Society
Secretary: Clive Down, 10 Glan-y-Llyn, Broadlands, North Cornelly, Bridgend County CF33 4EF.
Email: clivedown@btinternet.com
Meetings: Bridgend Bowls Centre, Bridgend. 2nd Friday, monthly, 7.30 p.m.

Bridgwater Astronomical Society
Secretary: Mr G. MacKenzie, Watergore Cottage, Watergore, South Petherton, Somerset TA13 5JQ.
Website: www.ourworld.compuserve.com/hompages/dbown/Bwastro.htm
Meetings: Room D10, Bridgwater College, Bath Road Centre, Bridgwater. 2nd Wednesday each month, Sept.–June.

Bridport Astronomical Society
Secretary: Mr G.J. Lodder, 3 The Green, Walditch, Bridport, Dorset DT6 4LB.
Meetings: Walditch Village Hall, Bridport. 1st Sunday each month, 7.30 p.m.

Astronomical Societies in the British Isles

Brighton Astronomical and Scientific Society
Secretary: Ms T. Fearn, 38 Woodlands Close, Peacehaven, East Sussex BN10 7SF.
Meetings: St John's Church Hall, Hove. 1st Tuesday each month, 7.30 p.m.

Bristol Astronomical Society
Secretary: Dr John Pickard, 'Fielding', Easter Compton, Bristol BS35 5SJ.
Meetings: Frank Lecture Theatre, University of Bristol Physics Dept., alternate Fridays in term time, and Westbury Park Methodist Church Rooms, North View, other Fridays.

Callington Community Astronomy Group
Secretary: Beccy Watson. *Tel:* 07891 573786
Email: enquiries@callington-astro.org.uk
Website: www.callington-astro.org.uk
Meetings: Callington Space Centre, Callington Community College, Launceston Road, Callington, Cornwall PL17 7DR. 1st Friday of each month, 7.30 p.m., Sept.–June.

Cambridge Astronomical Society
Secretary: Brian Lister, 80 Ramsden Square, Cambridge CB4 2BL.
Meetings: Institute of Astronomy, Madingley Road. 3rd Friday each month.

Cardiff Astronomical Society
Secretary: D.W.S. Powell, 1 Tal-y-Bont Road, Ely, Cardiff CF5 5EU.
Meetings: Dept. of Physics and Astronomy, University of Wales, Newport Road, Cardiff. Alternate Thursdays, 8 p.m.

Castle Point Astronomy Club
Secretary: Andrew Turner, 3 Canewdon Hall Close, Canewdon, Rochford, Essex SS4 3PY.
Meetings: St Michael's Church Hall, Daws Heath. Wednesdays, 8 p.m.

Chelmsford Astronomers
Secretary: Brendan Clark, 5 Borda Close, Chelmsford, Essex.
Meetings: Once a month.

Chester Astronomical Society
Secretary: John Gilmour, 2 Thomas Brassey Close, Chester CH2 3AE.
Tel: 07974 948278
Email: john_gilmour@ouvip.com
Website: www.manastro.co.uk/nwgas/chester/
Meetings: Burley Memorial Hall, Waverton, near Chester. Last Wednesday of each month except August and December at 7.30 p.m.

Chester Society of Natural Science, Literature and Art
Secretary: Paul Braid, 'White Wing', 38 Bryn Avenue, Old Colwyn, Colwyn Bay LL29 8AH.
Email: p.braid@virgin.net
Meetings: Once a month.

Chesterfield Astronomical Society
President: Mr D. Blackburn, 71 Middlecroft Road, Stavely, Chesterfield, Derbyshire S41 3XG. *Tel:* 07909 570754.
Website: www.chesterfield-as.org.uk
Meetings: Barnet Observatory, Newbold, each Friday.

Clacton & District Astronomical Society
Secretary: C. L. Haskell, 105 London Road, Clacton-on-Sea, Essex.

Cleethorpes & District Astronomical Society
Secretary: C. Illingworth, 38 Shaw Drive, Grimsby, South Humberside.
Meetings: Beacon Hill Observatory, Cleethorpes. 1st Wednesday each month.

Cleveland & Darlington Astronomical Society
Contact: Dr John McCue, 40 Bradbury Rd., Stockton-on-Tees, Cleveland TS20 1LE.
Meetings: Grindon Parish Hall, Thorpe Thewles, near Stockton-on-Tees. 2nd Friday, monthly.

Cork Astronomy Club
Website: www.corkastronomyclub.com
Email: astronomycork@yahoo.ie
Meetings: UCC, Civil Engineering Building, 2nd Monday each month, Sept.–May, 8.00 p.m.

Cornwall Astronomical Society
Secretary: J.M. Harvey, 1 Tregunna Close, Porthleven, Cornwall TR13 9LW.
Meetings: Godolphin Club, Wendron Street, Helston, Cornwall. 2nd and 4th Thursday of each month, 7.30 for 8 p.m.

Cotswold Astronomical Society
Secretary: Rod Salisbury, Grove House, Christchurch Road, Cheltenham, Gloucestershire GL50 2PN.
Website: www.members.nbci.com/CotswoldAS
Meetings: Shurdington Church Hall, School Lane, Shurdington, Cheltenham. 2nd Saturday each month, 8 p.m.

Coventry & Warwickshire Astronomical Society
Secretary: Steve Payne, 68 Stonebury Avenue, Eastern Green, Coventry CV5 7FW.
Website: www.cawas.freeserve.co.uk; *Email:* sjp2000@thefarside57.freeserve.co.uk
Meetings: The Earlsdon Church Hall, Albany Road, Earlsdon, Coventry. 2nd Friday, monthly, Sept.–June.

Crawley Astronomical Society
Secretary: Ron Gamer, 1 Pevensey Close, Pound Hill, Crawley, West Sussex RH10 7BL.
Meetings: Ifield Community Centre, Ifield Road, Crawley. 3rd Friday each month, 7.30 p.m.

Crayford Manor House Astronomical Society
Secretary: Roger Pickard, 28 Appletons, Hadlow, Kent TM1 0DT.
Meetings: Manor House Centre, Crayford. Monthly during term time.

Crewkerne and District Astronomical Society (CADAS)
Chairman: Kevin Dodgson, 46 Hermitage Street, Crewkerne, Somerset TA18 8ET.
Email: crewastra@aol.com

Croydon Astronomical Society
Secretary: John Murrell, 17 Dalmeny Road, Carshalton, Surrey.
Meetings: Lecture Theatre, Royal Russell School, Combe Lane, South Croydon. Alternate Fridays, 7.45 p.m.

Derby & District Astronomical Society
Secretary: Ian Bennett, Freers Cottage, Sutton Lane, Etwall.
Website: www.derby-astro-soc.fsnet/index.html
Email: bennett.lovatt@btinternet.com
Meetings: Friends Meeting House, Derby. 1st Friday each month, 7.30 p.m.

Doncaster Astronomical Society
Secretary: A. Anson, 15 Cusworth House, St James Street, Doncaster DN1 3AY
Website: www.donastro.freeserve.co.uk; *Email:* space@donastro.freeserve.co.uk

Meetings: St George's Church House, St George's Church, Church Way, Doncaster. 2nd and 4th Thursday of each month, commencing at 7.30 p.m.

Dumfries Astronomical Society

Secretary: Klaus Schiller, lesley.burrell@btinternet.com.

Website: www.astronomers.ukscientists.com

Meetings: George St Church Hall, George St, Dumfries. 2nd Tuesday of each month, Sept.–May.

Dundee Astronomical Society

Secretary: G. Young, 37 Polepark Road, Dundee, Tayside DD1 5QT.

Meetings: Mills Observatory, Balgay Park, Dundee. 1st Friday each month, 7.30 p.m. Sept.–Apr.

Easington and District Astronomical Society

Secretary: T. Bradley, 52 Jameson Road, Hartlepool, Co. Durham.

Meetings: Easington Comprehensive School, Easington Colliery. Every 3rd Thursday throughout the year, 7.30 p.m.

East Antrim Astronomical Society

Secretary: Stephen Beasant

Website: www.eaas.co.uk

Meetings: Ballyclare High School, Ballyclare, County Antrim. First Monday each month.

Eastbourne Astronomical Society

Secretary: Peter Gill, 18 Selwyn House, Selwyn Road, Eastbourne, East Sussex BN21 2LF.

Meetings: Willingdon Memorial Hall, Church Street, Willingdon. One Saturday per month, Sept.–July, 7.30 p.m.

East Riding Astronomers

Secretary: Tony Scaife, 15 Beech Road, Elloughton, Brough, North Humberside HU15 1JX.

Meetings: As arranged.

East Sussex Astronomical Society

Secretary: Marcus Croft, 12 St Mary's Cottages, Ninfield Road, Bexhill-on-Sea, East Sussex.

Website: www.esas.org.uk

Meetings: St Mary's School, Wrestwood Road, Bexhill. 1st Thursday of each month, 8 p.m.

Edinburgh University Astronomical Society

Secretary: c/o Dept. of Astronomy, Royal Observatory, Blackford Hill, Edinburgh.

Ewell Astronomical Society

Secretary: Richard Gledhill, 80 Abinger Avenue, Cheam SM2 7LW.

Website: www.ewell-as.co.uk

Meetings: St Mary's Church Hall, London Road, Ewell. 2nd Friday of each month except August, 7.45 p.m.

Exeter Astronomical Society

Secretary: Tim Sedgwick, Old Dower House, Half Moon, Newton St Cyres, Exeter, Devon EX5 5AE.

Meetings: The Meeting Room, Wynards, Magdalen Street, Exeter. 1st Thursday of month.

Farnham Astronomical Society
Secretary: Laurence Anslow, 'Asterion', 18 Wellington Lane, Farnham, Surrey GU9 9BA.
Meetings: Central Club, South Street, Farnham. 2nd Thursday each month, 8 p.m.

Foredown Tower Astronomy Group
Secretary: M. Feist, Foredown Tower Camera Obscura, Foredown Road, Portslade, East Sussex BN41 2EW.
Meetings: At the above address, 3rd Tuesday each month. 7 p.m. (winter), 8 p.m. (summer).

Greenock Astronomical Society
Secretary: Carl Hempsey, 49 Brisbane Street, Greenock.
Meetings: Greenock Arts Guild, 3 Campbell Street, Greenock.

Grimsby Astronomical Society
Secretary: R. Williams, 14 Richmond Close, Grimsby, South Humberside.
Meetings: Secretary's home. 2nd Thursday each month, 7.30 p.m.

Guernsey: La Société Guernesiasie Astronomy Section
Secretary: Debby Quertier, Lamorna, Route Charles, St Peter Port, Guernsey GY1 1QS. and Jessica Harris, Keanda, Les Sauvagees, St Sampson's, Guernsey GY2 4XT.
Meetings: Observatory, Rue du Lorier, St Peter's. Tuesdays, 8 p.m.

Guildford Astronomical Society
Secretary: A. Langmaid, 22 West Mount, The Mount, Guildford, Surrey GU2 5HL.
Meetings: Guildford Institute, Ward Street, Guildford. 1st Thursday each month except Aug., 7.30 p.m.

Gwynedd Astronomical Society
Secretary: Mr Ernie Greenwood, 18 Twrcelyn Street, Llanerchymedd, Anglesey LL74 8TL.
Meetings: Dept. of Electronic Engineering, Bangor University. 1st Thursday each month except Aug., 7.30 p.m.

The Hampshire Astronomical Group
Secretary: Geoff Mann, 10 Marie Court, 348 London Road, Waterlooville, Hampshire PO7 7SR.
Website: www.hantsastro.demon.co.uk; *Email:* Geoff.Mann@hazleton97.fsnet.co.uk
Meetings: 2nd Friday, Clanfield Memorial Hall, all other Fridays Clanfield Observatory.

Hanney & District Astronomical Society
Secretary: Bob Church, 47 Upthorpe Drive, Wantage, Oxfordshire OX12 7DG.
Meetings: Last Thursday each month, 8 p.m.

Harrogate Astronomical Society
Secretary: Brian Bonser, 114 Main Street, Little Ouseburn TO5 9TG.
Meetings: National Power HQ, Beckwith Knowle, Harrogate. Last Friday each month.

Havering Astronomical Society
Secretary: Frances Ridgley, 133 Severn Drive, Upminster, Essex RM14 1PP.
Meetings: Cranham Community Centre, Marlborough Gardens, Upminster, Essex. 3rd Wednesday each month except July and Aug., 7.30 p.m.

Heart of England Astronomical Society
Secretary: John Williams, 100 Stanway Road, Shirley, Solihull B90 3JG.
Website: www.members.aol.com/hoeas/home.html; *Email:* hoeas@aol.com

Meetings: Furnace End Village, over Whitacre, Warwickshire. Last Thursday each month, except June, July & Aug., 8 p.m.

Hebden Bridge Literary & Scientific Society, Astronomical Section
Secretary: Peter Jackson, 44 Gilstead Lane, Bingley, West Yorkshire BD16 3NP.
Meetings: Hebden Bridge Information Centre. Last Wednesday, Sept.–May.

Herefordshire Astronomical Society
Secretary: Paul Olver, The Buttridge, Wellington Lane, Canon Pyon, Hereford HR4 8NL.
Email: info@hsastro.org.uk
Meetings: The Kindle Centre, ASDA Supermarket, Hereford. 1st Thursday of every month (except August) 7 p.m.

Herschel Astronomy Society
Secretary: Kevin Bishop, 106 Holmsdale, Crown Wood, Bracknell, Berkshire RG12 3TB.
Meetings: Eton College. 2nd Friday each month, 7.30 p.m.

Highlands Astronomical Society
Secretary: Richard Green, 11 Drumossie Avenue, Culcabock, Inverness IV2 3SJ.
Meetings: The Spectrum Centre, Inverness. 1st Tuesday each month, 7.30 p.m.

Hinckley & District Astronomical Society
Secretary: Mr S. Albrighton, 4 Walnut Close, The Bridleways, Hartshill, Nuneaton, Warwickshire CV10 0XH.
Meetings: Burbage Common Visitors Centre, Hinckley. 1st Tuesday Sept.–May, 7.30 p.m.

Horsham Astronomy Group (was **Forest Astronomical Society**)
Secretary: Dan White, 32 Burns Close, Horsham, West Sussex RH12 5PF.
Email: secretary@horshamastronomy.com
Meetings: 1st Wednesday each month.

Howards Astronomy Club
Secretary: H. Ilett, 22 St George's Avenue, Warblington, Havant, Hampshire.
Meetings: To be notified.

Huddersfield Astronomical and Philosophical Society
Secretary: Lisa B. Jeffries, 58 Beaumont Street, Netherton, Huddersfield, West Yorkshire HD4 7HE.
Email: l.b.jeffries@hud.ac.uk
Meetings: 4a Railway Street, Huddersfield. Every Wednesday and Friday, 7.30 p.m.

Hull and East Riding Astronomical Society
President: Sharon E. Long
Email: charon@charon.karoo.co.uk
Website: http://www.heras.org.uk
Meetings: The Wilberforce Building, Room S25, University of Hull, Cottingham Road, Hull. 2nd Monday each month, Sept.–May, 7.30–9.30 p.m.

Ilkeston & District Astronomical Society
Secretary: Mark Thomas, 2 Elm Avenue, Sandiacre, Nottingham NG10 5EJ.
Meetings: The Function Room, Erewash Museum, Anchor Row, Ilkeston. 2nd Tuesday monthly, 7.30 p.m.

Ipswich, Orwell Astronomical Society
Secretary: R. Gooding, 168 Ashcroft Road, Ipswich.
Meetings: Orwell Park Observatory, Nacton, Ipswich. Wednesdays, 8 p.m.

Irish Astronomical Association
President: Terry Moseley, 31 Sunderland Road, Belfast BT6 9LY, Northern Ireland.
Email: terrymosel@aol.com
Meetings: Ashby Building, Stranmillis Road, Belfast. Alternate Wednesdays, 7.30 p.m.

Irish Astronomical Society
Secretary: James O'Connor, PO Box 2547, Dublin 15, Eire.
Meetings: Ely House, 8 Ely Place, Dublin 2. 1st and 3rd Monday each month.

Isle of Man Astronomical Society
Secretary: James Martin, Ballaterson Farm, Peel, Isle of Man IM5 3AB.
Email: ballaterson@manx.net
Meetings: Isle of Man Observatory, Foxdale. 1st Thursday of each month, 8 p.m.

Isle of Wight Astronomical Society
Secretary: J. W. Feakins, 1 Hilltop Cottages, High Street, Freshwater, Isle of Wight.
Meetings: Unitarian Church Hall, Newport, Isle of Wight. Monthly.

Jersey Astronomy Club
Secretary: Jodie Masterman, *Tel:* 07797 813681
Email: jodiemasterman@yahoo.co.uk
Chairman: Martin Ahier, *Tel:* (01534) 732157
Meetings: Sir Patrick Moore Astronomy Centre, Lex Creux Country Park, St Brelade, Jersey, 2nd Monday of every month (except August), 8.00 p.m.

Keele Astronomical Society
Secretary: Natalie Webb, Department of Physics, University of Keele, Keele, Staffordshire ST5 5BG.
Meetings: As arranged during term time.

Kettering and District Astronomical Society
Asst. Secretary: Steve Williams, 120 Brickhill Road, Wellingborough, Northamptonshire.
Meetings: Quaker Meeting Hall, Northall Street, Kettering, Northamptonshire. 1st Tuesday each month, 7.45 p.m.

King's Lynn Amateur Astronomical Association
Secretary: P. Twynman, 17 Poplar Avenue, RAF Marham, King's Lynn.
Meetings: As arranged.

Lancaster and Morecambe Astronomical Society
Secretary: Mrs E. Robinson, 4 Bedford Place, Lancaster LA1 4EB.
Email: ehelenerob@btinternet.com
Meetings: Church of the Ascension, Torrisholme. 1st Wednesday each month except July and Aug.

Knowle Astronomical Society
Secretary: Nigel Foster, 21 Speedwell Drive, Balsall Common, Coventry, West Midlands CV7 7AU.
Meetings: St George & St Theresa's Parish Centre, 337 Station Road, Dorridge, Solihull, West Midlands B93 8TZ. 1st Monday of each month (+/– 1 week for Bank Holidays) except August.

Lancaster University Astronomical Society
Secretary: c/o Students' Union, Alexandra Square, University of Lancaster.
Meetings: As arranged.

Layman's Astronomical Society
Secretary: John Evans, 10 Arkwright Walk, The Meadows, Nottingham.
Meetings: The Popular, Bath Street, Ilkeston, Derbyshire. Monthly.

Leeds Astronomical Society
Secretary: Mark A. Simpson, 37 Roper Avenue, Gledhow, Leeds LS8 1LG.
Meetings: Centenary House, North Street. 2nd Wednesday each month, 7.30 p.m.

Leicester Astronomical Society
Secretary: Dr P. J. Scott, 21 Rembridge Close, Leicester LE3 9AP.
Meetings: Judgemeadow Community College, Marydene Drive, Evington, Leicester. 2nd and 4th Tuesdays each month, 7.30 p.m.

Letchworth and District Astronomical Society
Secretary: Eric Hutton, 14 Folly Close, Hitchin, Hertfordshire.
Meetings: As arranged.

Lewes Amateur Astronomers
Secretary: Christa Sutton, 8 Tower Road, Lancing, West Sussex BN15 9HT.
Meetings: The Bakehouse Studio, Lewes. Last Wednesday each month.

Limerick Astronomy Club
Secretary: Tony O'Hanlon, 26 Ballycannon Heights, Meelick, Co. Clare, Eire.
Meetings: Limerick Senior College, Limerick. Monthly (except June and Aug.), 8 p.m.

Lincoln Astronomical Society
Secretary: David Swaey, 'Everglades', 13 Beaufort Close, Lincoln LN2 4SF.
Meetings: The Lecture Hall, off Westcliffe Street, Lincoln. 1st Tuesday each month.

Liverpool Astronomical Society
Secretary: Mr K. Clark, 31 Sandymount Drive, Wallasey, Merseyside L45 0LJ.
Meetings: Lecture Theatre, Liverpool Museum. 3rd Friday each month, 7 p.m.

Norman Lockyer Observatory Society
Secretary: G. E. White, PO Box 9, Sidmouth EX10 0YQ.
Website: www.ex.ac.uk/nlo/; *Email:* g.e.white@ex.ac.uk
Meetings: Norman Lockyer Observatory, Sidmouth. Fridays and 2nd Monday each month, 7.30 p.m.

Loughton Astronomical Society
Secretary: Charles Munton, 14a Manor Road, Wood Green, London N22 4YJ.
Meetings: 1st Theydon Bois Scout Hall, Loughton Lane, Theydon Bois. Weekly.

Lowestoft and Great Yarmouth Regional Astronomers (LYRA) Society
Secretary: Simon Briggs, 28 Sussex Road, Lowestoft, Suffolk.
Meetings: Community Wing, Kirkley High School, Kirkley Run, Lowestoft. 3rd Thursday each month, 7.30 p.m.

Luton Astronomical Society
Secretary: Mr G. Mitchell, Putteridge Bury, University of Luton, Hitchin Road, Luton.
Website: www.lutonastrosoc.org.uk; *Email:* user998491@aol.com
Meetings: Univ. of Luton, Putteridge Bury (except June, July and August), or Somerics Junior School, Wigmore Lane, Luton (July and August only), last Thursday each month, 7.30–9.00 p.m.

Lytham St Anne's Astronomical Association
Secretary: K. J. Porter, 141 Blackpool Road, Ansdell, Lytham St Anne's, Lancashire.
Meetings: College of Further Education, Clifton Drive South, Lytham St Anne's. 2nd Wednesday monthly Oct.–June.

Macclesfield Astronomical Society
Secretary: Mr John H. Thomson, 27 Woodbourne Road, Sale, Cheshire M33 3SY
Website: www.maccastro.com; *Email:* jhandlc@yahoo.com
Meetings: Jodrell Bank Science Centre, Goostrey, Cheshire. 1st Tuesday of every month, 7 p.m.

Maidenhead Astronomical Society
Secretary: Tim Haymes, Hill Rise, Knowl Hill Common, Knowl Hill, Reading RG10 9YD.
Meetings: Stubbings Church Hall, near Maidenhead. 1st Friday Sept.–June.

Maidstone Astronomical Society
Secretary: Stephen James, 4 The Cherry Orchard, Haddow, Tonbridge, Kent.
Meetings: Nettlestead Village Hall. 1st Tuesday in the month except July and Aug., 7.30 p.m.

Manchester Astronomical Society
Secretary: Mr Kevin J. Kilburn FRAS, Godlee Observatory, UMIST, Sackville Street, Manchester M60 1QD.
Website: www.u-net.com/ph/mas/; *Email:* kkilburn@globalnet.co.uk
Meetings: At the Godlee Observatory. Thursdays, 7 p.m., except below.
Free Public Lectures: Renold Building UMIST, third Thursday Sept.–Mar., 7.30 p.m.

Mansfield and Sutton Astronomical Society
Secretary: Angus Wright, Sherwood Observatory, Coxmoor Road, Sutton-in-Ashfield, Nottinghamshire NG17 5LF.
Meetings: Sherwood Observatory, Coxmoor Road. Last Tuesday each month, 7.30 p.m.

Mexborough and Swinton Astronomical Society
Secretary: Mark R. Benton, 14 Sandalwood Rise, Swinton, Mexborough, South Yorkshire S64 8PN.
Website: www.msas.org.uk; *Email:* mark@masas.f9.co.uk
Meetings: Swinton WMC. Thursdays, 7.30 p.m.

Mid-Kent Astronomical Society
Secretary: Peter Parish, 30 Wooldeys Road, Rainham, Kent ME8 7NU.
Meetings: Bredhurst Village Hall, Hurstwood Road, Bredhurst, Kent. 2nd and last Fridays each month except August, 7.45 p.m.
Website: www.mkas-site.co.uk

Milton Keynes Astronomical Society
Secretary: Mike Leggett, 19 Matilda Gardens, Shenley Church End, Milton Keynes MK5 6HT.
Website: www.mkas.org.uk; *Email:* mike-pat-leggett@shenley9.fsnet.co.uk
Meetings: Rectory Cottage, Bletchley. Alternate Fridays.

Moray Astronomical Society
Secretary: Richard Pearce, 1 Forsyth Street, Hopeman, Elgin, Moray, Scotland.
Meetings: Village Hall Close, Co. Elgin.

Newbury Amateur Astronomical Society (NAAS)
Secretary: Mrs Monica Balstone, 37 Mount Pleasant, Tadley RG26 4BG.
Meetings: United Reformed Church Hall, Cromwell Place, Newbury. 1st Friday of month, Sept.–June.

Newcastle-on-Tyne Astronomical Society
Secretary: C. E. Willits, 24 Acomb Avenue, Seaton Delaval, Tyne and Wear.
Meetings: Zoology Lecture Theatre, Newcastle University. Monthly.

North Aston Space & Astronomical Club
Secretary: W. R. Chadburn, 14 Oakdale Road, North Aston, Sheffield.
Meetings: To be notified.

Northamptonshire Natural History Society (Astronomy Section)
Secretary: R. A. Marriott, 24 Thirlestane Road, Northampton NN4 8HD.
Email: ram@hamal.demon.co.uk
Meetings: Humfrey Rooms, Castilian Terrace, Northampton. 2nd and last Mondays, most months, 7.30 p.m.

Northants Amateur Astronomers
Secretary: Mervyn Lloyd, 76 Havelock Street, Kettering, Northamptonshire.
Meetings: 1st and 3rd Tuesdays each month, 7.30 p.m.

North Devon Astronomical Society
Secretary: P. G. Vickery, 12 Broad Park Crescent, Ilfracombe, Devon EX34 8DX.
Meetings: Methodist Hall, Rhododendron Avenue, Sticklepath, Barnstaple. 1st Wednesday each month, 7.15 p.m.

North Dorset Astronomical Society
Secretary: J. E. M. Coward, The Pharmacy, Stalbridge, Dorset.
Meetings: Charterhay, Stourton, Caundle, Dorset. 2nd Wednesday each month.

North Downs Astronomical Society
Secretary: Martin Akers, 36 Timber Tops, Lordswood, Chatham, Kent ME5 8XQ.
Meetings: Vigo Village Hall. 3rd Thursday each month. 7.30 p.m.

North-East London Astronomical Society
Secretary: Mr B. Beeston, 38 Abbey Road, Bush Hill Park, Enfield EN1 2QN.
Meetings: Wanstead House, The Green, Wanstead. 3rd Sunday each month (except Aug.), 3 p.m.

North Gwent and District Astronomical Society
Secretary: Jonathan Powell, 14 Lancaster Drive, Gilwern, nr Abergavenny, Monmouthshire NP7 0AA.
Meetings: Gilwern Community Centre. 15th of each month, 7.30 p.m.

North Staffordshire Astronomical Society
Secretary: Duncan Richardson, Halmerend Hall Farm, Halmerend, Stoke-on-Trent, Staffordshire ST7 8AW.
Email: dwr@enterprise.net
Meetings: 21st Hartstill Scout Group HQ, Mount Pleasant, Newcastle-under-Lyme ST5 1DR. 1st Tuesday each month (except July and Aug.), 7–9.30 p.m.

Northumberland Astronomical Society
Contact: Dr Adrian Jametta, 1 Lake Road, Hadston, Morpeth, Northumberland NE65 9TF.
Email: adrian@themoon.co.uk
Website: www.nastro.org.uk
Meetings: Hauxley Nature Reserve (near Amble). Last Thursday of every month (except December), 7.30 pm. Additional meetings and observing sessions listed on website.
Tel: 07984 154904

North Western Association of Variable Star Observers
Secretary: Jeremy Bullivant, 2 Beaminster Road, Heaton Mersey, Stockport, Cheshire.
Meetings: Four annually.

Norwich Astronomical Society
Secretary: Dave Balcombe, 52 Folly Road, Wymondham, Norfolk NR18 0QR.
Website: www.norwich.astronomical.society.org.uk
Meetings: Seething Observatory, Toad Lane, Thwaite St Mary, Norfolk. Every Friday, 7.30 p.m.

Nottingham Astronomical Society
Secretary: C. Brennan, 40 Swindon Close, The Vale, Giltbrook, Nottingham NG16 2WD.
Meetings: Djanogly City Technology College, Sherwood Rise (B682). 1st and 3rd Thursdays each month, 7.30 p.m.

Oldham Astronomical Society
Secretary: P. J. Collins, 25 Park Crescent, Chadderton, Oldham.
Meetings: Werneth Park Study Centre, Frederick Street, Oldham. Fortnightly, Friday.

Open University Astronomical Society
Secretary: Dr Andrew Norton, Department of Physics and Astronomy, The Open University, Walton Hall, Milton Keynes MK7 6AA.
Website: www.physics.open.ac.uk/research/astro/a_club.html
Meetings: Open University, Milton Keynes. 1st Tuesday of every month, 7.30 p.m.

Orpington Astronomical Society
Secretary: Dr Ian Carstairs, 38 Brabourne Rise, Beckenham, Kent BR3 2SG.
Meetings: High Elms Nature Centre, High Elms Country Park, High Elms Road, Farnborough, Kent. 4th Thursday each month, Sept.–July, 7.30 p.m.

Papworth Astronomy Club
Contact: Keith Tritton, Magpie Cottage, Fox Street, Great Gransden, Sandy, Bedfordshire SG19 3AA.
Email: kpt2@tutor.open.ac.uk
Meetings: Bradbury Progression Centre, Church Lane, Papworth Everard, nr Huntingdon. 1st Wednesday each month, 7 p.m.

Peterborough Astronomical Society
Secretary: Sheila Thorpe, 6 Cypress Close, Longthorpe, Peterborough.
Meetings: 1st Thursday every month, 7.30 p.m.

Plymouth Astronomical Society
Secretary: Alan G. Penman, 12 St Maurice View, Plympton, Plymouth, Devon PL7 1FQ.
Email: oakmount12@aol.com
Meetings: Glynis Kingham Centre, YMCA Annex, Lockyer Street, Plymouth. 2nd Friday each month, 7.30 p.m.

PONLAF
Secretary: Matthew Hepburn, 6 Court Road, Caterham, Surrey CR3 5RD.
Meetings: Room 5, 6th floor, Tower Block, University of North London. Last Friday each month during term time, 6.30 p.m.

Port Talbot Astronomical Society (formerly **Astronomical Society of Wales**)
Secretary: Mr J. Hawes, 15 Lodge Drive, Baglan, Port Talbot, West Glamorgan SA12 8UD.
Meetings: Port Talbot Arts Centre. 1st Tuesday each month, 7.15 p.m.

Preston & District Astronomical Society
Secretary: P. Sloane, 77 Ribby Road, Wrea Green, Kirkham, Preston, Lancashire.
Meetings: Moor Park (Jeremiah Horrocks) Observatory, Preston. 2nd Wednesday, last Friday each month, 7.30 p.m.

Reading Astronomical Society
Secretary: Mrs Ruth Sumner, 22 Anson Crescent, Shinfield, Reading RG2 8JT.
Meetings: St Peter's Church Hall, Church Road, Earley. 3rd Friday each month, 7 p.m.

Renfrewshire Astronomical Society
Secretary: Ian Martin, 10 Aitken Road, Hamilton, South Lanarkshire ML3 7YA.
Website: www.renfrewshire-as.co.uk; *Email:* RenfrewAS@aol.com
Meetings: Coats Observatory, Oakshaw Street, Paisley. Fridays, 7.30 p.m.

Rower Astronomical Society
Secretary: Mary Kelly, Knockatore, The Rower, Thomastown, Co. Kilkenny, Eire.

St Helens Amateur Astronomical Society
Secretary: Carl Dingsdale, 125 Canberra Avenue, Thatto Heath, St Helens, Merseyside WA9 5RT.
Meetings: As arranged.

Salford Astronomical Society
Secretary: Mrs Kath Redford, 2 Albermarle Road, Swinton, Manchester M27 5ST.
Meetings: The Observatory, Chaseley Road, Salford. Wednesdays.

Salisbury Astronomical Society
Secretary: Mrs R. Collins, 3 Fairview Road, Salisbury, Wiltshire SP1 1JX.
Meetings: Glebe Hall, Winterbourne Earls, Salisbury. 1st Tuesday each month.

Sandbach Astronomical Society
Secretary: Phil Benson, 8 Gawsworth Drive, Sandbach, Cheshire.
Meetings: Sandbach School, as arranged.

Sawtry & District Astronomical Society
Secretary: Brooke Norton, 2 Newton Road, Sawtry, Huntingdon, Cambridgeshire PE17 5UT.
Meetings: Greenfields Cricket Pavilion, Sawtry Fen. Last Friday each month.

Scarborough & District Astronomical Society
Secretary: Mrs S. Anderson, Basin House Farm, Sawdon, Scarborough, North Yorkshire.
Meetings: Scarborough Public Library. Last Saturday each month, 7–9 p.m.

Scottish Astronomers Group
Secretary: Dr Ken Mackay, Hayford House, Cambusbarron, Stirling FK7 9PR.
Meetings: North of Hadrian's Wall, twice yearly.

Sheffield Astronomical Society
Secretary: Darren Swindels, 102 Sheffield Road, Woodhouse, Sheffield, South Yorkshire S13 7EU.
Website: www.sheffieldastro.org.uk; *Email:* info@sheffieldastro.org.uk
Meetings: Twice monthly at Mayfield Environmental Education Centre, David Lane, Fulwood, Sheffield S10, 7.30–10 p.m.

Shetland Astronomical Society
Secretary: Peter Kelly, The Glebe, Fetlar, Shetland ZE2 9DJ.
Email: theglebe@zetnet.co.uk
Meetings: Fetlar, Fridays, Oct.–Mar.

Shropshire Astronomical Society
Contact: Mr David Woodward, 20 Station Road, Condover, Shrewsbury, Shropshire SY5 7BQ.
Website: http://www.shropshire-astro.com; *Email:* jacquidodds@ntlworld.com
Meetings: Quarterly talks at the Gateway Arts and Education Centre, Chester Street, Shrewsbury and monthly observing meetings at Rodington Village Hall.

Sidmouth and District Astronomical Society
Secretary: M. Grant, Salters Meadow, Sidmouth, Devon.
Meetings: Norman Lockyer Observatory, Salcombe Hill. 1st Monday in each month.

Solent Amateur Astronomers
Secretary: Ken Medway, 443 Burgess Road, Swaythling, Southampton SO16 3BL.
Website: www.delscope.demon.co.uk
Email: ken@medway1875.freeserve.co.uk
Meetings: Communications Room 2, Oasis Academy, Fairisle Road, Lordshill, Southampton, SO16 8BY. 3rd Tuesday each month, 7.30 p.m.

Southampton Astronomical Society
Secretary: John Thompson, 4 Heathfield, Hythe, Southampton SO45 5BJ.
Website: www.home.clara.net/lmhobbs/sas.html
Email: John.G.Thompson@Tesco.net
Meetings: Conference Room 3, The Civic Centre, Southampton. 2nd Thursday each month (except Aug.), 7.30 p.m.

South Downs Astronomical Society
Secretary: J. Green, 46 Central Avenue, Bognor Regis, West Sussex PO21 5HH.
Website: www.southdowns.org.uk
Meetings: Chichester High School for Boys. 1st Friday in each month (except Aug.).

South-East Essex Astronomical Society
Secretary: C. P. Jones, 29 Buller Road, Laindon, Essex.
Website: www.seeas.dabsol.co.uk/; *Email:* cpj@cix.co.uk
Meetings: Lecture Theatre, Central Library, Victoria Avenue, Southend-on-Sea. Generally 1st Thursday in month, Sept.–May, 7.30 p.m.

South-East Kent Astronomical Society
Secretary: Andrew McCarthy, 25 St Paul's Way, Sandgate, near Folkestone, Kent CT20 3NT.
Meetings: Monthly.

South Lincolnshire Astronomical & Geophysical Society
Secretary: Ian Farley, 12 West Road, Bourne, Lincolnshire PE10 9PS.
Meetings: Adult Education Study Centre, Pinchbeck. 3rd Wednesday each month, 7.30 p.m.

Southport Astronomical Society
Secretary: Patrick Brannon, Willow Cottage, 90 Jacksmere Lane, Scarisbrick, Ormskirk, Lancashire L40 9RS.
Meetings: Monthly Sept.–May, plus observing sessions.

Southport, Ormskirk and District Astronomical Society
Secretary: J. T. Harrison, 92 Cottage Lane, Ormskirk, Lancashire L39 3NJ.
Meetings: Saturday evenings, monthly, as arranged.

South Shields Astronomical Society
Secretary: c/o South Tyneside College, St George's Avenue, South Shields.
Meetings: Marine and Technical College. Each Thursday, 7.30 p.m.

South Somerset Astronomical Society
Secretary: G. McNelly, 11 Laxton Close, Taunton, Somerset.
Meetings: Victoria Inn, Skittle Alley, East Reach, Taunton, Somerset. Last Saturday each month, 7.30 p.m.

South-West Hertfordshire Astronomical Society
Secretary: Tom Walsh, 'Finches', Coleshill Lane, Winchmore Hill, Amersham, Buckinghamshire HP7 0NP.
Meetings: Rickmansworth. Last Friday each month, Sept.–May.

Stafford and District Astronomical Society
Secretary: Miss L. Hodkinson, 6 Elm Walk, Penkridge, Staffordshire ST19 5NL.
Meetings: Weston Road High School, Stafford. Every 3rd Thursday, Sept.–May, 7.15 p.m.

Stirling Astronomical Society
Secretary: Hamish MacPhee, 10 Causewayhead Road, Stirling FK9 5ER.
Meetings: Smith Museum & Art Gallery, Dumbarton Road, Stirling. 2nd Friday each month, 7.30 p.m.

Stoke-on-Trent Astronomical Society
Secretary: M. Pace, Sundale, Dunnocksfold, Alsager, Stoke-on-Trent.
Meetings: Cartwright House, Broad Street, Hanley. Monthly.

Stratford-upon-Avon Astronomical Society
Secretary: Robin Swinbourne, 18 Old Milverton, Leamington Spa, Warwickshire CV32 6SA.
Meetings: Tiddington Home Guard Club. 4th Tuesday each month, 7.30 p.m.

Sunderland Astronomical Society
Contact: Don Simpson, 78 Stratford Avenue, Grangetown, Sunderland SR2 8RZ.
Meetings: Friends Meeting House, Roker. 1st, 2nd and 3rd Sundays each month.

Sussex Astronomical Society
Secretary: Mrs C. G. Sutton, 75 Vale Road, Portslade, Sussex.
Meetings: English Language Centre, Third Avenue, Hove. Every Wednesday, 7.30–9.30 p.m., Sept.–May.

Swansea Astronomical Society
Secretary: Dr Michael Morales, 238 Heol Dulais, Birch Grove, Swansea SA7 9LH.
Website: www.crysania.co.uk/sas/astro/star
Meetings: Lecture Room C, Science Tower, University of Swansea. 2nd and 4th Thursday each month from Sept.–June, 7 p.m.

Tavistock Astronomical Society
Secretary: Mrs Ellie Coombes, Rosemount, Under Road, Gunnislake, Cornwall PL18 9JL.
Meetings: Science Laboratory, Kelly College, Tavistock. 1st Wednesday each month, 7.30 p.m.

Thames Valley Astronomical Group
Secretary: K. J. Pallet, 82a Tennyson Street, South Lambeth, London SW8 3TH.
Meetings: As arranged.

Thanet Amateur Astronomical Society
Secretary: P. F. Jordan, 85 Crescent Road, Ramsgate.
Meetings: Hilderstone House, Broadstairs, Kent. Monthly.

Torbay Astronomical Society
Secretary: Tim Moffat, 31 Netley Road, Newton Abbot, Devon TQ12 2LL.
Meetings: Torquay Boys' Grammar School, 1st Thursday in month; and Town Hall, Torquay, 3rd Thursday in month, Oct.–May, 7.30 p.m.

Tullamore Astronomical Society

Secretary: Tom Walsh, 25 Harbour Walk, Tullamore, Co. Offaly, Eire.
Website: www.iol.ie/seanmck/tas.htm; *Email:* tcwalsh25@yahoo.co.uk
Meetings: Order of Malta Lecture Hall, Tanyard, Tullamore, Co. Offaly, Eire. Mondays at 8 p.m., every fortnight.

Tyrone Astronomical Society

Secretary: John Ryan, 105 Coolnafranky Park, Cookstown, Co. Tyrone, Northern Ireland.
Meetings: Contact Secretary.

Usk Astronomical Society

Secretary: Bob Wright, 'Llwyn Celyn', 75 Woodland Road, Croesyceiliog, Cwmbran NP44 2OX.
Meetings: Usk Community Education Centre, Maryport Street, Usk. Every Thursday during school term, 7 p.m.

Vectis Astronomical Society

Secretary: Rosemary Pears, 1 Rockmount Cottages, Undercliff Drive, St Lawrence, Ventnor, Isle of Wight PO38 1XG.
Website: www.wightskies.fsnet.co.uk/main.html
Email: may@tatemma.freeserve.co.uk
Meetings: Lord Louis Library Meeting Room, Newport. 4th Friday each month except Dec., 7.30 p.m.

Vigo Astronomical Society

Secretary: Robert Wilson, 43 Admers Wood, Vigo Village, Meopham, Kent DA13 0SP.
Meetings: Vigo Village Hall. As arranged.

Walsall Astronomical Society

Secretary: Bob Cleverley, 40 Mayfield Road, Sutton Coldfield B74 3PZ.
Meetings: Freetrade Inn, Wood Lane, Pelsall North Common. Every Thursday.

Wealden Astronomical Society

Secretary: K.A. Woodcock, 24 Emmanuel Road, Hastings, East Sussex TN34 3LB.
Email: wealdenas@hotmail.co.uk
Meetings: Herstmonceux Science Centre. Dates, as arranged.

Wellingborough District Astronomical Society

Secretary: S. M. Williams, 120 Brickhill Road, Wellingborough, Northamptonshire.
Meetings: Gloucester Hall, Church Street, Wellingborough. 2nd Wednesday each month, 7.30 p.m.

Wessex Astronomical Society

Secretary: Leslie Fry, 14 Hanhum Road, Corfe Mullen, Dorset.
Meetings: Allendale Centre, Wimborne, Dorset. 1st Tuesday of each month.

West Cornwall Astronomical Society

Secretary: Dr R. Waddling, The Pines, Pennance Road, Falmouth, Cornwall TR11 4ED.
Meetings: Helston Football Club, 3rd Thursday each month, and St Michall's Hotel, 1st Wednesday each month, 7.30 p.m.

West of London Astronomical Society

Secretary: Duncan Radbourne, 28 Tavistock Road, Edgware, Middlesex HA8 6DA.
Website: www.wocas.org.uk
Meetings: Monthly, alternately in Uxbridge and North Harrow. 2nd Monday in month, except Aug.

West Midlands Astronomical Association

Secretary: Miss S. Bundy, 93 Greenridge Road, Handsworth Wood, Birmingham.

Meetings: Dr Johnson House, Bull Street, Birmingham. As arranged.

West Yorkshire Astronomical Society

Secretary: Pete Lunn, 21 Crawford Drive, Wakefield, West Yorkshire.

Meetings: Rosse Observatory, Carlton Community Centre, Carleton Road, Pontefract. Each Tuesday, 7.15 p.m.

Whitby and District Astronomical Society

Secretary: Rosemary Bowman, The Cottage, Larpool Drive, Whitby, North Yorkshire YO22 4ND.

Meetings: Whitby Mission, Seafarers' Centre, Haggersgate, Whitby. 1st Tuesday of the month, 7.30 p.m.

Whittington Astronomical Society

Secretary: Peter Williamson, The Observatory, Top Street, Whittington, Shropshire.

Meetings: The Observatory. Every month.

Wiltshire Astronomical Society

Chair: Mr Andrew J. Burns, The Knoll, Lowden Hill, Chippenham, SN15 2BT; 01249 654541

Website: www.wasnet.co.uk; *Email:* anglcburns@hotmail.com

Secretary: Simon Barnes, 25 Woodcombe, Melksham, Wiltshire SN12 6HA.

Meetings: The Field Pavilion, Rusty Lane, Seend, Nr Devizes, Wiltshire. 1st Tuesday each month, Sept.–June. Viewing evenings 4th Friday plus special events, Lacock Playing Fields, Lacock, Wilsthire.

Wolverhampton Astronomical Society

Secretary: Mr M. Bryce, Iona, 16 Yellowhammer Court, Kidderminster, Worcestershire DY10 4RR.

Website: www.wolvas.org.uk; *Email:* michaelbryce@wolvas.org.uk

Meetings: Beckminster Methodist Church Hall, Birches Barn Road, Wolverhampton. Alternate Mondays, Sept.–Apr., extra dates in summer, 7.30 p.m.

Worcester Astronomical Society

Secretary: Mr S. Bateman, 12 Bozward Street, Worcester WR2 5DE.

Meetings: Room 117, Worcester College of Higher Education, Henwick Grove, Worcester. 2nd Thursday each month, 8 p.m.

Worthing Astronomical Society

Contact: G. Boots, 101 Ardingly Drive, Worthing, West Sussex BN12 4TW.

Website: www.worthingastro.freeserve.co.uk

Email: gboots@observatory99.freeserve.co.uk

Meetings: Heene Church Rooms, Heene Road, Worthing. 1st Wednesday each month (except Aug.), 7.30 p.m.

Wycombe Astronomical Society

Secretary: Mr P. Treherne, 34 Honeysuckle Road, Widmer End, High Wycombe, Buckinghamshire HP15 6BW.

Meetings: Woodrow High House, Amersham. 3rd Wednesday each month, 7.45 p.m.

The York Astronomical Society

Contact: Hazel Collett, Public Relations Officer

Tel: 07944 751277

Website: www.yorkastro.freeserve.co.uk; *Email:* info@yorkastro.co.uk

Meetings: The Knavesmire Room, York Priory Street Centre, Priory Street, York. 1st and 3rd Friday of each month (except Aug.), 8 p.m.

Any society wishing to be included in this list of local societies or to update details, including any website addresses, is invited to write to the Editor (c/o Pan Macmillan, 20 New Wharf Road, London N1 9RR or astronomy@macmillan.co.uk), so that the relevant information may be included in the next edition of the *Yearbook*.

The William Herschel Society maintains the museum established at 19 New King Street, Bath BA1 2BL – the only surviving Herschel House. It also undertakes activities of various kinds. New members would be welcome; those interested are asked to contact the Membership Secretary at the museum.

The South Downs Planetarium (Kingsham Farm, Kingsham Road, Chichester, West Sussex PO19 8RP) is now fully operational. For further information, visit www.southdowns.org.uk/sdpt or telephone (01243) 774400